This is the only book to trace the development of modern probability theory. It shows how in the first thirty years of this century probability theory became a mathematical science. The author also describes the development of probabilistic concepts and theories in statistical and quantum physics. There are chapters dealing with chance phenomena, as well as with the main mathematical theories of today and with their foundational and philosophical problems. Among the theorists whose work is treated at length are Kolmogorov, von Mises, and de Finetti.

The principal audience for the book comprises mathematicians and others concerned with probability and statistics, historians and philosophers of science, and physicists. The birth of modern probability is closely tied to the change from a determinist to an indeterminist world-view, and therefore the book will also interest anyone fascinated by twentieth-century scientific developments in general.

*Creating Modern Probability*

**Cambridge Studies in Probability, Induction, and Decision Theory**

*General editor:* Brian Skyrms



This new series is intended to be the forum for the most innovative and challenging work in the theory of rational decision. It focuses on contemporary developments at the interface between philosophy, psychology, economics, and statistics. The series addresses foundational theoretical issues, often quite technical ones, and therefore assumes a distinctly philosophical character.

Other titles in the series

Ellery Eells, *Probabilistic Causality*

Richard Jeffrey, *Probability and the Art of Judgment*

Robert C. Koons, *Paradoxes of Belief and Strategic Rationality*

Cristina Bicchieri and Maria Luisa Dalla Chiara (eds.), *Knowledge, Belief, and Strategic Interaction*

Patrick Maher, *Betting on Theories*

Cristina Bicchieri, *Rationality and Coordination*

J. Howard Sobel, *Taking Chances*

# *Creating Modern Probability*

Its Mathematics, Physics and Philosophy in Historical Perspective

*Jan von Plato*

CAMBRIDGE UNIVERSITY PRESS
Cambridge, New York, Melbourne, Madrid, Cape Town, Singapore, São Paulo

Cambridge University Press
The Edinburgh Building, Cambridge CB2 2RU, UK

Published in the United States of America by Cambridge University Press, New York

www.cambridge.org
Information on this title: www.cambridge.org/9780521444033



First published 1994
Reprinted 1995
First paperback edition 1998

*A catalogue record for this publication is available from the British Library*

ISBN-13 978-0-521-44403-3 hardback
ISBN-10 0-521-44403-9 hardback

ISBN-13 978-0-521-59735-7 paperback
ISBN-10 0-521-59735-8 paperback

Transferred to digital printing 2006

# Contents

# Preface

Sheer curiosity led me to read more and more of the old literature on probability. These explorations resulted in a number of papers during the 1980s, and at a certain point it occurred to me that I should try to write down the results of my efforts in a systematic way. No one had pursued the background of modern probability in any detail, so that I felt free to let my own particular interests act as my guide. As a result, the emphasis here is on foundational questions.

A historical–philosophical colleague may find the references to secondary literature few in number. This is due to the fact that literature on the development of modern probability is truly scarce, and it is also due to my insistence on consulting all the primary sources in the first place. I also wanted to combine a useful bibliography of primary sources with manageable length. There should still be enough indications to allow the uninitiated to begin reading the secondary literature.

I am indebted to several institutions for support during the period of writing the book. Most important, a fellowship of the Academy of Finland has secured the continuity of my researches. In 1982, I had the good fortune of being invited to join the project 'Probabilistic Revolution' at the Zentrum für interdisziplinäre Forschung at Universität Bielefeld. My contribution to *The Probabilistic Revolution* (MIT Press, 1987) was finished at the History of Science Department of Harvard University, during a Fulbright Scholarship in the fall of 1984. I started working on the present book during a Humboldt fellowship in 1988–1989, spent at the Institut für Geschichte der Naturwissenschaften of Universität München, and continued during the year 1989–1990 as fellow of the Center for the History and Philosophy of Science of the University of Pittsburgh. A pleasant stay at Princeton University's Department of Philosophy during the fall of 1992 was just the right thing for a final revision of the manuscript.

A great number of colleagues have generously helped me: Lorenz Krüger made me a member of his research project in Bielefeld, and I learned a lot also in the way of knowledge and skill from the other members of the group during the year. In Munich I was hosted by Ivo Schneider, in Pittsburgh by Gerry Massey and John Earman, and at Princeton by Dick Jeffrey. Suggestions, comments and other help gratefully acknowledged have come from, among others, Georg Antretter, Domenico Costantini, Persi Diaconis, Maria Carla Galavotti, Yair Guttmann, Michiel van Lambalgen, Isaac Levi, Eugenio Regazzini, Jürgen Renn, Romano Scozzafava, Glenn Shafer, Abner Shimony and Sandy Zabell; I also thank series editor Brian Skyrms for his invitation to contribute. A particular help has been provided by Timo Eirola, Ilkka Norros and Per Martin-Löf, who read and explained and sometimes even translated the Russian papers I could not have consulted otherwise. Petri Mäenpää and Aarne Ranta have offered convivial scientific companionship in my far-away home country. Finally, I thank Hanna and Milla for the understanding they have shown towards my efforts.

# 1

# *Introduction*

## 1.1 WHAT THIS BOOK IS ALL ABOUT

The concept of probability plays a vital role in many sciences, in the theory and practice of scientific inference, and in philosophy and the modern world view more generally. Recent years have brought a number of articles and books dealing with the emergence of probabilistic thinking in the last century and the earlier part of this century. Several studies of the history of statistics appeared in the 1980s. One also finds accounts of the development of statistical physics, of quantum theory and of fundamental questions such as determinism and indeterminism in modern physics. But nothing comparable exists on modern probability, the mathematical discipline that in some way or other is at the basis of any related studies. With the main focus on the shift from classical to modern probability in mathematics, I have attempted to combine in this book a historical account of scientific development with foundational and philosophical discussion.

Classical probability formed a chapter in applied mathematics. Its bearing on the larger questions of science and philosophy was limited. The shift to modern probability started around the turn of the century. By the late 1930s probability theory had become an autonomous part of mathematics. The developments of these early decades have been overshadowed by the nearly universal acceptance of modern axiomatic and measure theoretic probability as embodied in the classic work of Andrei Kolmogorov of 1933, *Grundbegriffe der Wahrscheinlichkeitsrechnung*. Following that work, a mathematician would answer the question of what is probability by saying: Anything that satisfies the axioms. Expressed in technical jargon, probability is a normalized denumerably additive measure defined over a $\sigma$-algebra of subsets of an abstract space. Something is lost with this answer, however. For if the space is finite, the answer also shrinks down to saying: Probabilities are numbers between 0 and 1 such that if two events cannot occur simultaneously, the probability of either one of them occurring is the sum of the probability of the first and the probability of the second. The mathematician's formalistic approach to the question does not address the meaning of probability, but only the difference in form between classical and modern probability. The shift to modern probability included changes in form as well as content. It was necessary to find a secured role for the concepts of chance and statistical law before an autonomous mathematical theory of probability could emerge. The quantum mechanical revolution of 1925–1927 is the clearest landmark in this change of ideas. It made the elementary processes in nature indeterministic, and it made probability an ineradicable part of the description of those processes. It must be said, though, that this landmark has somewhat overshadowed its surroundings. For example, well before any quantum mechanics was formulated, there was a very definite need for a clear mathematical foundation of probability within classical statistical physics.

The book aims at giving a readable view of the birth of modern probability in its different aspects, and of the foundational debates surrounding it, rather than encyclopaedic coverage. The presentation, with its combination of scientific history and philosophical reflection, can be followed with various interests in mind, and the chapters are designed so that they can be read, if wished, more or less independently of each other. The chapters divide into three parts: First, the establishment of connections between probability and pure mathematics in the first two decades of the century is dealt with in Chapter 2. During that time, the idea of modern probability was born. Second, topics from statistical physics, quantum theory, and dynamical systems are treated in Chapters 3 to 5. Third, the establishment of probability theory as an independent mathematical science is dealt with in the last three chapters of this book. These chapters cover the main approaches to probability theory and its foundations, the frequentist, the formal-mathematical, and the subjectivist one, portrayed through the works and views of von Mises, Kolmogorov, and de Finetti.

Each of the three approaches renounced classical probability. The restrictions of this concept became evident through the growing importance of statistics. It uses a concept of *statistical* probability not reducible to the classical one. Statistical or frequentist probability was the dominant intuitive idea in the developments leading to modern probability. The main representative of this concept was Richard von Mises, who in 1919 formulated a theory of probability based explicitly on the frequentist idea. That theory was very much in the foreground in the 1920s and early 1930s. It was followed by the more abstract approach based on set theory and measure theory, the one accepted by the vast majority of probability theorists from the later 1930s to our days. Kolmogorov himself followed a rather vague intuition of frequentist probability, an attitude most likely to be found shared today by probability theorists. A completely different approach to probability is *subjectivism*, or Bayesianism as it is called in statistical contexts. Probability is understood as a measure of 'degree of belief,' and probability theory as a systematic guide to 'behavior under uncertainty.' Although the subjectivist theory has gained support mostly since the 1950s, it was developed almost single-handedly by Bruno de Finetti in the late 1920s and carly 1930s. Even though it may be less clear at first sight, this view, too, meant a farewell to classical probability.

The period under study here belongs to a phase of science still dominated by developments in Europe. Some of the most representative theoretical achievements of that period include *logic*, *probability*, and *quantum mechanics*. The way 'quantum mechanics emerged miraculously from the European consciousness'[1] is known in great detail today. The development of logic and the foundations of mathematics also have received a lot of attention, while the birth of modern probability remains in the dark. It is hoped that my story of modern probability will draw a few more lines to the portrait of the European scientist of a special era. Since the turn of the century she had been facing great changes in all directions: First of all, there is science itself, with the new discoveries in physics perhaps in the forefront. The classical nineteenth century image of a ready-made mechanical-material world had been shaken by relativity. The essential discontinuity of the microworld of atoms and quanta finally dislodged the parts of that image. But overthrowing old images was not peculiar to the sciences only. In the arts, modernistic ideas made their breakthrough: Painters no longer depicted a 'given' reality, and proven ideas of harmony were dismissed within music. Logical empiricism dispensed with the philosophies of the old school, declaring their problems outright meaningless. The period under study here coincides with what has been felt to be the 'golden era' of European science. The year 1933, when Kolmogorov's monograph appeared also stands as a symbol for the beginning of the end of that era in political terms.

Erwin Schrödinger wrote in the fervent year he had mostly used for inventing wave mechanics: 'In the end, one has to keep it clear that all specialized knowledge has meaning and interest *only* within the great scheme of a world view.'[2] This statement seems to be saying that the 'flow' of meaning proceeds from without, from a cultural context to a specific scientific sphere of ideas. But there cannot be any complete

[1] From Matthews (1987, p. 199).

[2] From a letter to the philosopher Hans Reichenbach, October 25, 1926. (University of Pittsburgh archives.)

reduction of the individual acts of discovery into something common. The experience of the actor in question is rather of one who has to try to find that narrow path past all the dead ends, or worse still, past wrong leads that never close. She tends to take a view from within, from a perspective whose center is the sphere of private thought. That fact lies in the nature of almost any concentrated effort on difficult matters. Thus there is occasion for an approach from without as from within: Our understanding can proceed from an appreciation of an intrinsic role to the public reception of an idea or result, or it can proceed from the more commonly accessible goals and reasons toward the specific context of discovery. My account is mainly based on the existing research literature, that is, on the 'public image' that scientists wanted to give of their achievements at times when the printed word still carried primary importance. The literature on probability from the end of the last century and the early decades of this century is not unsurveyably large. I have also done archival research of unpublished sources when available, but the extent of such efforts is still limited by a lack of organized sources. To the circumstances of the individuals and institutions that contributed to the creation of modern probability I have payed little attention. Such detailed work has to wait for a general organization of historical studies on probability.

## 1.2 SHIFT FROM CLASSICAL PROBABILITY

Early probability theory was concerned with a finite number of alternative results of a trial. The rule for the computation of probabilities was very simple in principle. A composite event consists of several elementary events; its probability is the sum of the probabilities of the elementary events. When a finite number of repetitions was allowed, the computation usually became more complicated. The calculus of combinatorics was developed to handle these cases. But the principle remained the same: Each sequence of repetitions counted as one elementary event, and again probabilities of composite events came out as sums.

To determine the probabilities of composite events, the elementary events themselves must have some probabilities. Probability's earliest appearance was in connection with games of chance. Computational schemes were based on treating the elementary events as symmetric. This results in giving each of a number $m$ of elementary events the same probability $1/m$. Symmetry in the results of a game is not mere appearance, however. The very idea of *fairness* of a game includes that it is the gambling device which must not favor some results over others.

The assignment of equal probabilities usually applied also when a sequence of $n$ repetitions formed the elementary event. The easiest case is having probability $1/m$ for each single result, with independent repetitions. The probability of each sequence is $(1/m)^n$.

The classical limit theorems of probability theory maintained the finitary character of a calculus of probability. What today is called Bernoulli's theorem illustrates the case. Let us assume a simple repetitive event, represented as a sequence of 1's (success, occurrence of the event) and 0's (failure, nonoccurrence). Let the probability of success be 1/2, and let the repetitions be independent. With $n$ repetitions, the probability of each single sequence is $1/2^n$. What is the probability that in $n$ repetitions, the relative frequency of successes $k/n$ is within a small number $\varepsilon > 0$ of 1/2? The condition for the frequency describes here a composite event. The elementary events of which it consists are the single sequences of repetitions of length $n$ fulfilling the inequality $|k/n - 1/2| \leqslant \varepsilon$. Its probability depends on $n$ and $\varepsilon$. An elementary derivation shows that this probability is greater than $1 - 1/(4\varepsilon^2 n)$. If we let $n$ grow to infinity, the probability approaches 1 independently of the choice of $\varepsilon$. Results of this type are called *weak laws of large numbers.*

Continuous quantities appeared in at least two ways within the classical calculus of probability. The first was in connection with *geometric probabilities.* Their earliest appearance seems to be in Newton in 1665. A famous example is Buffon's needle problem of 1777. It seeks for the probability that a needle falls upon a line in a grid of parallels. In the derivation of the probability law for the needle, different choices of the continuous independent variable, the one considered 'drawn at random,' lead to different probabilities. The problem is, what choice, if any, is the right one? A second place where continuous quantities appear in probability theory is in connection with limiting distributions. The simplest example is the distribution of the sum (or average) of binomial random variables. It approaches a normal distribution law under wide conditions. The latter is used as an approximation formula, as in error theory. Continuous random quantities were handled through their distribution functions with elementary integration theory. In that theory one was not led to questions characteristic of modern probability. An extensive presentation of classical probability in precisely the sense I use the term can be found in Charles Jordan's book *Chapters on the Classical Calculus of Probability.*

The finitary classical calculus of probability is based on the 'classical interpretation of probability': There is supposed to be a finite number $m$ of 'possible cases.' They are judged 'equipossible,' hence equiprobable, if, as the saying goes, 'there is no reason to think the occurrence of one of them would be more likely than that of any other.' The condition is called 'Laplace's rule of insufficient reason,' the whole judgment an 'indifference argument.'

The real world does not possess the absolute symmetries of the classical theory's 'equipossible cases.' This was shown by the collection of statistical data, proving real dice to be unfair, having a boy not being as probable as having a girl, and so on. Classical probability is not sufficient for frequentist applications, so that a conceptual change away from the classical interpretation was required. Neither is it sufficient on a mathematical level, so that also a change away from its finitary scope was required. This change, from about 1900 on, transformed the classical calculus of probability into a mathematically deep subject. By mathematical depth I mean here that the theory began using mathematical infinities in an essential way, specifically, in the form of *infinitary events* and associated rules for computing their probabilities. There certainly were quite early contributions to probability theory that involved infinite limits. We saw above that this took place with Bernoulli's law of large numbers. In the limit, the probability that the frequency differs at all from the value 1/2 is 0. But this is not a statement about an event consisting of an infinite sequence of trials. Instead it just comes from the finitistic statement that for a given $\varepsilon$, an $n$ exists and can actually be determined, such that the probability of $|k/n - 1/2| > \varepsilon$ is below any preassigned level $\delta > 0$. Jordan's book is based systematically on finite sets and limiting procedures. Written in the 1940s, it mentions (p. 23) the possibility of a probability theory for infinite events as in Borel (1909a) and Kolmogorov (1933). But Jordan thinks that the choice of an element from an infinite set is not a well-defined notion.

Probability's entering the realm of pure mathematics was a new role for it, one without precedent. Classical probability theory had been taken as an applied field, providing formulas for error terms, say, and perhaps offering some mathematical interest in connection with the foundations of statistical physics. It shared the typical features of a chapter in applied mathematics. The concepts and methods were specific to the applications, and not much of a general, abstract structure was involved. But from the turn of the century on, more and more connections between pure mathematics and probability theory kept emerging. The appearance of these connections was due to changes in both. First of all, the applicability of the general theory of measure to probability was immediately evident. This theory had been created, at the instigation of Borel, by Lebesgue at the turn of the century. The measure theoretic study of sets of real numbers and of real functions, and the asymptotic properties of sequences of natural numbers are connected through the identification of a real number with such a sequence, as in a decimal expansion. Thus the probabilistic problem of the limiting behavior of relative frequency, for example, could be formulated as a problem about the measure of a set of real numbers. On the other hand, a probabilistic reading could be given to results traditionally belonging to a special branch of arithmetic. The earliest initiative here comes from a surprising direction. In 1888 the astronomer Hugo Gyldén was led to ask for the distribution of the integers in what is known as a continued fraction expansion of a real number:

$$a_0 + \cfrac{1}{a_1 + \cfrac{1}{a_2 + \cdots}}$$

The $a_i$ are integers. Gyldén saw, even if he did not have any strict mathematical argument for it, that a certain regularity in the distribution of the $a_i$ would emerge for 'randomly chosen' irrational numbers. *Gyldén's problem* is, what is the exact form of the limiting distribution of the $a_i$'s as $n$ grows? He became initially occupied with the problem in his studies of planetary perturbations. There continued fractions are used for finding the closest approximations to a real number in terms of rational numbers. Gyldén aimed at showing that 'the probability for the occurrence of divergence in the series by which one usually represents planetary perturbations is smaller than any given value' (1888a, p. 83).

At the heart of Gyldén's problem lies a question of the nature of real numbers. When is a property of real numbers special? Gyldén saw one such property: It is special that the integers of a continued fraction expansion of a real number should obtain large values, or at least that they should do it too often, because that would mean the divergence of the series he was calculating. Being an astronomer who uses continued fractions all the time in approximation calculations, he had come to this insight from daily experience, so to speak. *Rational* numbers would give a very special case of continued fractions, for their expansions terminate. The intuitive idea of rationals being very 'special' cases of real numbers was made more precise by Henri Poincaré in 1890 and 1896. In his study of the three-body problem, he proved a 'probability 1' result, later called Poincaré's recurrence theorem. Counterexamples to the theorem's recurrent motion are not impossible, but *exceptional.* In his book on the calculus of probability of 1896 we find another similar case. A real number's being rational is similarly an exceptional 'probability 0' property. Hence with an infinity of possible results, probability 0 does not always mean impossibility, and probability 1 not certainty.

Emile Borel's measure theory of 1898 provided a conceptual basis for Gyldén's and Poincaré's intuitions. That theory was in fact applied to probabilistic purposes very soon, in 1900–1901. The mathematician Torsten Brodén was the first to do this, in his study of Gyldén's problem. He was soon followed by his Lundian colleague Anders Wiman, who determined in 1900 measure theoretically the asymptotic distribution of the integers in a continued fraction. These works never had a wide influence directly, but they passed on the study of probabilistic distribution problems in continued fractions from which, via Borel's work, a path can be followed to modern probability.

Borel in 1909 was the first to enter modern probability seriously, in the sense used in this book. His work included a special case of what came to be known as the strong law of large numbers, and occupies our attention in Section 2.2. He was the first to consider explicitly and systematically the probabilities of events whose occurrence depends on an infinity of elementary events. Borel's law made its appearance in a very special context, namely, in his theory of 'denumerable probabilities.' It was supposed to be a branch of probability lying strictly between the finite and the continuous, or 'geometric.' In trying to create such a theory, Borel was motivated by a concern over the foundations of mathematics. Little attention was ever paid to that background, with the consequence that Borel's intentions have almost always been gravely misunderstood. But Borel's mathematical application of denumerable probability, the study of distribution problems in arithmetic sequences, managed to attract considerable interest from mathematicians. These included G. H. Hardy and J. E. Littlewood who were number theorists and analysts, and the set theorist Felix Hausdorff, among others. Borel's strong law of large numbers in its present general and purely probabilistic form is the fruit of Francesco Paolo Cantelli's efforts of 1916–1917.

The famous list of open mathematical problems that David Hilbert presented in 1900 has played a considerable role in the development of the mathematics of this century. One of his problems, the sixth, was to axiomatize probability and mechanics. Attempts in this direction were at first not very successful. Instead, the connections between probability and problems and theories of pure mathematics worked as the 'pathways to modern probability.' Decisive influences on the mathematics often came from problems in physics. That was the case with Gyldén's and Poincaré's influence on Borel. A very similar pattern can be found in Hermann Weyl's work on the distribution problems of real numbers. Such distribution problems in themselves are analogous to the arithmetic distribution problems Borel had studied, and they likewise originated in the perturbation calculations of planetary motions. That was around

1909–1910. A purely mathematical formulation was soon achieved, whose central result was: If you take a real number $x$, multiply it successively by $1, 2, 3 \ldots$, and take only the decimal part, you will get, with probability 1 a sequence of numbers *distributed uniformly* in the unit interval. Weyl's influential paper of 1916 on the equidistribution of reals mod(1) became the classic reference. Some fifty years later it was seen as a landmark in building connections between probability and pure mathematics, so that an entire conference was held in 1962 on the developments initiated by Weyl (Koksma and Kuipers, 1962). At the time Weyl was not so interested in the foundations of the topic, but rather in the mathematics proper. Among other things, this work contained something like the earliest positive results on a problem coming to mathematics from statistical physics. I refer here to the *ergodic problem*, or the problem of showing that the long-range behavior in time of a statistical mechanical system can be determined from its physical description in a certain way. This problem bears a close connection to the existence of mean motions in astronomical dynamics. As we shall see, the conceptual connections between arithmetic, probability, astronomy and statistical mechanics were fully appreciated by Weyl. The above scheme for producting a set of points on the unit interval is mathematically equivalent to drawing on a circle of unit circumference, in succession, angles of equal size with a corresponding arc length $x$. If $x$ is irrational, a dense set of points is produced; if rational, the sequence is periodic and only a finite number of separate points appear. The former situation leads to an equidistribution according to Weyl's central result. Weyl notes that there is a connection between such rotations and the epicyclic models of celestial motion of Ptolemy and others in antiquity.

The connection seen by Weyl was the object of a study of the medieval scholar and churchman Nicole Oresme. In an astonishing piece of mathematical insight and sophistication, he treated the occurrence of an event depending on an infinity: whether a certain number was rational or irrational. His aim was to prove that it is 'probable' that an irrational number is found. Consequently, it is equally probable that a dense set of points is produced on the circle. 'Probable' here – several centuries before the advent of the calculus of probabilities – is that which appears with a high relative frequency. Oresme's results, apparently, had no lasting influence. They present a delightful incident for a modern reader. In no sense is it a systematic part of the development discussed in this book, even if its substance has strong connections with the materials presented in the second chapter and in Section 3.2(a). It is included here as a supplement.

## 1.3 PHYSICS

The development of physics has had a profound influence on our ideas about probability. This influence stems from two sources, quantum mechanics and statistical physics. The latter has remained somewhat in the shadow of relativity and quantum mechanics, the two great theories of physics of the century. Relativity changed our ideas of time and space, of the structure of the universe in the large. Quantum mechanics destroyed the deterministic doctrine of classical physics; it professed a fundamental randomness in the behavior of the ultimate constituents of the world in the small, the elementary particles of matter and the quanta of radiation.

The conceptual changes statistical physics brought into the modern world view do not have the dramatic character associated with relativity and quantum mechanics. But it looks that way partly because of hindsight: In one sentence, statistical physics studies the consequences of the atomistic structure of matter. One is so used to this atomistic hypothesis that an alternative seems rather unthinkable. It has become a thing one needs to be reminded of, namely, that not all physicists have been atomists, not even at the turn of the century. Among the most notable ones to mention are Ernst Mach, Pierre Duhem, and the 'energeticist' Wilhelm Ostwald. The atomistic conception started gaining ground over competing theories around 1850, through the consequences of the identification of heat as molecular motion. But it faced serious obstacles on the way, some of them only to be removed by quantum physics. The difficulties included the explanation of irreversibility, the problem of specific heats, and the behavior of radiant heat. Together with these difficulties, a thermodynamical approach to heat, dispensing with molecules and based on directly observable phenomena instead, was gaining some recognition. Toward the end of the last century Ludwig Boltzmann, the chief advocate of the statistical-molecular theory on the Continent, had to greet a contribution by an adversary 'as the first proof that these works are receiving any attention in Germany at all' (1896, p. 567). But this trend soon turned, and it was through statistical physics that atomism gained its ultimate victory by the end of the first decade of our century. For antiatomists that change must have appeared much more dramatic than, say, the one relativity theory could deliver. It should be noted that there was also a milder form of resistance to atomism: not a denial but rather disbelief in its fruitfulness, as in Poincaré, Zermelo, and some others.

Statistical physics took its first steps in the 1850s with the formulation of the mechanical theory of heat. Probability never played the kind of clear role in statistical physics as it would very soon do in quantum mechanics. The main obstacle and point of debate was the mechanical basis. Consequently, there are rather different views to be explored. Amidst the contrary tendencies, conceptual and mathematical problems were met that directed the way to a deeper understanding of probability. There was always a tension between the classical mechanics that was supposed to be valid on the level of the atomic motions, and the macroscopic behavior of matter. Specifically, while mechanical processes are reversible and symmetric in time, heat processes obviously have a preferred direction, namely, toward the equalization of temperature differences. The problem of irreversibility became the crucial one for the kinetic theory. Probabilistic arguments were invented for reconciling the two levels with each other, that is, the levels of the mechanical molecular processes and of the macroscopic observable ones. These arguments, in turn, called for a more sophisticated probability mathematics.

The conceptual basis of physics in itself provided a natural habitat for some of the essential features of modern probability theory. It provided: 1. a continuous state space, and 2. continuous time. Nothing was more natural than to apply measure theory to the state space of a statistical mechanical system. This state space was a Euclidean space $R^n$ for which Borel and Lebesgue had created their measure theory. That theory was fairly soon put to use in statistical physics, where it provided a safer ground for the discussion of the foundational problems of statistical mechanics. The second concept provided by physics, continuous time, took a long time to be incorporated into a concept essential for modern probability theory. Whereas continuous random quantities had been studied in probability at least since Newton, random events following each other in continuous succession formed an entirely new concept. There were two lines of development, a physical and a mathematical one. The former started with Boltzmann's equation in 1872, and is highlighted by Albert Einstein's 1905 theory of Brownian motion, Marian von Smoluchowski's work, the Fokker–Planck diffusion equation of 1914–1917, and Chapman's equation of 1928. On the mathematical side, the first random processes systematically developed were Markov chains, where time is discrete. The Markov property, characterizing these chains, says that the probability of the next state, given the present state, is independent of previous history. This condition has a clear physical analogue. When the first systematic mathematical theories of random processes with continuous time were developed in the late 1920s, they originated directly from a physical model. The Markov property for these processes was a probabilistic rendering of the most characteristic feature of classical mechanics: Given the law of motion and the present state, the future evolution is determined. The probabilistic analogue was obtained by replacing an exact future state by a probability distribution. A general theory of continuous random processes soon evolved; it had long been overdue. At least part of the reason for the 'delay' seems to lie in the belief that the use of continuous time would lead to determinism. When the mathematical theory was developed, the physical literature on the topic first remained unknown, except for Einstein's papers on Brownian motion. The theory of *stochastic processes* was the most profound thing statistical physics gave to probability theory. Since the barest beginnings of the mechanical theory of heat in the 1850s, some seventy years had passed.

In the chapter on statistical physics I shall review the first uses of the concept of probability in physics, and the various interpretations given to it by Boltzmann, Maxwell, Gibbs and Einstein. The incorporation of irreversible processes into a mechanical theory was perhaps the central conceptual problem here. Boltzmann and Maxwell were both quick to recognize the essential role played by the concept of probability. In Boltzmann, this is connected to his attempt at finding a derivation of the second law of thermodynamics. In 1868 he introduced the *time average* notion of probability: If $A$ is a subset of the set of all possible states, the time average probability of $A$ is the limit of the relative time the system spends in the set of states $A$. In Maxwell, a probabilistic hypothesis was essential for the derivation of the Maxwellian distribution law of molecular velocities. Maxwell's name is associated with the *ensemble* view of probabilities in statistical physics, an approach actually invented by Boltzmann and further developed by Gibbs. An ensemble in itself is a technical notion that can be understood in different ways, as we shall see in the discussion of Gibbs.

A systematic development of the Boltzmannian time average interpretation of probability leads to ergodic theory. Originally meant as a foundation of statistical mechanics, it gained a more and more independent status as part of mathematics. The theory had started as a theory of classical mechanical systems with many degrees of freedom (that is, a great number of molecules). Under the crucial *ergodic hypothesis*, originally formulated by Boltzmann, it was able to establish a foundation for the calculation of average quantities in statistical mechanics. The aim of the ergodic hypothesis is, more particularly, to justify the introduction of a suitable probability distribution for calculating average values of physical quantities. What Boltzmann's ergodic hypothesis is supposed to be has been indelibly imprinted in subsequent literature. It reads: One mechanical trajectory fills all of the state space, rendering the time averages over the trajectory equal to averages over the state space. As we shall see, this hypothesis, while being untenable on very simple grounds, is also plainly incorrect historically as a reading of Boltzmann.

Whatever the correct and intended version of Boltzmann's ergodic hypothesis was, the mechanical assumptions underlying ergodic theory were physically incorrect. As became slowly known, starting in 1900, the world is quantum mechanical and not classical. The mutual effects of quantum theory and statistical physics make for a complex, twisted story, both historically and systematically. As far as Boltzmann's kinetic theory is concerned, the most important problem where quantum behavior mattered, concerned specific heats. The failure of statistical mechanics in this respect remained an anomaly, until an explanation was found in quantum theory. Other factors, in which the secrets of quantum behavior of matter and radiation were not relevant, played a clearer role in the development of statistical physics. That was so in Brownian motion and other fluctuation phenomena. Also, when Richard von Mises in 1920 started his attempts at formulating a purely probabilistic ergodic theory, freed of the classical mechanics, his motivations did not come from quantum theory. Instead, in the late 1910s he had created an original theory of probability in which randomness was a basic, undefined concept. The concept of probability in turn was defined through randomness, as a limit of relative frequency in a random sequence. Randomness, according to von Mises, is incompatible with mechanical motion, hence the need for a purely probabilistic physics. By the early 1930s the developments initiated by von Mises were brought into fulfilment in the creation of a theory of abstract dynamical systems. This shift from a classical mechanical to a purely probabilistic theory took place some years after modern quantum mechanics was created. It was felt that through the probabilistic formulation, ergodic theory was freed of the threat posed by quantum mechanics. Oddly enough, these developments were triggered by the application of the mathematical methods of quantum theory to classical dynamics. The application was due to von Neumann, who also was the chief architect of what is known as the Hilbert space formalism of quantum mechanics.

Physics has its theoretical as well as experimental side. Two specific phenomena from the latter occupy an important role as far as probability is concerned. These are *Brownian motion*, discovered under the microscope in 1827, and *radioactivity*, discovered in 1896. The first of these, the spontaneous random motion of microscopic particles in a fluid, was a source of many experimental studies. These led to a qualitative understanding of the phenomenon. But no theoretical explanation was forthcoming for a long time: The probabilistic law of Brownian motion remained unknown. It was only in 1905 that Albert Einstein was able to predict the occurrence of Brownian motion theoretically as caused by the fluctuations in the random impacts of the molecules of the fluid on the particle. His theory contained a simple differential equation whose solution was the continuous time probability law for Brownian particles. The formulation of a theory, soon repeated in a somewhat different form by von Smoluchowski, led to experimental studies which finally defeated the last antiatomists.

The explanation of Brownian motion proceeded from a classical statistical mechanics. Such was not the case for radioactivity. The statistical law of radioactive decay was soon found; it was a law known for the specific property that it pays no respect to the previous history of the process. But no explanation of radioactivity could be found before the advent of quantum mechanics. Therefore, from a historical, factual point of view, the study of radioactivity belongs to the sphere of ideas of classical statistical physics. The first hint at a connection to quantum theory is from the year 1916, but the real answer only started coming in 1928, when alpha decay was explained as quantum mechanical tunneling over a potential barrier.

When did physicists become indeterminists? The majority certainly around or after 1925–1927; the well-known events of these years started when modern quantum mechanics was invented by Werner Heisenberg. Max Born gave the method for deriving the probability laws governing quantum mechanical processes in 1926. He already saw the indeterminism, but could not argue that it would be a final state of matters in the theory of quantum mechanics. Indeterminism in the small was instead established through the Heisenberg uncertainly relation in 1927. Both of these aspects of quantum mechanics were rather soon, if not universally, so generally accepted. It is often said that the scientific community is basically conservative, wishing to maintain its old edifices. How could it be that there was such an immediate acceptance of what now seem like dramatically new ideas? The scientific authority of persons such as Bohr certainly played a role, but the main reason must lie in the power of quantum mechanics to lead to physically significant results. The first remarkable event in this direction was Pauli's determination of the spectral lines of hydrogen early in 1926. Probability and indeterminism came along with the successes. But probabilistic methods and ideas had already penetrated classical physics through the kinetic theory, through Brownian motion, molecular statistics, radioactivity, radiation in general, and so on. They were familiar in the 'old' quantum theory that preceded quantum mechanics in 1900–1925. Then, suddenly, probability became part of the basic stuff of things. In looking at the development leading to the acceptance of a probabilistic and indeterministic quantum mechanics, it is good to make a distinction between giving up classical mechanics as the basis of physics and accepting indeterminism. The old quantum theory, from Plank to 1925, dispensed with classical mechanics in a way, but that was still not enough to make physics indeterministic. Einstein, for example, gave up the hope of finding a mechanical explanation of quantized radiation around 1908, and introduced probability in the description of 'quantum jumps' in 1916. But he never became an indeterminist in the Heisenberg–Born sense.

Some opposition notwithstanding, a certain view on quantum mechanical probability evolved soon after that probability was discovered by Born in June of 1926. This view is concisely exposed in von Neumann's 1932 book on the mathematical foundations of quantum mechanics. There we read (p. 2) that according to the traditional view, every probability judgment stems from the incompleteness of our knowledge. Accordingly, it was suggested that the statistical character of quantum mechanics also should be explained on the basis of 'hidden parameters.' 'Explanation through hidden parameters has reduced many a seemingly statistical way of behavior back to the causal foundation of mechanics' (p. 109). Enter von Neumann's proof of the impossibility of incorporating any such hidden variables into the formalism of quantum mechanics: It was meant to show that quantum mechanics is an *irreducibly acausal* theory of elementary processes. (To what extent it did, is still debated.) As will be seen, irreducible character of quantum mechanical probability was already familiar, at least since Heisenberg's uncertainty relation in 1927.

Von Neumann's view of probability in physics has become standard. In classical physics probabilities are basically nonphysical, epistemic additions to the physical structure, a 'luxury' as von Neumann says, while quantum physics, in contrast, has probabilities which stem from the chancy nature of the microscopic world itself. Epistemic probability is a matter of 'degree of ignorance' or of opinion, if you permit. The quantum mechanical probabilities, instead, are computed out of the $\psi$-function so that no place seems to be left over at which the knowing subject could inject his ignorance. The standard view is that these two kinds of probabilities, the classical–epistemic and the quantum mechanical–objective, exhaust all possibilities. This view is known, after Popper, by its popular name as the *propensity interpretation* of probability. Whether his is a viable view of quantum mechanical probabilities can be discussed. But as an account of objective probability in general, the propensity theory raises two fundamental questions: 1. Why should quantum mechanics be the explanatory ground for all objective probabilities? and 2. Why should a mechanical basis turn statistical behavior into something apparent only, into a mere 'scheinbar statistisches Verhalten,' as von Neumann puts it? Was the reality of statistical law not precisely one of the lessons to be learned from classical statistical physics?

The accounts of what was going on beyond the quantum mechanical revolution, or in preparation of it, have not led to any final result. One of the central questions, namely, the 'break' from an epistemic notion of probability, has been variously located. I shall steer a rather strict course here, mainly based on a handful of key papers on quantum mechanics from the 'professional' physics literature. Here the developments are well known. What is clear is the decisiveness of the break with the old quantum theory that preceded modern quantum mechanics in 1900–1925. The old theory tried to resolve the puzzles of quantized radiation and of the spectral lines of atoms. It was based on the Rutherford model of the atom, where electrons orbit around the atomic nucleus according to the laws of classical mechanics. Corresponding to the characteristic discrete lines in the spectrum of an atom, there is assumed to be a discrete sequence of energy levels. In transitions energy is emitted or absorbed in discrete proportions. Bohr (1913) tried to derive the laws of spectral lines by a combination of classical trajectories of the electrons around the nucleus, of specific 'quantization rules' and of the 'correspondence principle.' The rules state that the frequency of the light is determined from the energy difference in the 'quantum jump' divided by Planck's quantum of action $h$. The correspondence principle requires that classical laws be regained as a limiting case from the quantal ones. By the 1920s the application of the old quantum rules had become very complicated. The lack of a good theory was felt as a 'crisis,' well testified by contemporary opinions.

Probability was a familiar concept in the old quantum theory. It had a twofold basis: on the one hand in the derivation of Planck's radiation law itself, on the other in connection with the transitions of energy states that atoms are experiencing. These were treated by explicit probabilistic methods by Einstein. In his crucially important work of 1916–1917 Einstein studies the atoms of a gas, introducing the *probability* $W_n$ of an atom's being at the $n$th energy level. If the gas receives energy through radiation, there will be jumps into higher energy levels $m$ in addition to the emissive jumps into lower levels. The probabilities $W_{nm}$ of such jumps, in a given interval of time, can likewise be introduced through a condition of statistical equilibrium. That condition enabled Einstein to derive Planck's radiation law. He also got Bohr's quantization rules for the differences of energy levels. The $W_{nm}$ are the *transition probabilities* between the different states of an atom. They reappear in the very first paper on quantum mechanics, that of Heisenberg in 1925.

After Heisenberg's opening, quantum mechanics emerged in a few years, years that belong to the most intense in the development of physics and scientific thought in general. Heisenberg had made a break with the past by explicitly creating a theory based on 'observable concepts' only, and by declaring meaningless the classical conceptions of continuous trajectories of particles. The general formulation of his theory was given in the form of *matrix mechanics* in 1925. Soon there followed a quite independent development: Schrödinger's wave mechanics in the first half of 1926. Schrödinger, though having earlier been an indeterminist in his philosophy, reverted for a while to a 'continuum' conception of the microworld. Then the concept of probability in quantum mechanics arose through Born's work in the middle of 1926, from the soil of Einstein's and Heisenberg's transition probabilities, de Broglie's matter waves, Einstein's gas theory, and Schrödinger's wave function. Heisenberg's uncertainty relation from early 1927 showed that quantum mechanics does not permit the kind of exact concept of physical state as classical mechanics, thereby certainly reinforcing the indeterminism of elementary processes Born had envisaged in 1926. In 1927 the quantum mechanical building of the new physics was approaching completion in the works of Paul Dirac. The systematic mathematical formulation of the theory was being developed in Weyl's 1927 work on group theory and quantum mechanics and in the general formulation in terms of Hilbert spaces and their operators by von Neumann. The philosophy of quantum mechanics was dominated in the early years by the 'Copenhagen spirit of quantum theory.' Its central doctrine was Bohr's 1928 idea of complementarity, a philosophical generalization of the uncertainty relations. Part of the Copenhagen philosophy is the relativity of the division into the observer and the observed. The interference phenomena of quantum mechanics pointed at the 'quantum entanglement,' or the nonlocal character of the quantum mechanical concept of state.

The explanation of quantum phenomena led to the acceptance of probability and chance as belonging to nature itself. For good or bad, the classical theory was never able to convince more than a minority in this respect. In Chapter 5 I shall discuss one further theme connecting probability and physics, namely, the attempts at continuing the explanation of probability and chance along classical lines. It is a part of foundations of probability almost forgotten today. The reason for the neglect certainly lies in part in the dominance of ideas from quantum theory.

As I have emphasized, without the changes brought about by physics, modern probability theory could not have obtained the remarkable position it today enjoys in scientific and philosophical thinking. But the impact of quantum physics on probability is, under the period discussed here, largely restricted to the indeterminism–probabilism of Heisenberg–Born as defined above. For in quantum theory, the set of events typically becomes discrete because of quantization, in contrast to the continuous state spaces of classical statistical mechanics. From the latter, modern probability got its continuous time random processes, while quantum mechanics gave no technical contribution of comparable magnitude. From these technical and conceptual points of view, modern probability owes more to classical statistical mechanics than to quantum theory. From a foundational and philosophical point of view, it seems to be the other way around.

## 1.4 THE FINAL STAGE, 1919–1933

Modern probability in our present measure theoretic sense was a creation of the late 1920s and early 1930s. Its definitive formulation is Kolmogorov's axiomatization of probability in his *Grundbegriffe der Wahrscheinlichkeitsrechnung* in 1933. It is the kind of theory in which a mathematician of today would be trained. His other training would include a set theoretic way of thinking about mathematical existence. But we must keep in mind that during most of the period under discussion in this book, notions of mathematical existence underwent heavy debates. Traditionally, mathematics had been pursued *constructively*. Borel and some other French mathematicians, for example, required all mathematical objects to be defined 'by a finite number of words.' A sequence of integers, say, was defined by defining a function that for each $n$ gives the $n$th member of the sequence. That function represented the mathematical law of the sequence. It was part of the notion of function that it had to be *computable*. (This was hardly spelled out, for no one contemplated the existence of noncomputable functions.) A sequence due to chance does not follow a mathematical law at all. Thus it was a great problem for Borel to find a way of representing chance mathematically. To this end, he had to widen the requirement of finite definability.

The conceptual situation traditional mathematics met while trying to incorporate chance, was analogous to the situation in physics. Mathematical laws were as strict as the deterministic laws of physics, the main difference being that they dealt with objects that did not have any immediate correspondence in the physical world.

The old sense of mathematical existence prevailed until the turn of the century. Then came David Hilbert with his idea of existence as consistency. Let chance produce an indefinitely, or even infinitely, long sequence of numbers. Existence as consistency says that *there is* some mathematical law that the sequence follows. Knowing which law, is an entirely different matter. The limits of mathematical existence were soon raised even much higher by set theory's nondenumerable infinities. Debate about mathematical existence was not settled with that, but rather opened. Remarkably, many of the most important contributors to probability discussed in this book had reservations about the formalist or set theoretic foundations of mathematics. That was the case with Borel and Weyl, the latter the best known supporter of the intuitionist Brouwer. But the same holds even for the foremost of the probability theorists, Kolmogorov himself. Although not well known, he remained a convinced intuitionist all his life, with a special 'license' to pursue also set theoretic classical mathematics.

The above reservations about mathematical existence are best kept in mind when studying the frequentist theory of probability of von Mises that preceded the Kolmogorovian measure theoretic approach. The former is of special importance for the more abstract and general measure theoretic probability. The frequentist theory formed a large part of the intuitive background of Kolmogorov. It was a theory that aimed at 'the closest connection between the mathematical theory and the empirical origin of the concept of probability' as Kolmogorov himself puts it (1933, p. 2). In 1919 von Mises published two long works on probability. The first one was a survey of the mathematics of probability. He deplored its low state, saying it was lagging behind other parts of mathematics in the formulation as well as in the proof of its results. Specifically, he wanted to have more rigorous conditions and derivations of the central limit theorem. This challenge was met in a few years. Von Mises' second work of 1919 was concerned with the foundations of probability. He felt that probability theory was also here lagging behind other parts of mathematics. His own foundational system is based on the following ideas. There is a sample space of possible results, each represented by a number. An experiment is repeated indefinitely. The resulting sequence of numbers is called a *collective* if the following two postulates are satisfied:

I. Limits of relative frequencies in a collective exist.
II. These limits remain the same in *subsequences* formed from the original sequence.

Probability is a concept that applies only to collectives. It is a *defined* notion, the limit of relative frequency. The second condition above is a postulate of *randomness.* The subsequent development of the theory of collectives has centered on the proper definition of randomness. The following argument was repeatedly presented: Let us suppose we have a binary collective. If the limit of the relative frequency of 0's in the collective differs from 0 and 1, there exist subsequences containing only 0's, say. Surprisingly many people thought this trivial observation shows the impossibility of defining randomness. From von Mises' expositions it is clear that the nonexistence of subsequences in the randomness postulate is intended in a way different from the unlimited notion of existence such as one has in set theory. Several ideas were pursued towards clarifying the principles of choice of subsequences that can be considered admissible. For von Mises, the randomness postulate was an expression of his indeterminism. He said that in classical physics there exists an 'algorithm' for computing, at least in principle, the future course of events. As was noted in Section 1.3, on the basis of this conviction, he formulated in 1920 a program toward developing a purely probabilistic theory of statistical physics. He assumed that there is a finite set of discernible macroscopic states, with probabilistic laws of transition between them. The states together with their transition probabilities form what later became to be called *Markov chains.* Their characteristic property is that the probability of transition from state $i$ to state $j$ does not depend on previous history of the process. Von Mises' approach to statistical physics contributed remarkably to the development of the mathematical theory of Markov chains. On the other hand, it was limited to treating only discrete time processes.

Another kind of criticism of von Mises' ideas objected to the use of infinite sequences as collectives. These criticisms should be set against the background of von Mises' philosophical convictions. He was one of the founders of logical empiricism, and considered mathematical infinity an idealization that is accepted in probability theory in the same way as in other parts of mathematics. Thus he would say, for example, that one must not assume a continuous mechanical process behind a discrete observable one if the former is not simple enough to be scientifically useful. It would be similar in other cases: No infinity could claim empirical reality directly, but only as a useful idealization.

For a broader perspective on the criticisms on von Mises' theory, it is useful to compare the theory with the idea of probability as a measure of 'proportion' in a population. Such an approach postulates the existence of a class of cases, or a population, finite or infinite, and of subclasses representing events; say, the population of human beings and the subclass of females. Probability is a concept which applies to an event resulting from the *random choice* of an individual from the population. In von Mises' theory the random choice is repeated and therefore ordered in time. Its results can thus be represented as a sequence of numbers. The randomness postulate has the same content as the condition of random sampling in statistics in general. These two share the same problems. For example, one may ask if a property appears as truly randomly distributed in a class, or if a different frequency of occurrence could be obtained by a more informed way of sampling. Certainly, there exist in some sense subpopulations with different frequencies. These remarks are mere variants to the objections toward von Mises' notion of collective, such as the claim that there exist subsequences with different limits of frequencies.

Kolmogorov's first paper on measure theoretic probability appeared in 1929. It was published in Russian and was programmatic in character. At the time, Kolmogorov viewed measure theoretic probability as a way of incorporating probability theory into pure mathematics: 'One gains the impression that the formulas of the calculus of probability express one basic group of mathematical laws of the most general kind,' he wrote in the paper. The mathematical line of development behind this view lay in the measure theoretic study of asymptotic properties of sequences of natural numbers. The first result here was Borel's strong law of large numbers. Kolmogorov himself had started his probabilistic career with a joint paper with Alexander Khintchine in 1925. They gave a sufficient condition for the convergence of a sum of random variables in the probability 1 sense. The exceptions have Lebesgue measure 0.

Very soon Kolmogorov's motivations changed. In 1929 he had not been sure that it would be worthwhile to work out in detail a measure theoretic approach to probability. In a long paper of 1931 on continuous time random processes, measure theoretic probability is needed for the treatment of problems of statistical physics. That is also the stated main motivation in Kolmogorov's *Grundbegriffe*. It is sometimes said that the maturity of a field of mathematics is measured by the degree to which it can 'forget' about its history. Measure theoretic probability forms an outstanding example in this respect, after it was given a compact logical–mathematical formulation by Kolmogorov in 1933. Here it is, essentially in the form given to it by Kolmogorov. First, there is an abstract *space* $\Omega$. Second, there is a *Boolean algebra* $\mathscr{F}$ of subsets of $\Omega$, defined by the following clauses:

I. $A \in \mathscr{F}$ and if $-A$ is its complement relative to $\Omega$, $-A \in \mathscr{F}$.
If $A \in \mathscr{F}$ and $B \in \mathscr{F}$, $A \cup B \in \mathscr{F}$ and $A \cap B \in \mathscr{F}$.

II. $\Omega \in \mathscr{F}$. It follows that the empty set $\emptyset \in \mathscr{F}$.

There is a function $P:\mathscr{F} \rightarrow [0, 1]$ such that:

III. $P(\Omega) = 1$
IV. If $A \cap B = \varnothing$, $P(A \cup B) = P(A) + P(B)$.

The preceding conditions axiomatize the notion of a finitely additive probability space. Next, denumerable unions and intersections of sets are included, which extends the Boolean algebra into a *σ-algebra*:

V. If for $i = 1, 2, 3, \ldots$ $A_i \in \mathscr{F}$, $\bigcup_{i=1}^{\infty} A_i \in \mathscr{F}$. It follows that also $\bigcap_{i=1}^{\infty} A_i \in \mathscr{F}$.

The finitely additive measure $P$ is finally assumed also *denumerably additive*:

VI. If $A_i \cap A_j = \varnothing$ whenever $i \neq j$, $P(\bigcup_{i=1}^{\infty} A_i) = \sum_{i=1}^{\infty} P(A_i)$.
This is the condition of $\sigma$-additivity.

Probability theory in this formulation is the theory of normalized $\sigma$-additive measures on abstract spaces. It would be of no particular interest were it not for the special structure given to the basic space $\Omega$, and for the intuitive meanings given to the theory's basic notions. These latter are revealed by the special terminology suggested already by Kolmogorov. The space $\Omega$ is the set of *elementary events*. Members of $\mathscr{F}$ are *events*, or *random events*, and $\Omega$ is the *certain event*, $\varnothing$ the *impossible* one. $P(A)$ is the *probability* of the event $A$. Kolmogorov himself in 1933 saw the notion of *probabilistic independence* and its generalizations as a characteristic mark of probability spaces and measures. In the simplest case, the space $\Omega$ is of the form of a product $\Omega' \times \Omega'$. A product measure $P$ over $\Omega$ is defined from a measure $P'$ over $\Omega'$ by the stipulation $P(A) = P'(A')P'(A'')$, where $A = (A', A'')$.

As its name indicates, Kolmogorov's treatise was concerned with the *basic concepts* of probability. Its two mathematical novelties were the theory of conditional probabilities when the condition has zero probability, and the introduction of an infinite product space for the treatment of limit theorems and random processes in general. Both novelties grew straight from the ground of statistical physics, as we shall see. In elementary probability, a conditional probability is defined as follows: If $P(B) > 0$, the *conditional probability* $P(A|B)$ of event $A$ given event $B$ is the number $P(A \cap B)/P(B)$. A conditional probability measure is a probability measure over a new space we can call a conditional probability space. This is one place where Kolmogorov demonstrates the special interest of probability measures through his definition of a conditional probability also for the case of $P(B) = 0$. He found such cases in his study of systems of statistical physics that have a continuous state space $S \subset R^n$. The system is assumed to be in an exact state $x_t$ at time $t$, represented by a point of $S$. The events to be considered belong to a conditional probability space corresponding to a given value of $x_t$, and a conditional probability measure is given over that space. A second place where Kolmogorov gives a specific structure to the basic space $\Omega$ of probability theory is in connection with *random processes*. If $T = N$, $\Omega^T = \Omega \times \Omega \times \Omega \times \cdots$ is the space of infinite realizations of a discrete time random process. If $T = R$, $\Omega^T$ is the space of continuous time trajectories of a process. Measure theory provides a way of extending a probability defined over a finite product $\Omega_1 \times \cdots \times \Omega_n$ (or $\Omega_{t_1} \times \cdots \times \Omega_{t_n}$ with $t_i \in R$) into a probability measure over the space of infinite realizations, discrete or continuous. It then becomes possible to answer probabilistic questions about the class of all possible realizations. For example, what is the probability that the limiting frequencies of a discrete process exist? Or, given that a Brownian particle is at point $x_0$ at time zero, what is the probability that it is at a distance $< d$ from $x_0$ at time $t$? With the formulation of the general notion of a random or stochastic process, two lines of development toward measure theoretic probability became united. With the notion of a discrete time process, the limit laws of probability theory could be formulated as 'probability 1' theorems about the space $\Omega^N$. The continuous time case, on the other hand, became the most characteristic object of study in modern measure theoretic probability.

After measure theoretic probability became the accepted standard, little understanding was nor has it ever been shown toward the frequentist theory. Still, it is of interest for several reasons. First, for its obvious historical role. Toward the end of the 1920s, several alternative formulations of the frequentist theory started appearing. These theories, and the one by von Mises in particular, were the object of remarkable debates on foundations of probability. Another reason for interest in the frequentist theory is its revival in the 1960s. This development was initiated by no one else than Kolmogorov himself. To many it must have seemed a step back. But reading Kolmogorov's 1933 book, one notices how greatly it was influenced by von Misesian ideas with regard to the meaning of probability. In the 1950s Kolmogorov writes that in 1933 he did not state how probability is applied because he did not know it. His survey (1956, Engl. 1963a) contains a frequentist approach to the application problem of probability. The law of large numbers states that the probability and the relative frequency of an event are, with a sufficient number of trials, with a very high probability close to each other. The second probability in this statement, however, cannot be interpreted frequentistically according to Kolmogorov (1956). When in the 1960s he started developing anew the old von Mises concept of a *random sequence*, he thought it would lead to a solution of the application problem of probability: The conclusions of his measure theoretic probability would be applicable to a single empirical sequence of results if that sequence could be proved random in the modified von Mises sense. (This is a doubtful idea, for the randomness of a sequence does not seem to be a property that could ever be definitively proved.) Empirical sequences are of necessity finite, so that a concept of a finite random sequence was needed. The new Kolmogorovian definition of randomness for finite sequences has led to an extensive literature on 'algorithmic information theory,' extending far beyond the general time bounds of this book. But the topic is so essential for the understanding of Kolmogorov's works on probability that it could not be passed by.

Coming to the last chapter of the book, there is a smaller group of probability theorists and statisticians, often called 'Bayesians,' who do not share the formalistic approach that is prevalent in probability mathematics today. In the Bayesian philosophy, chance and probability have a position different from what my discussion in Section 1.3 of the importance of statistical physics and quantum mechanics for probability would suggest. Still, their position is not one of derivative notions. They belong to the basic furniture, not of the world itself, but of the *system of knowledge*. The Bayesian subjectivist view of probability is not tied to old-fashioned determinist views of science, either. The purest form of Bayesianism, that of Bruno de Finetti, was developed as a response to the change toward indeterminism in scientific development in the late 1920s. I shall discuss it in the final Chapter 8. De Finetti's work is remarkable in many respects. Being very original and prodigious as well as somewhat isolated, he created a philosophy of probability of his own, and probability mathematics to accompany that philosophy. His best known piece of work concerns the notion of exchangeability. He wanted to deny the reality of objective statistical probabilities, and to replace them by exchangeability that he thought would have immediate intuitive meaning: It is the judgment that in a sequence of trials, the order of successes is irrelevant for the assessment of probabilities. The 'reduction' of unknown objective probabilities to subjective ones was based on a technical result of de Finetti's, his famous representation theorem of 1928. In 1929 he independently developed the theory of continuous time random processes, where the changes in the process at nonoverlapping intervals of time are probabilistically independent. De Finetti's work on continuous time processes was immediately recognized and made him well known among the small circle of experts in modern probability. His foundational ideas remained less known until the

Bayesian statistics of L. J. Savage (1954) and the development of game theory and decision theory revived interest in his contributions.

De Finetti published his subjectivist program for the foundations of probability in the early 1930s. Probability was to be interpreted as a primitive concept in an account of human behavior under uncertainty. Its basic nature is qualitative. But it can also be measured numerically, through the betting ratios one would be willing to accept for the occurrence of the event one is uncertain about. Corresponding to these ideas, de Finetti created a theory of *qualitative probability*, and showed how the usual properties of numerical probability can be derived from the notion of *coherent bets*. There are four axioms for the qualitative probability relation $E \succeq E'$, to be read as 'the event $E$ is at least as probable as the event $E'$.' The events $E$ and $E'$ are defined to be identically probable, $E \cong E'$, if $E \succeq E'$ and $E' \succeq E$. If $E \succeq E'$ but not $E \cong E'$, one writes $E \succ E'$. Finally, $E + E'$ is the event $E$ or $E'$. The axioms are:

I. $E \succeq E'$ or $E' \succeq E$ for any events $E$ and $E'$.
II. $A \succ E \succ B$ for $A$ certain, $B$ impossible, and $E$ neither of these.
III. $E \succeq E'$ and $E' \succeq E''$ implies $E \succeq E''$.
IV. If $E_1$ and $E_2$ are both incompatible with $E$, $E + E_1 \succeq E + E_2$ if and only if $E_1 \succeq E_2$. Specifically, $E_1 \cong E_2$ if and only if $E + E_1 \cong E + E_2$.

The fourth axiom is the essential one. It states that the qualitative probability of a composite event $E + E_1$ remains the same if $E_1$ is replaced by a qualitatively identically probable event $E_2$.

De Finetti considers the idea of coherent bets as a convenient though somewhat arbitrary and simplified way to introduce numerical probabilities. One assumes a 'banker' who has to accept bets for an event $E$ for any sum $S$ chosen by a bettor. The banker has the right to choose the *probability* $p$ of the event $E$. The *bet* is defined as the random quantity $G$, the *gain* of the bettor,

$$G = (|E| - p)S$$

where $|E| = 1$ if $E$ occurs, $|E| = 0$ if $E$ fails to occur. If bets for several events $E_1, \ldots, E_n$ are proposed simultaneously,

$$G = (|E_1| - p_1)S_1 + \cdots + (|E_n| - p_n)S_n.$$

*Coherence* requires that the banker be never led into a betting scheme with *sure loss*, that is, with $G$ positive for at least one choice of the sums $S_i$, under all possible results of the events $E_i$. De Finetti proves that coherence implies that the numbers $p_i$ associated to the events $E_i$ fulfil the axioms of finitely additive probability. Central among de Finetti's concerns about the mathematical formulation of probability theory was the question of denumerable additivity. He found no justification for it from his interpretation of probability, and left it as a property that is sometimes fulfilled, sometimes not.

Probability's coming of age meant its becoming an autonomous part of mathematics. Mathematicians, with their preference for a timeless image of their subject matter of study, tend to think the abstract mathematical structure of current probability theory calls for no external justification. But let us try to imagine, for the sake of the argument, a view of the world where *chance* and *statistical law* are understood as derivative notions rather than as part of the basic furniture of the world. This brings us right back to the previous century's dominating world view – to a world governed by the iron laws of classical physics, a world where variation is deviation, chance fortuitousness, and the reign of statistical law a mystery. This is to say, without the change in the conceptual appropriation of the world, witnessed through almost all of science and philosophical thinking as a 'probabilistic revolution,' a mathematically sophisticated probability theory would have little conceptual autonomy. Against this background, the shift from classical to modern probability appears as part of a great movement, the very change from classical to modern science itself.

# 2

# Pathways to modern probability

## 2.1 FIRST STEPS IN MEASURE THEORETIC PROBABILITY. AXIOMATIZATION

### *2.1(a) Gyldén's problem in continued fractions*

Measure theory originated at the end of the last century from problems encountered mainly in mathematical analysis, the theory of trigonometric series, and integration theory. Measure first was a generalization of geometric measure in Euclidean space. Current measure theory originated as an abstraction from making the concepts independent of real numbers and real spaces. This abstract kind of measure theory was first given in Fréchet (1915).[1]

In Borel (1898) a generalization of length on the real line was proposed which is now called the *Borel measure.* The definition is repeated in Borel's first paper on probability (1905b): First *measurable sets* are defined as consisting of closed intervals, finite or denumerable unions of closed intervals, and complements relative to a given measurable set. The *Borel measure* of an interval $[a, b]$ is $b-a$, that of a denumerable set of pairwise disjoint closed intervals the sum of the lengths of the intervals, and the measure of a complement $E-F$ the measure of $E$ minus that of $F$. If an arbitrary set $E$ is contained in a measurable set $A$ of measure $\alpha$ and contains a measurable set $B$ of measure $\beta$, its measure $m$ is less than or equal to $\alpha$ and greater than or equal to $\beta$. Finally, following Lebesgue's extension of the original Borel measure, if one can prove that the measure of a set $E$ is, for any $\varepsilon$, less than or equal to $m+\varepsilon$ and greater than or equal to $m-\varepsilon$, its measure is $m$ (Borel, 1905b, pp. 987–88). One idea with defining measure in this way was to make the measure of any measurable set actually computable with any desired degree of accuracy.

Poincaré's *Calcul des probabilités* (1896, p. 126; 1912 ed., p. 148) had treated as an example the question, what is the probability that an

[1] Hawkins (1970) is a history of measure and integration theory exploring in depth the nineteenth century mathematical background.

arbitrary $x \in [0, 1]$ is rational? Equivalently, what is the average of the indicator function $I_Q$ of the set of rationals $Q$ over the unit interval?[2] The intuitive answer is that this average should be 0. Poincaré's famous work of 1890 on the three body problem treats a related question. He proves his *recurrence theorem* in the course of which he defines probability as relative volume in the bounded state space of a dynamical system. On this basis he shows that the set of initial conditions not eventually returning arbitrarily close to themselves, has probability 0 (1890, pp. 69–72). Borel's and Lebesgue's novel integration theory provided the mathematics for these intuitions and integrations and more. Its applicability to probability became immediately evident.

Two years before Poincaré's work appeared, the astronomer Hugo Gyldén published two papers in the *Comptes rendus* dealing with a probabilistic distribution problem stemming from planetary theory. These can hardly have gone unnoticed to Poincaré.[3] The problem itself arose from convergence questions in the approximate computation of planetary motions. One central problem of mathematical astronomy concerns the long-term behavior of motions of bodies. The question is, whether there exists an asymptotic mean motion. As we shall see, probability entered such problems through the use of continued fraction expansions in Gyldén's works of 1888.[4] The French papers Gyldén (1888b), and (1888d) are shorter versions of two Swedish papers Gyldén (1888a), and (1888c).

Gyldén was led to ask for the probability of having the value $a_n = k$ in a continued fraction representation of a real number of the interval $[0, 1]$. This is yet another way of representing real numbers. They are not very handy for arithmetic operations, but have other useful properties not possessed by decimal expansions. The continued fraction representation of a real number $x$ is obtained as follows: First take the integer part $[x] = a_0$ of it, whence $x = a_0 + x_1$ for some $x_1 \in [0, 1]$. Next form $[1/x_1] = a_1$, with $1/x_1 = a_1 + x_2$ and $x_2 \in [0, 1]$. Then go on applying the formulas $[1/x_i] = a_i$ and $1/x_i = a_i + x_{i+1}$, where $x_{i+1} \in [0, 1]$. The

[2] Defined by $I_Q(x) = 1$ if $x \in Q$, $= 0$ otherwise.

[3] The three-body paper mentions in its introduction Gyldén in connection with convergence problems of celestial mechanics. The introduction says that the paper had been completed by June 1, 1888. Gyldén's paper was communicated to the French Academy some days later. The longer Swedish version was presented to the Swedish Academy in February 1888. On the basis of these data, it seems impossible to tell who comes first. Buchholz (1904), however, tells that Poincaré's famous essay was printed twice, in 1889 and in revised form the following year.

[4] The first probabilistic problem in continued fractions stems from Gauss. Cf. Kuzmin (1928) for the problem and its solution.

conventional expression for a continued fraction is

$$a_0 + \cfrac{1}{a_1 + \cfrac{1}{a_2 + \cdots}}. \tag{2.1}$$

Assuming now $a_0 = 0$ so that $x \in [0, 1]$, write $x = (a_1, a_2, \ldots)$. The $n$th *convergent* is the rational number $x_n = (a_1, \ldots, a_n, 0, 0, \ldots)$. If $x$ is rational, there is an $a_n$ such that $x = (a_1, \ldots, a_n, 0, 0, \ldots)$. This is equal to the convergent $x_n$, and one writes $x_n = p_n/q_n = (a_1, \ldots, a_n)$. For irrational $x$ the fraction is genuinely infinite. Continued fractions are useful for finding *best rational approximations* for reals. The number $a/b$ is such an approximation if for any other approximation $c/d$ with $0 < d \leqslant b$, $|x - a/b| < |x - c/d|$. These best rational approximations to a real number $x$ are its convergents $p_n/q_n$.[5] Generally speaking, any value is possible for the $a_i$.

*Gyldén's problem*, as I shall call the question of the limiting distribution of integers in a continued fraction, was prompted by a question in the perturbation theory of planetary motions. One asks if there is a *mean motion* of a variable $\omega$ describing planetary motion (motion of the perihelion length, for example). By definition, there is such a motion if $\omega$ can be given as a multiple of time $ct$ plus a bounded function of time $\chi$. Dividing by $t$ one gets $\omega/t = c + \chi/t$. When $t \to \infty$, $\omega/t \to c$, this being the mean motion. In the perturbation calculations of planetary motions, the motion is approximated by a finite combination of uniform circular motions. Any motion behaving stably can be arbitrarily well approximated in that way. The combination can be represented in symbols as a finite sum of the form

$$\sum_{j=1}^{n} r_j e^{i(\lambda_j t + \alpha_j)} \tag{2.2}$$

where $i = \sqrt{-1}$. The linear term in parentheses in the exponent describes a uniform motion, and the transformation makes it circular, with $r_j$ the amplitude. One obtains equivalently, by the law $e^{i\omega} = \cos\omega + i\sin\omega$, a finite trigonometric series. In the limit $n \to \infty$, the series should converge. The terms in such a series contain as factors the successive convergents

[5] See, for example, Hardy and Wright (1960, Sec. 10.15). This book gives a good introduction to continued fractions and related matters. Koksma's 1936 *Diophantinsche Approximationen* offers a survey of the number theoretic connections of the topics under discussion. It has a remarkably extensive and useful bibliography.

of a continued fraction representation of some real number $x \in [0, 1]$.[6] One sees by examples that for some real numbers, one obtains a convergent series, for others a divergent one. The way it goes seems to depend on how large the terms $a_n$ get in the representation of $x$, and how often they do get large. Gyldén (1888a) aims at proving the improbability of the case of divergence. His argumentation is that of an astronomer rather than a mathematician. But anyway, he aims at showing that the probability of a value $k$ for the $n$th number $a_n$ in a continued fraction is 'more or less' inversely proportional to $k$. Therefore 'the probability for the occurrence of divergence in the series by which one usually represents planetary perturbations, is smaller than any given value' (1888a, p. 83). He assumes that the digits of $x$ 'do not follow any law, but can each be taken as completely random' (p. 79). He further assumes a uniform probability distribution over $[0, 1]$. Then he derives approximate values for the probability that $a_n$ exceeds a given number. Next, given two successive convergents $x_{n-1} = p_{n-1}/q_{n-1}$ and $x_n = p_n/q_n$, he considers the ratio $q_{n-1}/q_n$ of their denominators. Note that $x$ is chosen randomly so that the denominators, and consequently also their ratio, are random quantities. There is then, as one now says, an expected value for the ratio. Gyldén obtains the right limiting value $\sqrt{2} - 1$ for the expectation as $n \to \infty$. A comparison of theoretical values with computations from a number of observed planetary motions shows 'complete accordance' (1888a, p. 83). Another paper of the same year, Gyldén (1888c) studies 'the probability of obtaining great numbers in the continued fraction development of irrational decimal expansions.' Here Gyldén's problem acquires a life of its own, independent of its astronomical origins. Gyldén makes an approximate determination of the distribution of the values $a_n = 1$, $a_n = 2, \ldots$ in a continued fraction for randomly chosen irrationals. Gyldén describes a system of urns for the choice (p. 354). To test the agreement between his theoretical values and actual distributions of the $a_n$'s, he takes several kinds of data, including observed approximate planetary mean motions, logarithms chosen blindly from a table, and numbers chosen by a person (p. 354). The agreement is so good that Gyldén judges the assumptions leading to the theoretical result, and the derivations themselves, to be 'at least approximately correct' (p. 355).

As mentioned above, two years after Gyldén, Poincaré used probabilistic arguments in planetary theory in his famous essay on the three-body problem. Gyldén's papers lay otherwise dormant until the turn of the century, when Torsten Brodén, from Lund, Sweden, took them up

[6] This goes back to Lagrange.

'for revision.'[7] His first paper (1900) was followed by one by his Lundian colleague Anders Wiman (1900), who finally gave an exact determination of the limiting distribution law of $a_n$ as $n$ grows, from the assumption of uniform distribution over the unit interval. This determination uses measure theory, though that is not so evident in Wiman (1900). But it can be seen as follows. For $x \in [0, 1], a_1 = k$ holds exactly when $x$ is between $1/(k+1)$ and $1/k$. The probability of $a_1 = k$ is the length of the interval, $1/k - 1/(k+1) = 1/(k(k+1))$. The probability of $a_2 = k$ is the sum over lengths of nonoverlapping intervals corresponding to $a_1 = 1$ and $a_2 = k, a_1 = 2$ and $a_2 = k, a_1 = 3$ and $a_2 = k$, and so on into infinity:

$$\sum_{i=1}^{\infty} \left| \frac{k}{k+i} - \frac{k+1}{ik+i+1} \right|. \tag{2.3}$$

Similarly, the probability of $a_n = k$ is an $(n-1)$-fold infinite sum. The correct law for the probability of having $a_n = k$ as $n \to \infty$ can now be determined (Wiman 1900, p. 835):

$$\frac{1}{\log 2} \log\left( \frac{1 + 1/k}{1 + 1/(k+1)} \right). \tag{2.4}$$

A privately published hefty debate, conducted in German, was created between the two colleagues. Brodén's (1901a) 'Remarks on set theory and probability theory prompted by a paper by Mr. A. Wiman' was followed by Wiman's (1901) 'Remarks on a probability problem suggested by Gyldén.' Wiman's main criticism in his (1900, p. 832) was that Brodén did not keep apart the notions of average value and expected value. Both Brodén and Wiman were mathematicians following continental developments in the theory of functions, the latter having been in contact with Borel, as one can judge from some publications. As is acknowledged in Borel (1905b), the papers referred to were the first ones noticing the applicability of measure theory in probability. For the very first time, such a suggestion seems to appear in Brodén (1900, p. 243).[8] But Wiman (1900) had actually put measure theory into use for obtaining the probability law (2.4). The next year his suggestions were more explicit (1901, p. 18): 'We are quite decidedly of the opinion that if one should want to develop probability theory in the sense of the

[7] However, Gyldén's properly astronomical work of the 1890s was widely known, as is witnessed by Poincaré (1905), Bohl (1909), and so on. For his work in general, see Bohlin (1897). There we read that Gyldén himself valued especially his work on probability in continued fractions (p. 403).

[8] This same paper contains the result that the probability of having in the sequence $a_1, a_2, \ldots$ at least one $a_n$ with value $k$ or greater than $k$ is 1 (p. 256).

modern theory of sets, one should above all make use of the Borelian notion of content.'[9]

### 2.1(b) *Hilbert's sixth problem: the axiomatization of probability*

The very same year 1900 that Brodén and Wiman made their first contributions, another very definitive initiative toward the development of a mathematical probability theory was taken; this time by no one less than David Hilbert. At the international congress of mathematicians in Paris in 1900, Hilbert presented his famous list of unsolved problems. It has played a considerable role ever since.[10] The list begins with the famous continuum problem. A less well-known problem is the sixth: following the example of the *Grundlagen der Geometrie*, 'to treat axiomatically those physical disciplines in which mathematics plays a predominant role.' These are in the first place the calculus of probability and mechanics. Hilbert adds that it would be desirable to have, together with the logical investigation of the axioms of probability theory, a rigorous and satisfying development of the methods of determining averages in physics, especially in the kinetic theory of gases (Hilbert 1900, p. 306). Hilbert refers to lectures Georg Bohlmann had given on the mathematics of insurance earlier, in the spring of 1900 in Göttingen.[11] These lectures appear in Bohlmann (1901), which does not do much more than call some of the basic properties of probability calculus by the name of axioms. After Hilbert's call, a couple of inconclusive early attempts were made at founding probability theory on set theory and measure theory. In the dissertation of Rudolf Laemmel (1904), set theoretical concepts are used, but measure theory appears only rudimentarily. 'To ascertain to probabilities a greater epistemological value, one has to try to replace the intuitive or empirical procedure by a determination of probability through a process of hypotheses.' (Laemmel 1904, p. 359). His hypotheses call for a set of possible cases, the set of those cases falling under some predicate, and the giving of weights to each of the cases. A notion of 'content' is introduced by a fourth hypothesis. A content would usually be given as a length, area and so

[9] I have been unable to find a copy of Brodén (1901a). The final word in the debate was another 11-page private publication of the same year, also by Brodén.

[10] See, for example, the entertaining review *Mathematical Developments Arising from Hilbert Problems.* (Proceedings of Symposia in Pure Mathematics, vol. 28, 1976).

[11] See Schneider (1988, pp. 354–55) for more details on this.

on. It remains an intuitive nineteenth-century notion, and no contact is made with the measure theory of Borel and Lebesgue. But Laemmel's axioms contain the principle of denumerable additivity (p. 365).

An axiomatization of probability based on the most recent theory of measure, Lebesgue's (1904), is given by Ugo Broggi in a dissertation in Göttingen in 1907, in which Hilbert was involved. Broggi wants to put up a system for which to prove, following the model of Hilbert's geometry, consistency, completeness, and the mutual independence of the axioms. Referring to advice of Hilbert's assistant Zermelo, he assumes it 'objectively decided' whether an arbitrary element of a set $M$ has a property $A$ or not.[12] Since it is possible to assume an $x \in M$ given without saying anything on whether it has property $A$ or not, probability can be applied in spite of the fixity of all properties. Its applicability is based on our ignorance (1907, p. 369). Broggi aims at defining probability as the relative measure of those $x$ which have property $A$. His axioms, however, are put generally: Probability is a nonnegative function of events such that the certain event has value 1, and such that it further has the additivity property. Broggi further claims to derive denumerable additivity from finite additivity.[13] He goes on to prove consistency by noting that the definition of probability as relative Lebesgue measure fulfils his axioms (p. 372). His claim that the axiomatization is complete (or categorical, that is, no two systems of different structure can fulfil the axioms) seems to assume that probabilities apply to the unit interval, or its $n$-dimensional generalization. He assumes further that 'equal extensions (length, area, volume,...) are equally probable..., without this, determinations of probability are senseless' (p. 372). The categoricity property becomes the same as the uniqueness of Lebesgue measure, it seems (p. 374). For the case of a denumerable set $M$ of possible events, Broggi calls for an ordering of its elements. He defines probability as the limit, supposing it exists, of relative frequency in ever-growing subsets delivered from $M$ by the ordering of its elements.

Laemmel and Broggi's attempts at Hilbert's sixth problem seem rather groping, even if they start from set theory, and the latter even from measure theory. But there is another aspect to Hilbert's problem: his concern about the foundations of statistical mechanics. He wanted a firm mathematical basis for the determination of average values

[12] We should not be surprised that the father of Zermelo–Fraenkel set theory sees the properties of objects as corresponding to fixed sets.

[13] Steinhaus (1923) refutes this in passing by noting that there are essentially finitely additive measures.

there.[14] These average values are of course determined with the help of a probability distribution for the quantity considered, so the problem became one of justifying probabilistic assumptions. Let us briefly review some of the developments up to Hilbert.

It is very natural in statistical physics to identify probability with relative geometric volume, or its measure theoretic generalization. This identification goes back to the early works of Maxwell and Boltzmann. In Maxwell a differential element $dxdydz$ of the space of positions of molecules enters in the determination of the Maxwellian velocity distribution. His probabilistic assumption in his (1860) amounts to the spherical symmetry of a probability density over the direction of motion of molecules. In a general Hamiltonian formulation of statistical mechanics, the normalized geometric measure, as defined by a differential element of the Hamiltonian phase space, is introduced for the computation of averages. It is singled out as the measure that is preserved in the motion of an isolated system. The crucial *ergodic problem* consists in finding and justifying hypotheses to the effect the geometric measure of a set of states $A$, or phase average, as it is called, would coincide with the limit of relative time the system spends in the set $A$. The latter is also an average, namely, the *time average probability* of $A$, as introduced by Boltzmann around 1870. Much closer to Hilbert, the time-invariant geometric or volume measure over phase space was the essential tool in the proof of Poincaré's recurrence theorem of 1890. Now Hilbert's assistant Zermelo had been working on the mathematical foundations of statistical physics. He presented the famous recurrence objection in 1896, based on Poincaré's theorem, and was led to debates with Boltzmann on the foundations of statistical mechanics.[15] In 1900 he published another paper in which the introduction of a stationary probability over the state space of statistical mechanical systems was defended.

These facts should give enough indication of the direction of Hilbert's concern about the foundations of statistical mechanics at the time of his problem list. It took a number of years before anything mathematically definitive was proposed. In Borel's paper on measure theoretic probability (1905b, p. 990) application to statistical mechanics is suggested, but the first essential use of measure theoretic arguments to overcome

[14] Ch. 22 of Hilbert's 1912 book on integral equations is devoted to the kinetic theory. See also Brush (1972) for commentary. Hilbert lectured on statistical mechanics at least in 1913, but there is nothing foundational about these lectures. They are physically oriented, as one can see from notes taken by Hans Reichenbach, now at the University of Pittsburgh archives. For a lecture of 1919 see Bernays (1922).

[15] To be discussed in detail in Sec. 3.1(b).

problems in statistical mechanics is only due to Plancherel in 1913. I shall return to these developments in the discussion of the role of probability in statistical mechanics in Chapter 3.

Hausdorff's book *Grundzüge der Mengenlehre*, 1914, takes up probability as an example and application of measure theory.[16] If $P$ is a subset of $M$ and if both are measurable, the measure of $P$ divided by that of $M$ can be defined to be the probability that a point of the set $M$ belongs to the set $P$. Further, Hausdorff says that probability 0 is not impossibility in this definition (1914, pp. 416–17). In this connection, he says that his definition is 'on the whole quite arbitrary.' Previously Hausdorff says that 'many theorems on the measure of point sets take on a more familiar appearance when expressed in the language of probability calculus' (p. 416). In Hausdorff it is the measure (normalized, of course) that is defined to be a probability, while later understanding of the relation would be rather the other way around, with probability defined formally as a measure. Why Hausdorff sees his definition as arbitrary is presumably because, proceeding in his direction, the normalized measure need not have anything to do with probability in any proper sense. In conclusion, while Hausdorff does not *identify* mathematical probability with measure, he does point out that a normalized measure has all the formal properties of mathematical probability. His book was for a long time the standard reference for set theory; therefore the connection between probability and measure theory can be considered well established in the mathematical literature by 1914. In Section 2.3 many more early examples of the identification of probability with measure will be found.

It is not difficult to display a whole array of early measure theoretic approaches to probability. The applicability of Lebesgue's powerful integration theory was also obvious to many. Why did success not come at once, instead of twenty or thirty years later? In part, such a question seems a result of hindsight, one connected with Kolmogorov's *Grundbegriffe der Wahrscheinlichkeitsrechnung*. Its mathematical success had such a dramatic effect as to make many people question the reasonableness of anything that preceded it. On the other hand, the mathematical requirements of applied work on probability are not usually on any level that would require proper use of measure theory.[17] What is more, measure theoretic probability proper is needed only for

[16] Hausdorff had published in 1901 a paper on the normal law of errors.

[17] In the preface to Nelson (1987) it is suggested that the measure theoretic strictures on probability never were much more than a way of salvaging our mathematical consciences, that is, they were not important for uses of probability theory.

handling infinite sets and sequences of events. Outside the strictly infinitistic domain, to say that probability is a normalized measure is just to say that probabilities are numbers between 0 and 1 that have to be added in a special way.

## 2.2 BOREL AND THE INTRINSIC PROPERTIES OF REALS

Probability theory occupies only a small part of the mathematical work of Emile Borel.[18] His main contributions are in analysis and set theory. The book on modern analysis, *Leçons sur la théorie des fonctions*, is especially well known.[19] Modern means based on measure theory, and this is one of the works that started it all. We have met already his first paper on probability, Borel (1905b), where he notices the applicability of measure theory to the formulation of probability. But his fame as probability theorist comes from a result now bearing his name, Borel's strong law of large numbers.

Borel wrote several books on probability mathematics, including *Eléments de la théorie des probabilités* (1909), and expositions intended for a broader range of readers, such as the book *Le hasard* (1914). Later he began to edit a series of monographs on probability, collectively called the 'Traité du calcul des probabilités et ses applications.' It included his *Applications à l'arithmétique et à la théorie des fonctions* (1926) and *Mécanique statistique classique* (1925) and ended in 1939 with his *Valeur pratique et philosophie des probabilités*, a book still quite readable today. In 1928 Borel was instrumental in the establishment of the Institut Henri Poincaré, a research institution devoted to probability theory and mathematical physics. Several well-known lecture series were held at the institute in the 1930s, including de Finetti's *La prévision* (1937).[20]

Probability theory and mathematical physics were the titles of the chair of Henri Poincaré. Borel became the successor. He was a successor of Poincaré also in an intellectual sense, already in the early work on complex analysis that paved the way to measure theory, and later in probability theory. The two, and more or less only two, remarkable books on probability of recent origin were Joseph Bertrand's *Calcul des probabilités* (1888) and Poincaré's with the same title, first edition in 1896, second in 1912. Bertrand belongs still to a somewhat older

[18] Some 8 percent of his *Oeuvre*, to be precise.

[19] First edition 1898; second edition, with appendices including *Les probabilités dénombrables*, in 1914.

[20] Cf. *Annales de l'Institut Henri Poincaré*.

tradition. With Poincaré, French probability assumed its somewhat eclectic character. While still admitting the subjective role of probability, as stemming from ignorance of the true course of events, it also proclaimed the objectivity of statistically stable phenomena in nature. This is apparent in Poincaré (1902), one of his popular expositions of scientific and philosophical matters. He discusses a game of chance which, when observed for a long time, leads us to judge 'that the events are distributed in conformity with the calculus of probability, and here we have what I call *objective probability*' (1902, pp. 218–19). Thus the teachings of the calculus of probability are identified with a law of large numbers. Poincaré also offered some of the earliest explanations of statistical regularity outside statistical physics proper, as we shall see in Chapter 5.

Borel followed Poincaré's ideas on the double aspect of probability. The tendency of the French to accept both a subjective and an objective significance to probability is further witnessed by Paul Lévy's *Calcul des probabilités* (1925). Lévy was one of the foremost figures in mathematical probability theory from the 1930s on. After it became acceptable to do purely mathematical and abstract work on probability, his writings paid only little attention to foundational issues.

### *2.2(a) Foundations of mathematics*

Around 1900 the foundations of mathematics were confronted by a whole array of paradoxes. These included Cantor's paradox on the cardinality of the class of cardinal numbers, Russell's paradox on the class of all classes not containing themselves as elements, and Richard's paradox.[21] The last mentioned can be put as follows: Consider all real numbers of the unit interval definable by a finite number of words. These definitions can be organized into a sequence, according to length and within the same length alphabetically. Now change the first decimal of the first real number, say, if it is less than or equal to five, increase it by one, if greater than five, decrease it by one. Then do the same for the second, and so on, and form a decimal of these new numbers. These last two sentences define a real number of the unit interval in 50 words. But this number is different from each and every single real number of the unit interval definable by a finite number of words. The similarity to Cantor's famous diagonal argument for the nondenumerable infinity of the reals is striking.

[21] A good selection of original texts and commentaries can be found in van Heijenoort (1967).

In the constructive tradition of foundational philosophy of mathematics, mathematical objects are taken to exist as products of thought. This is the tradition Borel shared. In an exposition of his scientific work up to 1912, he wrote that 'the mathematical entities we study are real for us only because we think of them' (1912, p. 128). He admitted that this is philosophically an idealist position.[22] Borel's 1903 paper on the arithmetic analysis of the continuum starts with the claim: 'All of mathematics can be deduced from the sole notion of an integer; here we have a fact universally acknowledged today' (1903, p. 1439). The arithmetization of analysis is the achievement of Weierstrass. But, Borel continues, one can require a more strict arithmetic basis such as Kronecker had used in algebra and as Jules Drach had attempted in analysis. The requirement is that definitions in mathematics must be based on 'a limited number of integers' (p. 1440). This principle relates directly to Borel's remarkable solution to Richard's paradox (1908b, pp. 1271–76): He holds firmly that the only mathematical objects we might ever encounter are all definable in a finite number of words. However, we are not *able* to decide in general whether something is a definition of a real number in finite words. For example, we may make the suggested definition depend on the solution of an unsolved mathematical problem. In a suggestive terminology, Borel says there is a distinction between denumerable infinity in the classical sense, and *effective enumerability*.[23] An example of a denumerable and also effectively enumerable sequence is obtained from the 40 characters in the (French) alphabet, periods and so on included: Put all finite strings of these characters in a lexicographical order. But the set of real numbers definable by a finite number of words is *not* effectively enumerable. This is the conclusion one has to take from Richard's paradox. An effectively enumerable set may have subsets not effectively enumerable.

Borel's views have been echoed in the thought of constructivist mathematicians at various times. Making counterexamples from the assumed inexhaustibility of unsolved mathematical problems was a favorite move of Brouwer's.[24] The banishment of the higher Cantorian transfinities, and the resolution of the diagonal argument as showing some infinite sets not to be effectively enumerable, is the position in attempts at basing

[22] At some other times Borel used the terms idealist and realist in exactly the opposite way.

[23] This was three decades before any systematic theories of effective procedures as developed by Gödel, Church, Kleene, Post, Turing and others.

[24] It is not easy to say what kind of relation Borel took to Brouwer's intuitionistic mathematics. On the other hand, Borel's ideas had a direct influence on Brouwer. See Heyting (1934) and Troelstra (1982). There is a previous unpublished note of 1979 by Jervell, 'From choice sequences to the axiom of choice – a historical note.'

mathematics on the notion of effective computability or recursiveness. Borel however did not create any precise theory of what it means for a procedure to be *effective*. His notion of calculability was the intuitive one of following a finite prescription. In his 'Le calcul des integrals définies' (1912b) we find some of the basic concepts of computable analysis, and some of Borel's insights into the mathematics that follows from these notions. A real number is calculable if one knows how to obtain arbitrarily good rational approximations to it (1912b, p. 830). He says that a representation of a real number as a decimal, for example, is not theoretically important. That particular representation is not 'invariant' under arithmetic operations (ibid.). It seems that Borel here hints at the impossibility of obtaining decimal expansions for all calculable reals. It was only later definitely noticed that an argument such as the one in Richard's paradox assumes real numbers to have decimal expansions. The insight that this would not always be the case is Brouwer's (1921). Say, one can calculate arbitrarily long an expansion $a = 0.000\ldots$ without knowing whether some time a nonzero number appears. For the number $0.5 - a$, one cannot determine even the first decimal before $a$ is expanded into infinity. Borel (1912b) goes on to define a real function as calculable if its value is calculable for a calculable argument. Calculability means that one knows how to obtain arbitrarily good approximations to the function value, given good enough approximations of the argument. Therefore calculable functions have to be *continuous* (p. 834). Here we have in fact the basic intuition of Brouwer's famous 1924 'uniform continuity theorem.' The contrary to a calculable function is an *asymptotically definable* function. It is one whose value does not depend on any finite part of the development of the argument (say, in decimals or rational approximations). A prime example is the indicator function of the irrationals. It was Borel's belief in his crucial paper on probability (1909a) that 'the day when these *undefinable* elements are really set apart and when one does not let them intervene in a more or less implicit way, there certainly follows a great simplification in the methods of analysis' (1909a, p. 1079). But if we are to follow the recursive analysts of the latter years, it certainly proves the case that mathematical analysis becomes both complicated and unaesthetic if practiced with recursively definable reals only.[25] The 'calculable' analysis Borel was outlining seems more reminiscent of the constructivism of Errett Bishop. This is an approach where, we could say, a systematic 'prevention' of trying the infinitely difficult is built into the concepts. As a result, to give an example, one is not committed to

[25] See Bridges and Richman (1987, Ch. 3) for illustration of this point.

deciding whether two reals share the same decimals into infinity. It may be the case, for all we know, that the only way to find this out is to compare each individual pair of decimals separately. But this form of constructivity, too, is more complicated than Classical Analysis. One cannot get more rewards for less, is what constructive analysis teaches us, despite Borel's hopes to the contrary. More specifically, one cannot keep books on what is finitary and what infinitary without paying a price in the form of more complicated definitions. The reward is that if something can be shown to exist, it can also be computed.

If the Borelian Continuum is taken to consist of calculable real numbers only, it proves to be too meager for many purposes. A great problem, one that led Borel to consider widening his criteria of mathematical definability, was where to put the sequences describing results from random experiments. They would have their proper place in an arithmetic continuum of the classical kind. But that was not available for Borel. In his paper (1912b) outlining the basic concepts of computable analysis, he says that 'a noncalculable value can only be conceived as defined by chance; the properties of the function are represented by coefficients of probability' (p. 836).[26]

Borel's constructive ideas (1912b) extend also to the introduction of Borel sets and of the Borel measure, discussed in Section 2.1(a). A set is *well defined* if its indicator function is calculable. That is the case for the interior points of a closed interval (p. 838). Its measure is its length. (Obviously, the end points have to be calculable, though this is not mentioned.) The measure of the complement of an interval relative to another is obtained by a simple subtraction. The measure of a denumerable set of disjoint intervals is the sum of the measures of the intervals. In the definition of Lebesgue measure, a set is measurable if it contains, and is contained in, well-defined sets whose measures can be brought arbitrarily close to each other. The consequence Borel aims at is that a measure should become a calculable function (p. 844).

The French 'semi-intuitionists' – Borel, Lebesgue, and Baire – had doubts about the validity (or constructive meaningfulness) of Zermelo's axiom of choice. Borel felt it legitimates 'an arbitrary choice without a law' (1912, p. 133). Such a choice is not, as Borel objected in a short note prompted by Zermelo's paper (1904) on his axiom, a legitimate concept for nondenumerably infinite sets (Borel 1905a, p. 1252). A correspondence concerning the axiom, 'Cinq lettres sur la théorie des ensembles,'

[26] In contrast, Bishop's constructive analysis is a 'nominalistic' theory. There is no place for a thing like 'sequences chosen arbitrarily' in it. This is made more than clear in Bishop (1967, p. 6).

was published by Borel in 1905. Hadamard believed in the legitimacy of arbitrary nondenumerable choice if the choices are 'independent of each other.' Lebesgue, on the other hand, said he does not believe in the notion of an absolutely nondenumerable set. For him, Baire and Borel, existence was not to be inferred from mere consistency, but from a definition, or from *naming* a characteristic property. Baire further limited the naming to refer only to a finite number of previously defined objects.[27] In 1908 Borel held the infinities transcending the denumerable to be 'purely negative notions.' In particular, the classical nondenumerably infinite continuum was such a purely negative concept. Instead, the continuum should be based on a geometric intuition. For a complete arithmetization of the continuum it would suffice to admit a *denumerable infinity* of successive choices. Its legitimacy is debatable, says Borel, but it is still essentially different from a nondenumerable choice which is 'totally devoid of sense' (1908a, p. 1268). With denumerable choice, one would be sure that any single choice is made in a finite time, so that the difference of two systems of choice is borne out finitely. This is a typical position of a constructivist who accepts only potential, not actual infinities. 'The continuum is never given as a whole arithmetically. Every one of its members can be defined. (Or at least there is not one of which we could admit that it could not be defined.)' (p. 1269.)[28] But the set of effectively defined reals is always denumerable.

We have now reviewed Borel's position in the foundations of mathematics at some length. It can be seen that by 1908, his faith in the requirement of finite definability had started faltering. I shall come back to how he came to accept the extension of mathematical existence through the idea of denumerable choice. Appreciation of that conceptual change is quite crucial for a proper understanding of his theory of denumerable probabilities, the topic of Section 2.2(c).

### *2.2(b) Philosophy of probability*

Before going to Borel's philosophy of probability proper, let us first look at his views on certain related philosophical topics. A paper of his on radioactivity and determinism, of the year 1920, identifies determinism with predictability. It should be remembered that there was no general view about ultimate indeterminism in radioactivity before the late

[27] This correspondence (Borel 1905c) is reprinted in Borel's *Oeuvres* pp. 1253–65. An English translation is contained in Moore (1982), a book on the history and philosophy of the axiom of choice.

[28] Borel comes here close to inventing the distinction between a proposition and its double negation.

1920s.[29] Borel's strongly empiricist philosophy is untouched by the possible existence of objective chance. It says instead that there is indeterminism whenever Nature's actions are unpredictable. Borel gives elsewhere at least two cases where we may infer this kind of indeterminism. In 1912 he discusses mechanical systems with a very great degree of freedom, such as $10^{24}$, the magnitude of the number of molecules in a macroscopic system. He concludes that the integrals of motion, besides total energy, are not effectively determinable (1912, pp. 181–82). In that case it is impossible to determine the system's motion. It is unclear whether he thought this impossibility could be explained by some precise concept of effectiveness, or whether he meant a more practical sort of impossibility. The second alternative is followed when, in another paper of 1912, he says that 'the theoretical necessity of definitions with a finite number of words transforms, in the field of physics, into the practical necessity of definitions with a small number of words: only these can be considered the scientific ones' (1912c, p. 2134). He says that with a colossal number of variables, physical properties are usually sufficiently determined by averages and the statistical mean deviations from these (pp. 2134–35), a rather traditional kind of reason for the introduction of statistics and probability.

Borel had found another, very impressive case for indeterminism alias unpredictability in 1906. It is based on the effect of *gravitational instability*, a phenomenon that renders mechanical determinism purely hypothetical.[30] Borel had come to such ideas through the study of Maxwell and the kinetic theory of gases, in combination with Poincaré's definition of chance as instability.[31] In the study of molecular collisions, one tries to determine conditions under which all directions of motion of a molecule become equally probable. Borel notes that the convexity of surfaces tends to disperse the trajectories in collisions (1906a, pp. 1687–88, note). What are the given data in a problem, contains a necessary indetermination, for example, through the variation in attraction of the stars (p. 1685). One can compute the effect of such a very minor disturbance on the trajectory of a molecule. The result is that it *grows exponentially* in collisions. Borel liked working out incredibly small and large numbers on this exponential basis: How small can a perturbing mass be and how far and how much or little should it be moved in order for the trajectory of a given molecule to become completely indeterminate in some very short time such as a fraction of

[29] This matter will be discussed in detail in Sec. 3.5.

[30] This kind of instability was rediscovered in the 1970s.

[31] To be discussed in detail in Ch. 5.

a second? Answers to such questions included bodies of the mass of a fraction of a molecule, at a distance of billions of light-years, moved by a fraction of a millimeter, and so on.[32] These studies of Borel's never denied the theoretical possibility of determinism. They denied its scientific meaningfulness instead. It was still in the style of the times that the idea of *spontaneous* chance in nature did not occur to him at this stage. His philosophy of probability reflects this state of affairs.

As we have seen, the human impossibility of prediction marked the entrance of indeterminism for Borel. Obviously, the door was opened for probability at the same time. The concept of probability bore some traces from its entrance; it related partly to what we are able to do and what not, what we take to be equivalent to practical certainty, and so on. Borel's first paper on probability starts with the observation, bearing a close resemblance to one in Poincaré (1896, pp. 98–99; 1912, pp. 121–22), that the solution of probability problems with continuous quantities must be based on a *convention.* Following Bertrand, Borel says the probability that $x \in [0, 1]$ is between 0 and 1/2 is the same as that $x^2$ is between 0 and 1/4. 'This would be absurd if one supposed that each of these probabilities has an *intrinsic* value, defined objectively independent of any convention' (1905b, p. 985).

Borel saw probability theory as having a scientific and a practical value, both illustrated in two follow-up papers to his first contribution to probability (1905b). The scientific value is displayed in his (1906a) application of probability to physics, statistical mechanics especially, to be dealt with in Section 3.2(a). The practical value is the topic of Borel (1906b). He speaks there in terms of games of chance, but clearly the conclusions bear more general validity. The probability calculus helps one compute the *fair value* of a game by determining the expected value of gain. This gives us a rule of conduct, to wit, one of paying no more than the fair value for the game (p. 992). However, being offered a game with one chance in two for winning 1000 francs, some might still choose a sure gain of 400 instead of the game. Others might be willing to pay 600 for the right to play the game. Borel concludes that it may not always be possible to measure probability by certainty (that is, by determining an equally preferred certain gain; p. 994). Instead one can reduce complicated situations to ones where the probability is given in a simple way. I may choose from two lotteries proposed the one which has greater expected value, or the one which gives the greatest chance of getting a sum exceeding a predetermined bound. 'In either case, the

[32] From Borel (1948, pp. 133–36), a new edition of Borel (1914) that was not at hand. One of Borel's calculations assumes a minute mass to be displaced on Sirius.

calculus of probability dictates my decision' (p. 998). In his 1909 book Borel introduces, somewhat belatedly considering it was proposed already by Daniel Bernoulli and discussed by Bertrand (1888), the notion of utility ('commercial value') to improve the outlooks of the rule of expected utility.

Borel follows Poincaré in distinguishing between subjective and objective probability. The latter accepts the 'objective reality' of stable frequency phenomena as incontestable. Borel makes a very precise distinction: In many cases we have to be content with acting on the basis of probabilities and gains, through the evaluation of expected utility. But sometimes a probability is so close to 1 that it is practically the same as certainty, and we act as if it were a certainty (1906b, p. 1002). The difference between subjective and objective probability is for Borel one of *degree*. Although neither Poincaré nor Borel make it very explicit in this context, their examples indicate the following as a typical situation of practical certainty. There would be some event repeated a great number of times in regular circumstances, with the consequence of a statistically stable behavior of frequencies. In such a situation it is practically certain that the frequencies coincide with the probabilities (pp. 1000–1002). Borel's early concept seems striking, in view of the meaning given today to subjective probability. Subjective probabilities are degrees of belief of someone, and their values can be freely assessed. But in Borel it is not the values of the probabilities that are subjective, contrary to what one might assume. Instead, his example of coin tossing requires perfect physical symmetry of the coin, in order for *equal chances* to prevail for the two results (1906b, p. 992; 1909, pp. 21–22). Then if there is a situation where the probabilities are far away from 1 and 0, probability has a subjective value in the sense that some action has to be taken which it is up to the actor to choose. The topic of 'practical certainty' itself is one of those older ideas on probability, found already in Bernoulli. What is called 'the rule of high probability' says that a sufficiently high probability can be considered 'practically certain,' and a sufficiently low correspondingly 'practically impossible.' What is a probability low enough to be considered an impossibility depends on the scale of things. Borel's liking for big numbers made him define four such scales: the first human one ignores events whose probabilities are less than $1/10^6$, and the last universal one those with $1/10^{1000}$ Borel's notion of humanly impossible events reflects a factual aspect of behavior: the negligence of risks on the order of magnitude of one in a million (1906b, p. 1004). At the other end, an event with a universally negligible probability would be an accident more rare than could be reasonably expected within known cosmic dimensions.

We find a developed notion of subjective probability in Borel's essay (1924) on Keynes' *Treatise on Probability*. This notion bears close resemblance to the theory of subjective probabilities de Finetti developed around 1930. Borel accepts Keynes' basic idea which is that a probability is always relative to a corpus of knowledge. However, he says that in a case like probability of radioactive decay, probability is treated by physicists just like any other physical constant. It is equally objective, even if nothing ever is *absolutely* objective (1924, p. 2173). Borel now accepts genuine subjective probabilities in cases like betting. 'The moment one accepts the subjective character of probability it is not possible to try to make perfect or more precise whatever is imperfect or imprecise in its definition; for a modification in the system of knowledge... has as consequence a modification of this probability.' One can only 'substitute for it *another*, entirely different probability' (p. 2174). He explicitly renounces what is today known as 'temporal coherence,' or the updating of subjective probability through Bayesian conditioning, saying one can change one's subjective probabilities through mere reflection (p. 2174). If probabilities are interpreted subjectively, it can be questioned whether their numerical values can be determined with an arbitrary precision. Borel thinks there is no difference of principle between the numerical precision of subjective probabilities and physical constants. In each case an infinity of decimals remains undetermined. In fact, what Borel proposes is theoretically *better* than physics can offer; for in the latter, there is no uniform method for obtaining arbitrary accuracy, contrary to what is the case for his suggested *method for measuring* subjective probability. He says 'it seems that the method of betting permits, in most cases, a numerical evaluation of probabilities having precisely the same character as the evaluation of prices by a proposed exchange' (p. 2180). When the bid goes up, a sale is reached, and when the price goes down, a buy. A measurement of subjective probability is obtained by a qualitative comparison to standard bets in games of chance. Paul is offered a choice of 100 francs if his prediction of rain tomorrow turns out true, or 100 francs if a dice gives 5 or 6. If he prefers the former, his probability of rain is more than a third. 'The method applies to all verifiable judgments. It permits a numerical evaluation of probabilities with a precision comparable to that in evaluating a price' (p. 2181).[33]

The last section of Borel's essay on Keynes discusses cases where doubt is cast on the uniqueness of subjective probabilities. There are countries where results of elections are the subject of betting, with

[33] Some ideas of this kind can be found in Bertrand's 1888 book, cf. below, p. 271.

published odds. An elector who has placed a bet for a certain candidate may decide to vote for the candidate. (Naturally, the group of electors has to be fairly small for any effect.) Another similar case is when your opponent raises high in poker, making you think he has a good hand. In such cases, 'the fact that a bet is proposed modifies the probability of judgment to which the bet leads' (1924, p. 2182). In these problems, as Borel puts it, 'chance combines with the abilities of players' (p. 2183). Such questions properly belong to the theory of games. Borel published three papers dealing with zero-sum games, the last one (1927) being referred to in von Neumann (1928).[34] Borel suggests erroneously that what became to be known as the von Neumann minimax theorem, stating there is an optimal strategy in such games, would not be generally valid.

### *2.2(c) Borel's denumerable probabilities*

In 1909 Borel published his seminal paper on denumerable probabilities and their arithmetic applications. Denumerable probability was to be a species between the classical case of a finite probability space and that of a continuous one. This paper has been the source of numerous misunderstandings, though there is nothing especially confusing about its mathematics. It is rather the conceptual situation, as also seen by Borel, that is far from clear. A proper evaluation of the paper has to take into account Borel's constructive ideas. Otherwise, the whole point about having a separate species of denumerable probability is lost. It would reduce to measure theoretic probability of the unit interval.[35] On the other hand, if all real numbers (or denumerable sequences) have to be given by some finite prescription, where shall chance find its mathematical representation? That is a very difficult question for anyone who endorses constructivist ideas on mathematics.

Borel's introduction of denumerable probability between the finite and continuous cases is an attempt to create a theory that would be faithful to his view of the continuum as 'a purely negative notion,' as a 'transitory instrument' that is only 'a way to study denumerable sets, these being the only reality accessible to us' (1909a, pp. 1055–56). The theory of denumerable probability requires an extension of mathemati-

[34] Borel's papers are printed in English translation in *Econometrica 6* (1953) together with Fréchet's comments and a note by von Neumann.

[35] This reduction is taken for granted in the historical study of Borel by Barone and Novikoff (1978). Their neglect of Borel's relevant works leads to such strange claims as that Borel thought binary reals denumerable. See, however, Knobloch (1987) which takes Borel's ideas in their right context.

cal existence from finite definability, an extension that Borel had admitted as a possibility by this time. A reader of 'Les probabilités denombrables' is puzzled because the matter is not explicitly explained in it. But according to Borel (1908a, p. 1268) existence would not necessarily be the same as finite definability; a denumerable infinity of *successive and arbitrary choices* would be allowed. That way the continuum would be exhausted. Heyting's well-known summary of the situation in the foundations of mathematics in 1934, while explaining the background of intuitionism, comments as follows on the matter. (He speaks of noncomputable functions where at this point a sequence not admitting finite definition would be in place.) 'Noncomputable functions are not exactly excluded. They are subsumed under the calculus of probability; a noncomputable number cannot be given individually. It can only be thought of as being determined by chance' (1934, p. 6). Thus he has no difficulty in understanding Borel's intentions. But they remained largely incomprehensible to those admitting unlimited existence of mathematical objects in the set theoretic sense.

I shall now proceed to outline the probability mathematics of Borel's paper proper. The first part introduces denumerable probabilities. The second applies them to decimal expansions, the third, following the line starting with Gyldén's problem, to the theory of continued fractions. Some of Borel's moves in the setting up of the general theory become motivated only when he comes to the applications. Specifically, he focuses on the case of a denumerable sequence of events with probabilistic independence, but a different success probability $p_n$ for each event. As it turns out, the $p_n$ reappear later as the probabilities that the relative frequency in $n$ trials is within certain bounds. The use made of the theorems of the first part of the paper is, in the first place, to show the strong law of large numbers for a simple event of probability 1/2, such as coin tossing. Second, they are used to characterize the behavior of the successive denominators in a continued fraction.

Three categories of denumerable probability are distinguished: a finite number of possible results in infinite repetition, an infinity of results in finite repetition, and infinity of results with infinity of repetitions. Taking in the first the simple special case of success and failure, Borel asks for the probability that there is never a success (1909a, problem I, p. 1056). He assumes, without mentioning it, that the repetitions are independent, with $p_n$ the success probability in trial $n$. Then 'the application of the principle of composite probability' gives the sought for probability $a_0 = (1 - p_1)(1 - p_2)\dots$. Under the condition that each $p_n < 1$, if the sum $\Sigma p_i$ is finite, $a_0$ is well defined and has a value strictly between 0 and 1, and if the sum is infinite, $a_0 = 0$. The event of never having a success depends on an infinity. Its probability is determined by Borel's extension of the product rule to denumerable infinity. Borel's problem III (p. 1058) is, what is the probability $a_\infty$ for the success to appear an infinite number of times? If the success probabilities $p_n$ approach 0 as fast as to make their sum converge, the probability of an infinity of successes is 0. If the series diverges, the probability of an infinity of successes is 1 (pp. 1056–59). There are two essential points: how to justify the computation of the probability $a_0$ as an infinite product, and how to interpret the probability of a genuinely infinitary event. Borel tackles these as follows. Assume the series converges. Taking the $n$ first events, classical probability answers to the question, what is the probability that there is no success (problem I). When $n$ grows, the probability approaches a limit so that no conceptual novelty is involved (p. 1057). But consider the case of divergence in problem III. One cannot ask for the probability $a_\infty$ in $n$ trials and let $n$ grow. We encounter here *a genuinely infinitistic event.* Borel suggests instead the following: choosing a fixed number $m$, one asks for the probability of getting more than $m$ successes in $n$ trials. What is the limit of this probability as $n$ grows? The answer is: it is 1, independently of the choice of $m$. Borel tries to give a betting interpretation for this probability. It is advantageous to bet any sum against a mere franc for the event of having (eventually) more than $m$ successes for any fixed $m$ (p. 1059). It does not seem to have bothered Borel that a bookie accepting such a bet would have to have a more intimate relation with infinity than the bettor. At any stage, the latter either wins or goes on playing, whereas the bookie only wins after an infinity of trials.

In treating problems with an infinity of possible results, Borel briefly considers the possibility of having a probability over a denumerable set that is not denumerably additive. One obvious candidate would be a uniform distribution over the natural numbers, which would represent the choice of an arbitrary natural number.[36] Borel says of genuine finite additivity that 'such a hypothesis does not seem logically absurd to me, but I have not encountered circumstances where its introduction would have been advantageous' (p. 1060).

Borel applies his results to decimal expansions and continued fractions. There one asks for the asymptotic distribution of integers in denumerable sequences. Givc a real number of the unit interval, its *decimal expansion to the base $q$* is the sum of the form $\Sigma(a_i/q^i)$, where

[36]Each number has probability 0 of being chosen. Consequently the denumerable sum is also 0. But the probability of some number being chosen is 1. See Sec. 8.5 for more on this.

$0 \leqslant a_i < q$. With $q = 10$ we have decimal numbers proper, with $q = 2$ binary expansions. In both cases the common notation for the decimal number is $0 \cdot a_1 a_2 a_3 \cdots$. Borel asks for the probability that a decimal, that is, one of the $a_i$, belongs to a given set. He makes two hypotheses of a probabilistic character: 1. the decimals are independent. 2. Their probabilities are $1/q$. He says these hypotheses are inexact since a decimal expansion always has to be given by a law.

Borel says his hypotheses are 'easily justified if one takes not the logical point of view, but the geometric one' (p. 1066). A decimal expansion is a point of the unit interval, and the probability that it belongs to a segment is the same as the length of the segment. Borel says, with his (1905b) paper on measure theoretic probability in mind, that 'one can interpret and verify from this geometric point of view the results we are going to obtain' (p. 1066). But this was a point of view Borel would not take because of his constructive concept of the continuum.

A modern reader equates at once the space of denumerable sequences with the unit interval. Denumerable probability becomes continuous probability on $[0, 1]$. How should one interpret a continuous probability? Usually an idea is transported from the discrete case: A sufficiently large sample will approximate the continuous distribution. Naturally, the sample is discrete, even finite, so that it has to be made precise in what sense it approximates the continuous distribution. The very idea of sampling from a continuous population (here $[0, 1]$) is, however, purely metaphorical. For each sample point is a real number, and even Borel himself very explicitly denied that it makes any sense to think that choosing even *one* number $x \in [0, 1]$ is *ever* completed, not to speak of an infinite sequence of such choices.[37] The classical mathematician can easily accommodate sequences due to chance in his continuum, as well as infinite sequences of real numbers, each chosen by chance, and so on. On the other hand, if only finitely definable mathematical objects are accepted, a very difficult problem is met. For there does not seem to be any way of representing sequences due to chance mathematically. Borel on the contrary says that (individual) decimal expansions have to be given by a mathematical law; this according to him would be contrary to the assumption that the digits follow a probability law.

The simplest case of binary expansions results from setting $q = 2$. By classical arguments, distribution of success frequency approaches a normal limiting distribution. It gives for the probability that the number of successes in $2n$ trials is between $n - \lambda\sqrt{n}$ and $n + \lambda\sqrt{n}$, the limiting

[37] See, Borel (1905a, p. 1252) or (1908a, p. 1268).

expression

$$\Theta(\lambda) = \frac{2}{\sqrt{\pi}} \int_0^\lambda e^{-\lambda^2} d\lambda. \tag{2.5}$$

Set $\lambda = \lambda_n = \log n$ and call a success the event that the success count in $2n$ trials is within the corresponding limits $n - \log n\sqrt{n}$ and $n + \log n\sqrt{n}$. The probability of success $p_n = \Theta(\lambda_n)$ comes from the above integral. The probability of failure $q_n$ is $1 - p_n$, and the series $\Sigma q_i$ converges. Therefore, by the solution to problem III, *the probability of an infinite number of failures is* 0. There is a gap in Borel's argument, for the success probability $p_n$ is not simply given by the normal approximation $\Theta(\lambda_n)$ of (2.5), and hence the convergence of $\Sigma(1 - p_i)$ is not guaranteed by the convergence of the series formed from the $\Theta(\lambda_n)$.[38] But the gap can be filled.

One version of the weak law of large numbers says: for any $\varepsilon$ and $\delta$ there is a number of trials $n$ such that the probability of deviating from the limit frequency 1/2 by more than $\varepsilon$ is less than $\delta$. Borel's strong law says more: Let us divide the above limits by the number of trials $2n$, so that the deviations from the limit frequency 1/2 are

$$\pm \frac{\log n}{2\sqrt{n}}. \tag{2.6}$$

These two numbers approach zero as $n$ grows. Borel's result says that there is a probability 1 that from some value $n$ on, these limits *always* apply. A weak law does not state any convergence, just a probable proximity, whereas the strong law says relative frequency converges in the ordinary sense to a limit if a set of cases of probability 0 is excluded. But for the weak law, the limits $\varepsilon$ and $\delta$ effectively determine a number of trials $n$. In the strong law, there is only a proof of existence of a number.

Borel calls a decimal number *normal* if each digit has the limit frequency 1/10. His theorem implies, if we put $q = 10$, that any number is with probability 1 normal. In Borel such a property concerns the set of all numbers that are given as decimals, as one sees from what he says about rationals (1909a, p. 1070).[39] These latter correspond to periodic decimal expansions. Is it possible to define the probability that the

[38] This remark was made repeatedly in the subsequent literature. Borel did not care or remember, as his booklet (1926), a sort of extended version of the article under discussion, shows.

[39] As was said, Borel seems to have at least doubted that there would always exist a decimal expansion of a real number.

period is of length $k$, Borel asks. The problem is of second category, with an infinity of possible results. A uniform probability would have to violate denumerable additivity: Each of the possible lengths $k$ would have to have zero probability, so that the infinite sum of these probabilities would also be zero. But it is certain, hence has probability 1, that one of the possible lengths is obtained. Further, the more restricted assumption that (at least) the period of length 1 has zero probability, is unacceptable for Borel. No answer is suggested; Borel only says that 'whatever the procedure adopted for defining an arbitrary periodic fraction, it is certain that the probability for having a one-digit period cannot be regarded zero; for such is surely produced a certain number of times in a finite number of experiments.'

Borel calls a number *absolutely normal* if not only the single digits, but also any finite sequence of digits appears with the right limit frequency. The limit is $1/10^n$ for a sequence of length $n$. He considers the possibility that such numbers could not be constructed. There might not exist a definition in a finite number of words for any one of these. Nevertheless, the probability that a number is absolutely normal, is 1 (1909a, p. 1069).

The third section of Borel's paper treats continued fractions. The way probability enters the theory of continued fractions reveals some of his conceptual uncertainties. He begins with a formal notation for probability: For $x \in [0, 1]$, $p_{i,k}$ denotes the probability that in the continued fraction expansion of $x$ one has $a_i = k$. There is no mention of any probabilistic assumption about $x$, it is only stated that the notation designates the probability 'in a general way' (p. 1072). The sum of these probabilities over the different values of $k$ equals 1. 'A priori, one can make arbitrary hypotheses on the $p_{i,k}$' (ibid.). While discussing decimal expansions Borel said that his probabilistic hypotheses are inexact. He thought decimal expansions must be given by an arithmetic law, so that the strict application of probabilistic laws would be excluded. His two probabilistic hypotheses were that the decimals appear independently and with equal probabilities. Then he mentioned that these hypotheses can be easily verified from the geometric (that is, measure theoretic) point of view. In connection with continued fractions, he directly takes this latter way of determining probabilities. It was the same application of measure theory to probability that Wiman had made in 1900. Then $a_1 = k$ is an event which appears precisely when $x$ is between $1/k$ and $1/(k+1)$. 'The geometric probability that $x$ appears in this interval is equal to its length; so that if we adopt this geometric probability, we have $p_{1,k} = 1/k(k+1)$' (pp. 1072–73). Next let $P_{n,k} = p_{n,1} + \cdots + p_{n,k}$. This sum gives the probability that $a_n \leqslant k$, so $1 - P_{n,k}$ gives the probability of the converse inequality. Measure theory is needed when one sums over the first index $n$, as we saw in connection with Wiman. The essential result following from Borel's choice of Lebesgue measure of the unit interval as probability is this: Let first $\varphi$ be a nondecreasing function of $n$ such that the series

$$\sum_{n=1}^{\infty} \frac{1}{\varphi(n)} \tag{2.7}$$

converges.[40] It follows that the series

$$\sum_{n=1}^{\infty} (1 - P_{n,\varphi(n)}) \tag{2.8}$$

also converges. From the solution to problem III, the probability of having infinitely often $a_n > \varphi(n)$ is *zero*. Or, one has with probability 1 after a finite number $n$ always $a_n \leqslant \varphi(n)$. Borel thought the way the function $\varphi$ relates to this probability, that is, $\varphi$'s having to grow just that bit faster than $n$ as to make the series (2.7) converge, was the most interesting result of his paper (p. 1077).

In his concluding section IV, Borel briefly treats further interesting cases of denumerable probability, such as the probability that a given Diophantine equation has rational roots.[41] He introduces a notation for the probability that a coefficient $a$ has the value $m$, and so on, and he starts deducing relations among such probabilities. To answer probabilistic questions in Diophantine equations, one has to make precise hypotheses about these probabilities. He mentions the possibility of a choice similar to the uniform geometric probability he had used with continued fractions, but stresses that 'above all, it would be interesting to study hypotheses naturally imposed by the study of concrete problems' (p. 1079).

Borel calls for a comparison of the theory of denumerable probability with the results obtained by the theory of continuous or geometric probability, once the former has been developed. Cantor's continuum contains undefinable points; this is what the diagonal argument shows in his opinion. These elements should be put aside, which according to Borel should lead to 'a great simplification in the methods of Analysis' (1909a, p. 1079). Borel's thoughts seem to be changing. It is clear that in 1909 he had an idea of 'sequences due to hazard' in mind, as produced by a denumerable random choice, even though his several suggestions

[40] Remember that $\Sigma 1/n = \infty$.

[41] A Diophantine equation says that a polynomial with integer coefficients has value zero.

and remarks seem rather contradictory. On the one hand, he would remove undefinable points out from the continuum. On the other hand, genuine random sequences would not be among the finitely definable ones. Very little earlier, in his Rome address (1908a), the possibility of denumerable choice is contemplated. In the paper on the paradoxes of set theory, Borel (1908b, p. 1275) says one cannot indicate a way of fixing on the line a unique and well-determined point which does not belong to the set of effectively definable points: 'The proposition saying *there are* such points is true or false according to whether one admits or does not admit the possibility of a denumerable infinity of arbitrary choices; but here we have a *metaphysical* question, in the sense that a positive or negative answer would never have an influence on the development of Science' (p. 1275). In 1909 Borel had thought it possible that the notion of absolute normality could be incompatible with the finite definability (constructibility) of a sequence. This specific possibility we refuted by Sierpinski's (1917) construction of an absolutely normal number (See also Lebesgue 1917). Had it rather gone the way Borel surmised, a strictly mathematical condition for the definition of the *set* of random sequences would have been found, even if *single* random sequences would have remained inconstructible.

Let us now recapitulate some of the essentials of Borel's procedure. He starts by considering probabilities *hypothetically*, not worrying that their introduction might be counter to our intuitions: The objects considered supposedly follow arithmetic laws, alien to the notions of probability and chance. A formal notation is put down for the probabilities, and their relations are studied. There is the possibility of a geometric point of view which requires the idea of a random choice of a real number. Borel stresses in his discussion of measure theoretic probability (1905b) that continuous probabilities must necessarily be based on a *convention*, there is no *intrinsic*, *uniquely right* way of assigning probabilities. Above we saw a good example: We may adopt the convention that gives the probability numbers $p_{i,k}$ for continued fractions from Lebesgue measure of the unit interval. But that uniform measure is only one possibility. In his conclusions to the paper on denumerable probability Borel says explicitly that a probability question concerns properties of numbers of some given class *chosen arbitrarily* from that class. It must be kept in mind that the completed choice of a real number is inapplicable. Therefore the above measure theoretic determination of probabilities is not really acceptable for Borel, even though he does use it. The theory of denumerable probability tries to resolve the problem of what it means to choose a real number at random. With the help of probabilistic hypotheses, the class of mathematical objects we can consider is extended. The new objects, arbitrarily chosen infinite sequences, differ from the finitely definable ones in that we can only consider them as a totality, not individually.

In his paper (1912d) Borel imagines one thousand people each choosing a digit at random, this being repeated and repeated. Still one cannot *define* an unlimited decimal number by any analogous process. 'I don't think anyone has ever dreamed of asking an *infinity* of persons to each write a digit at random'. Contrasting his views to those of Zermelo and Hadamard, he says the definition of a number in this way 'uses an infinity of words.'

I for my part consider it possible to pose *probability problems* on *those* decimal numbers which would be obtained by choosing digits either completely at random, or imposing some restrictions, restrictions leaving part of the choice made at random. But it is impossible to talk of *one* of these numbers. (1912d, p. 2130.).[42]

Borel's attempt at a mathematical probability theory more restricted in its assumptions than the measure theoretic one, received little understanding. Perhaps it is fair to say that he had been himself ambiguous about denumerable probabilities. He does start with probabilistic hypotheses defined over finite sequences of events. But later he uses the measure over the unit interval. A second weak point is the justification of his infinitistic product rule, used in deriving the strong law of large numbers. Later studies tried to dispense with denumerable probabilities in one way or another. In Steinhaus (1923) the objective was to build an axiomatic theory of denumerable probability in terms of measure theory that would 'allow once and for all to pass from one interpretation to another in researches of this type.' These two alternatives are the Borelien denumerable probabilities and the measure theoretic ones (p. 286). On p. 307 Steinhaus gives the geometric representation of an *arbitrary* probability of a simple event. Let $p$ be the probability of success and $q = 1 - p$ that of failure. One divides the unit interval into $[0, q)$ and $[q, 1]$. Letting $x$ be an arbitrary point of $[0, 1]$, if $x$ falls in the first interval, we have failure, if in the second, success. Next the two intervals are again divided in the same proportion as before and one asks to what part $x$ falls. A sequence of simple events is created, with a law of large numbers stating that the limiting relative frequency of successes is $p$ for almost all sequences. Here $p$ is the probability of the event of having the point in the interval $[q, 1]$. It is obtained from a

[42] Borel's notion of a denumerable sequence of arbitrary choices was the direct ancestor of Brouwer's notion of choice sequence. See note 24 above.

uniform (Lebesgue) probability on [0, 1] representing 'arbitrary choice' of a point. Ending his paper, Steinhaus claims that in Borel (1909a) and Bernstein (1912a) 'probability' is only an expression for 'measure' (Borel or Lebesgue measure) or else does not appear at all.[43] That was to become the accepted view of denumerable probabilities, even though it fails to represent correctly Borel's approach. For him, the right way of studying 'sequences due to hazard' was through probabilistic hypotheses, instead of the 'geometric point of view,' that is, measure theory. Further misunderstandings are revealed by Paul Lévy's commentary on Borel's paper.[44] Lévy tells us that a denumerable sequence is represented by a point of the unit interval. The probability of any property of a sequence is 'nothing else but the measure of the set of points $t$ corresponding to the property; this remark, very simple but so important, has been made independently by many authors, notably P. Lévy [1], B. Jessen [1], [2], H. Steinhaus [1], [2], and A. Denjoy [1].' (Lévy 1940, p. 222).[45] Lévy's remark is, almost *verbatim*, the 'geometric point of view' Borel considers in his (1909a). Lévy goes on saying that the notion of limiting frequency that Borel introduced could bring alive the attempts at founding probability on an 'empirically' introduced frequency. 'I have often underlined the inherent contradiction in these tentatives which admit, as it were, as an axiom that which they should explain; this is also the view supported by M. Borel.' (1940, p. 222). Lévy refers here implicitly to von Mises' frequentist theory of probability. Borel can hardly have been against von Mises' ideas. As late as 1937 Borel says (p. 1159) his negative view on the possibility of 'imitating' chance 'does not diminish the interest of recent researches on the notion of collectives (von Mises) or admissible numbers (Copeland).' By 1940 the success of Kolmogorov's measure theoretic probability had already made people see differently what went previously, a familiar phenomenon ever since. Later commentators of Borel do not fare much better. Barone and Novikoff (1978), for example, state as the aim of their paper, to gather 'crucial evidence against asserting that Borel realized that "denumerable probability" *was* measure theory in [0, 1]' (p. 132).

A little hindsight will be helpful in settling the matter of denumerable versus measure theoretic probability – a kind of hindsight that would

[43] Interesting remarks of a historical character can be found in Steinhaus (1938).

[44] Published in a 1940 *Festschrift* for Borel. I quote from Borel's *Oeuvres*, vol. 1, where it is reprinted.

[45] The paper of Steinhaus is his (1923), Lévy's [1] is his (1929). Comparing Steinhaus (1938, p. 65, note 14) with Wiener (1924, p. 569, notes 1 and 3), one sees that the discovery was not independent.

have been in the reach of Lévy and others who since 1933 knew Kolmogorov's *Grundbegriffe*. There you find the following: For the sake of simplicity, assume integer-valued random variables $x_1, x_2, \ldots$. Then a system of *finite dimensional distributions* $p_i$ is defined by the equations

$$\Sigma_i p_1(x_1 = i) = 1$$

$$\Sigma_i p_{n+1}(x_1, \ldots, x_n, x_{n+1} = i) = p_n(x_1, \ldots, x_n). \tag{2.9}$$

The first assures normalization, the second consistency in the sense that the lower dimensional probabilities are marginal distributions of the higher ones. What Kolmogorov in his 1933 book called his *Hauptsatz*, also called the *extension theorem*, says that a consistent system of finite dimensional distributions has a unique extension into a measure over the infinite dimensional space of events. The simplest case is when $x_i = 0$ or $x_i = 1$, with $p_1(x_i = 0) = p_1(x_i = 1) = 0.5$ for all $i$, and with independent repetitions. The extension gives the Lebesgue measure on [0, 1]. In Borel (1909a), we have seen more complicated ways of assigning the finite dimensional distributions. While for Borel these made sense as probabilities, referring to finitely verifiable events as they did, he did not find this to be the case in general for the measure over the limit of infinite sequences of events.[46] No mathematical trick can overcome his philosophical refusal of the actually infinite here.

Before entering into the further developments relating to Borel's strong law of large numbers of 1909, let us have a look at his book *Eléments de la théorie des probabilités* from that same year. It proves to be what the title promises. There are three parts, devoted to discrete probability, continuous or geometric probability, and the probability of causes, the latter comprising statistical inference. The discrete part uses the example of coin tossing throughout and concentrates on working out many numerical examples. Probability is there based on the idea of equipossibility. Borel says that though this kind of approach is not comprehensive, there are many examples which show how it works. And it is sufficient for a mathematical theory that there exist cases to which it is applicable (p. 22). Only the weak law of large numbers is treated, while denumerable probability does not appear at all in this elementary textbook.[47]

[46] Kolmogorov in turn thought the infinitary measure his extension theorem gives would also be conceptually a harmless extension of the finitary case; see Sec. 7.4(a).

[47] A much later edition of 1950, English translation in Borel (1965), adds some remarks on denumerable probability.

Borel's strong law of large numbers seems to have caught mathematicians by surprise. It was thought to give a strange property of real numbers and it therefore aroused the curiosity of analysts: The measure of binary decimals with a limiting frequency of 1's different from 1/2, is zero.[48] According to Steinhaus (1923, p. 286), this property of real numbers was known as 'Borel's paradox.' For Borel himself there was no unclarity. As we just noted, he saw the 'probability question' receiving an answer through a convention: that which makes precise the sense in which a number is drawn arbitrarily from a set. Thus his theorem did not state any *intrinsic property* of the reals. Instead, it stated a consequence of his two hypotheses: having equal probabilities half and half for 0 and 1, with independence of consecutive digits.

From 1910 on Borel's theorem appears in a variety of contexts. The convergence with probability 1 is obtained through different measure theoretic arguments, while the limits for convergence are improved in a series of investigations. Faber (1910) constructs a continuous function $f$ for which he shows the following. The set of points $x$ at which $f$ does not have a derivative has Lebesgue measure 0. Further, with $n(1)$ the number of 1's and $n(0)$ the number of 0's in the $n$ first binary digits of $x$, if $\liminf(n(1)/n(0)) < 1 - \varepsilon$ or if $\limsup(n(1)/n(0)) > 1 + \varepsilon$, there is no derivative.[49] It follows that the set of $x$ for which $\lim(n(1)/n(0)) = 1$ has measure 1. The proof of this statement may be taken as the first correct one of Borel's law. A problem Faber poses shows that he was not yet quite as sure as Borel about the probabilistic assumptions underlying the law. He asks, 'is the probability that... a number belongs to a definite given set of measure 0, always equal to 0? And conversely: is a set always of measure 0 if the probability that a point belongs to it is 0? (p. 400). That is to say, he posed the question whether probability and Lebesgue measure have to be mutually continuous with respect to each other.

The next person to discuss Borel's strong law of large numbers was Felix Bernstein (1912a). This work uses measure theoretic probability without even bothering to mention that it is a matter of some novelty. Bernstein's paper connects the line of research on distribution problems in continued fractions, that is, form Gyldén to Borel, to continuous

[48] As in Hausdorff (1914, p. 420), to be discussed shortly.

[49] The meaning of $\limsup(n(1)/n(0)) > 1 + \varepsilon$ is as follows: Take for each natural number $k$ the supremum (least upper bound) of the sequence of numbers $n(1)/n(0)$, where $n = k$, $n = k + 1, \ldots$. Take the infimum (greatest lower bound) of these suprema as $k \to \infty$. This number, also written $\inf\sup(n(1)/n(0))$ is $> 1 + \varepsilon$. $\liminf(n(1)/n(0))$ is $\sup\inf(n(1)/n(0))$. A probabilistic reading will be given soon.

distribution problems, as first studied by Bohl (1909). I shall therefore discuss it in Section 2.4(a).

Hausdorff's book on set theory (1914) also gives without further ado a measure theoretic reading to Borel's strong law. This takes place when he wants to show as an application of measure theory how it handles probabilistic questions, the example being the one we met above: It is shown that the set of binary expansions of the reals of the unit interval for which the limit of relative frequency of 1's is 1/2, has measure 1. By his interpretation of measure as probability, Hausdorff gets that 'there is probability 1 for having, in the dyadic expansion of $x$, asymptotically as many 0's as 1's (pp. 419–20). Hausdorff, contrary to Borel, sees here an intrinsic property of real numbers:

This theorem is strange. On the one hand, it looks like a plausible extension of the 'law of large numbers' into infinity; on the other hand, however, the existence of a limit for a sequence of numbers, and of a prescribed limit in addition, is a very special case one should consider a priori extremely improbable.

Let us consider Borel's theorem in the following from: Putting as above $n(1)$ for the number of 1's in the binary expansion up to $n$, $n(1)/n \to 1/2$ as $n \to \infty$, except for a set of measure 0. After proving the theorem of Borel, Hausdorff goes on to study the asymptotic behavior of the oscillation of frequency. This oscillation is measured by the number $|n(1) - n/2|$, the deviation of the number of occurrences of 1 from the average $n/2$. Borel had obtained the limits $\pm \log n\sqrt{n}$. Hausdorff mentions that one can continue and prove that, in effect, $n^{1/2+\varepsilon}$ is an upper bound, with probability 1 (p. 421). In other words, there is probability 1 that the frequency always stays within $n^{1/2+\varepsilon}$ of $n/2$ after some finite number $m$. Hausdorff concludes by discussing the application to continued fractions, displaying nicely the connection of the growth of Borel's function $\varphi$ to the probability problem.

The improvement of the bounds of oscillation of frequency was continued by Hardy and Littlewood (1914) in connection with their generalization of what is known as Kronecker's approximation theorem. The very simplest special case of Kronecker's result says that the decimal parts of a sequence $x, 2x, 3x, \ldots$ form a dense set in the unit interval in case $x$ is irrational. Kronecker's theorem in its general form says the following[50]: Let $x_1$ to $x_k$ be linearly independent irrational numbers, that is, ones for which there are no integers $a_1, \ldots, a_k$ such that $a_1x_1 + \cdots + a_kx_k = 0$. Let $0 \leqslant \alpha_1, \ldots, \alpha_k < 1$. There is an increasing sequence of natural numbers $n_i$ such that, with the brackets denoting

[50] I shall turn to this matter in Sec. 2.4(a).

the decimal parts, $(n_i x_1) \to \alpha_1, \ldots, (n_i x_k) \to \alpha_k$ as $i$ grows. Hardy and Littlewood prove in this number theoretic setting the following. Taking the binary expansion of $x$, with $n(1)$ the sum of 1's in the first $n$ digits,

$$\frac{|n(1) - n/2|}{\sqrt{n \log n}} \to 1 \quad \text{as } n \to \infty \tag{2.10}$$

except for a set of $x$ of measure 0 (1914, pp. 185, 215). The value $\sqrt{n \log n}$ is an upper bound for the deviation or oscillation of frequency $|n(1) - n/2|$ in the lim sup sense. They also show that the bound $\sqrt{n}$ is too strict, that is, that lim inf $|n(1) - n/2| > \sqrt{n}$ (p. 187). Working in number theory, they do not use the probability reading of measure 0. Instead they say a property holding except for a set of measure 0 is 'almost always true' (pp. 189–90). In this terminology they state their results as: 1. 'It is almost always true that the deviation from the average in the first $n$ places is not of the order exceeding $\sqrt{n \log n}$.' 2. 'It is almost always true that the deviation in both directions is sometimes of order exceeding $\sqrt{n}$' (p. 191).

The upper bound of Hardy and Littlewood was improved by Khintchine (1923) to $\sqrt{n \log \log n}$ in what is known as the law of iterated logarithm. This result was formulated in number theoretic language. One year later, this time with an explicitly probabilistic formulation, Khintchine (1924) showed that this bound cannot be improved. He considered a simple event with arbitrary success probability $p$, and showed that there is a function $\chi(n)$ such that for any $\varepsilon$ and $\delta$, there is an $n_0$ for which, with a probability $> 1 - \delta$ we have, for *all* $n > n_0$, the inequality

$$\left| \frac{n(1) - n/2}{\chi(n)} \right| < 1 + \varepsilon \tag{2.11}$$

and with the same probability we have, for *some* $n > n_0$, the inequality

$$\left| \frac{n(1) - n/2}{\chi(n)} \right| > 1 - \varepsilon. \tag{2.12}$$

The solution gives, with $q = 1 - p$, for the function $\chi(n)$ the asymptotic expression $\sqrt{2pqn \log \log n}$. A whole series of generalizations followed, by Khintchine and Kolmogorov (1925).

A general probabilistic formulation of a strong law of large numbers for real random variables was obtained by Francesco Paolo Cantelli in

1916–1917. This was a line of development completely different from Borel's. Cantelli (1917) considered the limiting behavior of relative frequency for a simple event with any success probability $p$. He says some authors claim that the relative frequency behaves as if it were converging to the limit $p$ in the ordinary mathematical sense (pp. 39–40). Earlier (1916b, p. 192) he had already stated that such a claim leads to contradictions. The argument leading to the Cantelli law in (1917) starts from noting that for an unlimited sequence of events $i_1, i_2, \ldots$ the probability of the first $n+1$ events (that is, successes) cannot be greater than that of the first $n$ events. Therefore the unlimited sequence of probabilities of ever-growing sequences is bounded from above and must have a limit $l$, which limit equals the probability of a simultaneous occurrence of all the events in the unlimited sequence. Cantelli concludes that $l \geqslant 1 - \Sigma p_e$, where $p_e$ is the failure probability for the $i$th event. Next, in order to use a result from (1916a), he generalizes by the introduction of random variables. The average of $n$ independent real random variables $X_1, \ldots, X_n$ is denoted by $X_{(n)}$.[51] Assuming that $E(X_n)$ converges with $n$ to some value $M$, Cantelli constructs a sequence of intervals for the random variables $M - X_{(n)}, M - X_{(n+1)}, \ldots$. The intervals are so constructed that the *measure* of these intervals goes to 0 when $n$ grows. Then he denotes with $l_{(n)}$ the probability that the random variables $M - X_{(n)}$ belong from $n$ on to the intervals. This probability has the limit 1, which comes from the sum of the failure probabilities converging to 0 with $n$, with the above inequality for $l$. Therefore the probability that the random variables $X_{(1)}, X_{(2)}, \ldots$ converge to the limit $M$ in the ordinary sense is as close to unity as desired (p. 44). Thus were established the results now carrying the name of Borel–Cantelli lemmas. The most general results in this direction are due to Kolmogorov from the late 1920s. Cantelli's earlier paper (1916a) shows acquaintance with the work of Chebyshev, Lyapounov, as well as Markov's 1908 generalization of the law of large numbers to dependent events but he never mentions Borel in these years.[52]

At the end of his paper Cantelli (1917) turns back to the special case of simple events (that is, two-valued random variables) with arbitrary success probability $p$. As mentioned, this generalization of Borel's strong law was obtained several years later also by Steinhaus (1923).[53]

[51] Cantelli's argument needs the assumption that these have finite fourth moments.

[52] See also Regazzini (1987) for details and information on Cantelli.

[53] See the end of Sec. 2.2(c). In Steinhaus the strong law of large numbers is connected with the name of Bernoulli. Referring to that paper, he tells us later (1938) to read this as Cantelli.

## 2.4 CONTINUOUS DISTRIBUTION PROBLEMS. WEYL'S VIEW OF CAUSALITY

### *2.4(a) The equidistribution theorem of reals mod(1)*

If you grant yourself the ability to choose an *arbitrary* real number $x$, then multiply it by 2, 3, 4,... and take only the decimal parts, you get, with probability 1 a dense set that is *uniformly distributed* in the unit interval. Mathematics, a science organized economically, calls this 'the equidistribution of reals mod(1).' Hidden behind that catchphrase is a complex story of simultaneous discovery; apparently, three separate problems led to the same result. These were concerned with the mean motion of the planets, the thermal conductivity of metal rings, and, least surprisingly, number theory. What was distilled out of these studies by the astronomer P. Bohl, the young universalist Hermann Weyl, and the set theorist Waclav Sierpinski, respectively, namely, distribution results about sequences of real numbers, is analogous to the distribution problems of integers we have met in previous sections. That there is a connection between the existence of mean motion and probabilistic problems in continued fractions was realized already by Gyldén. Bohl (1909) does not bring this forth, even though he refers to other works of Gyldén. Instead, Bernstein's (1912a) 'application of set theory to a problem stemming from the theory of secular perturbations' was the work that connected the researches of Gyldén, Brodén, Wiman and Borel on the distribution of integers in continued fractions with those of Bohl, Sierpinski and Weyl on the distribution of reals.

Bernstein uses a measure theoretic approach to probability throughout. Borel's law for continued fractions follows from a quite neat theorem he proves. The notation $[x]$ denotes the greatest integer less than or equal to $x$. Bernstein's theorem says that if $n_i$ is an increasing sequence of natural numbers, the set of those reals $x \in [0, 1]$ for which $\lim(n_i x - [n_i x]) = 0$, has measure 0. That is, even choosing a subsequence of multipliers, the decimal parts of the sequence of real numbers $n_i x$ cannot be made to converge to zero except for a set of Lebesgue measure 0 (1912a, p. 421). As a consequence of this theorem, Bernstein shows that the set of $x \in [0, 1]$ whose continued fraction developments $(a_1, a_2, \ldots)$ are from some $n$ on bounded from below ($a_n \geqslant k > 1$) or from above ($a_n \leqslant k$) has measure 0 (p. 428). Given a value $k$, there is a probability 1 that in an arbitrary continued fraction, one eventually has $a_n = k$. In particular, $a_n = 1$ infinitely often with probability 1.

Nothing of the above connection was known to Bohl when in 1909 he studied the *problem of mean motion* in planetary astronomy. One starts from the assumption that the planetary orbits are ellipses. For the Sun and one planet, say the Earth, six parameters are enough to fix the motion along a particular ellipse. The first five of these are: the longitude of the ascending node, inclination of the orbit, the longitude of perihelion, eccentricity and the greater half-axis.[54] The gravitational disturbances caused by other planets tend to perturb the motion from its Keplerian ideal orbit. The main theoretical question to study in perturbation theory is, whether the motion remains within finite bounds, or whether planets could eventually escape into space or hit the Sun. Obviously, the latter two possibilities could not occur if the planets possessed truly constant average periods, distances, and so on. The existence of these averages is the problem of mean motion. In general terms, it is known as the problem of the *stability of the solar system*, one with a tradition going a long way back in time. Some of the greatest names in the history of celestial mechanics each believed to have succeeded in the proof. In the 1880s King Oscar II of Sweden offered a prize for the one who solves the problem of stability. Poincaré's long paper on the three-body problem showed instead that no proof can be found with the known methods. That gave him the prize.[55] Probabilistic concepts appeared in this context for the first time in Gyldén in 1888 and in Poincaré, as we have seen in Section 2.1(a).

In preparation of the following discussion, let us repeat the mathematical description needed for the problem of mean motion. First we have uniform rectilinear motions as given by the equations.

$$x_j = \lambda_j t + \alpha_j, \qquad j = 1, 2, 3, \ldots. \tag{2.13}$$

By taking the Fourier transforms $e^{2\pi i x_j}$ of these, with $i = \sqrt{-1}$, the motions are made into uniform circulations which are summed together into a combined motion:

$$z = \sum_j a_j e^{2\pi i x_j}. \tag{2.14}$$

The parameters in (2.13) and (2.14) are: $a_j$ are the amplitudes of the individual rotations, $\lambda_j$ their frequencies, and $\alpha_j$ the initial phases. If the frequencies are linearly dependent, the motion is periodic; in the contrary case it is aperiodic. Mean motion can be shown to exist under special assumptions. Lagrange showed that if one of the amplitudes $a_j$

[54] If one adds the angular direction of the planet at some time, the motion becomes completely determined through all time.

[55] See Abraham (1967) for these matters. Buchholz (1904) reports curious details on Poincaré's essay and the debates surrounding the problem of the stability of motion.

exceeds the sum of all the others, there is mean motion. If we think of the terms of the sum (2.14) above as vectors, putting the longest in the origin, Lagrange's assumption means that it kind of overrides the others.

Studying a system of three planets, Bohl was able to prove the existence of mean motion. That meant that there is a bounded function of time $\chi$ so that variables like perihelion length could be given in the form $ct + \chi(t)$. Let $\lambda_1, \lambda_2$, and $\lambda_3$ be the frequencies or circulation times of the planets. Consider the number $(\lambda_2 - \lambda_1)/(\lambda_3 - \lambda_1)$. Bohl (1909) showed that if it is commensurable (rational), there will exist a mean motion. But if it is incommensurable, this would not always be the case. Instead, the appropriate quantities could exceed any bounds. Both kinds of motions are dense, and Bohl concluded (p. 198) that it is impossible to make inferences to the existence of mean motion on an experimental basis. For within any experimental accuracy, both possibilities are contained. Bernstein examined in his (1912a) the question from a new point of view: the connection between measure theory and the observable world. He proposes (p. 419) a general principle which, he says, 'seems to belong to the foundations of theoretical physics and astronomy':

**Axiom.** When one relates the values of an experimentally measured quantity to the scale of all the reals, one can exclude from the latter in advance any set of measure 0. One should expect only such consequences of the observed events which are maintained when the observed value is represented by another one within the interval of observation.

It follows that 'only quantities with finite extension have real meaning.' Bernstein calls his principle the axiom of the limited arithmetizability of observations (p. 420). When considering parameters depending continuously on the directly observable quantities, any set of measure 0 can be ignored.[56] This approach opens a new possibility for answering the problem of mean motion: The values for which there is no such motion could very well form a set of measure 0 and be therefore physically meaningless according to Bernstein's axiom.

In Section 2.3 we met Kronecker's approximation theorem: If $x$ is irrational, the sequence of decimal parts of $x, 2x, 3x, \ldots$ is dense in $[0, 1]$.[57] The first one to make the step from the density of a quantity to its distribution was Bohl in 1909. He imagined a ray, with points on it following each other at distance $r$, to be 'rolled around a circle of

[56] Comparing with Borel's constructivist ideas, it is no wonder that Bernstein's axiom appealed to him; cf. Borel (1912a, p. 1090).

[57] The density, attributed to Kronecker (1884) in a general formulation, was known in connection with continued fractions at least by 1842, judging from Dirichlet (1842, p. 635).

circumference 1' (1909, p. 224). A part of length $s$ of the circumference is taken. With $n$ points, let $\varphi(n)$ be the number of points lying in that part. Then if $r$ is irrational,

$$\lim \frac{\varphi(n)}{n} = s. \tag{2.15}$$

Such is the first proved version of the *equidistribution* for irrational $x$ of the decimal parts of $nx$, $n = 1, 2, 3, \ldots$. The equidistribution theorem was discovered almost simultaneously also by Sierpinski (1910) and Weyl (1910). It can be formulated as saying that the dense fractional multiples of a number $x \in [0, 1]$ are *uniformly distributed* except for a set of measure 0. This result is today associated with Weyl's remarkable paper (1916). 'Über die Gleichverteilung von Zahlen mod. Eins.' Out of the three discoverers of the equidistribution theorem, he was able to formulate the result in mathematical terms that permitted generalization. He also saw its connections to mathematical astronomy, statistical mechanics, and probability theory. A detailed historical discussion of Weyl's paper, of the contributions of Bohl and Sierpinski, and of some of the lines of mathematical research initiated by Weyl (1916) can be found in Hlawka and Binder (1986).[58] Sierpinski's note, reprinted in Hlawka and Binder, is in number theoretic terms. When delivering his Gibbs lecture in 1950, Weyl recalled how he arrived at his version of the equidistribution theorem in his 1910 paper, 'Die Gibbssche Erscheinung und verwandte Konvergenzphänomene.' He was considering the behavior of a metal ring having two parts of different thermal conductivities $\alpha$ and $\beta$. With $\alpha + \beta = 1$ and a great temperature difference between the parts, he wanted to determine how the difference equalizes in time (1910, p. 403). To handle the case of $\alpha$ irrational, he used a number theoretic result which implies the equidistribution (p. 407). He heard from Bernstein about the connection to the problem of mean motion and to Bohl's equidistribution theorem (1950, p. 117).

Even though the equidistribution theorem had such diverse historical origins, it soon became pure mathematics. That meant generalization; one line was suggested by Kronecker's theorem. Its special case, referred to above, says that the decimal parts of the sequence $x, 2x, 3x, \ldots$ form a dense set in the unit interval in case $x$ is irrational. Since by irrationality, $x \neq m/n$, $nx - m \neq 0$ for any integers $m, n$. The proof of density can be reduced to showing that there are an infinity of $m, n$ such that $nx - m < \varepsilon$

[58] A whole conference, edited by Koksma and Kuipers (1962), was held on the influence of Weyl's paper.

for any positive $\varepsilon$.[59] This is to say, the fractional part at times becomes arbitrarily small. Bohl (1909) had given a simple geometric way of looking at the situation: Since considering reals mod(1) is the same as 'rolling' the real line around a circle of unit circumference, the subsequent points are separated by a constant angle on that circle. For rational $x$, the angle is commensurable; for irrational $x$ it is incommensurable.[60] The general form of Kronecker's theorem applies to an arbitrary number of dimensions $k$: Let $x_1, \ldots, x_k$ be linearly independent irrationals, that is, ones for which are no integers $a_1, \ldots, a_k$ different from 0 such that $a_1 x_1 + \cdots + a_k x_k = 0$. It follows that the points $(nx_1, \ldots, nx_k)$, where $n = 1, 2, 3, \ldots$ and only the decimal part is taken, are dense in the $k$-dimenstional unit cube. Weyl (1914a) generalizes the equidistribution theorem of 1909–1910 to this case. He formulates the one-dimensional case in a way reminiscent of Bohl: To consider reals mod(1) is the same as 'to wrap the real line around a circle of unit circumference or, analytically speaking, to adjoin the complex number $e^{2\pi i x} = e(x)$ to the real $x$' (Weyl 1914a, p. 488). These complex numbers are, through the relation $e(x) = \cos 2\pi x + i \sin 2\pi x$, radius vectors of the circle of unit circumference. Equidistribution of a denumerable sequence of reals mod(1) is defined as follows: Let $A$ be an arc of the circle. Let $n_A/n$ be the relative number of members falling in arc $A$ for the $n$ first members of the sequence. The sequence is equidistributed if the limit of this relative number is the same as the length of arc $A$, for any arc. To prove the equidistribution property, Weyl invented the following very ingenious condition, subsequently going under the name of *Weyl's criterion*: A necessary and sufficient criterion for the equidistribution mod(1) of a sequence of reals $x_1, x_2, x_3, \ldots$ is that

$$\lim_{n \to \infty} \frac{1}{n} \sum_{i=1}^{n} e(mx_i) = 0 \tag{2.16}$$

whenever $m \neq 0$. Each member of the sum is a plane vector of length $1/2\pi$. The sum of these vectors, divided by $n$, has to approach the origin for equidistribution and vice versa. The criterion obviously generalizes to several dimensions; say, for sequences of pairs $(x_i, y_i)$ in two dimensions, one sums radius vectors of a ball. Then the corresponding equidistribution theorem follows under the conditions of Kronecker's density theorem, as above (Weyl 1914a, pp. 491–92, 1916, p. 318).

[59] See the Supplement for more details and background concerning these matters.

[60] This is the way Oresme handles the problem. As explained in the Supplement, he arrives at considering the iterated application of an angle by taking two successive conjunctions of two planets.

Weyl moves easily from his criterion to a much more general level. Instead of considering a denumerable sequence of points in a $k$-dimensional unit cube, he considers the *linear continuous motion* of a point in a $k$-dimensional manifold. This kind of motion had been studied by König and Szücs (1913). They prove, on the basis of Kronecker's approximation theorem as Weyl notes (1916, p. 320), that the linear motion of a point inside a cube, with geometric collisions at the walls, produces a dense trajectory. It is immediate to ask for the distribution of the moving point. Weyl shows that if a point moves linearly with constant speed in a $k$-dimensional manifold, its *relative sojourn time* in a given part of space is the same as the (relative) volume of that part of space. The condition for this result is that the linear equations of motion for the $k$ components $x_i$ between collisions, $x_i = a_i + b_i t$, have linearly independent directions $b_i$. Relative sojourn time is defined, with $t_A$ the time the point is in part $A$ of the space in time period $t$, as the limit $\lim(t_A/t)$ as time goes to infinity. Next Weyl notes that the relative volume of $A$ can be considered the *a priori probability* for finding a randomly chosen point in $A$. The result above says that (Weyl 1916, p. 320)

$$\text{relative sojourn time} = \text{a priori probability.} \qquad (*)$$

With effortless superiority, Weyl has proved the first *ergodic theorem.*[61] This took place in the middle of a mathematical paper and is an aspect of his work that perhaps has not received the attention it should deserve. Weyl himself wrote a more popular, clear account (1914b) of his mathematical results and their bearing on the foundations of statistical mechanics and the perturbation theory of planetary motions. The equidistribution theorem is given in several formulations belonging to rather different fields of inquiry. The first one, stating the equidistribution of reals mod(1), belongs properly to analytic number theory. With a formulation for the case of linear motion, such as billiards in the two-dimensional case, Weyl expresses the theorem as the principle (*) above. This leads him straightaway to the foundations of statistical mechanics. He notes that the example of billiards motion is the simplest one which satisfies what is known as the *ergodic hypothesis* (1914b, p. 459). The hypothesis was (supposedly) formulated by Boltzmann some forty years earlier, in an attempt at giving a foundation for the calculation of average values in statistical mechanics. Finding a proof

[61] Weyl (1914a, p. 490) refers to Rosenthal's paper on the ergodic hypothesis of the year 1913. For the latter paper cf. p. 105.

for it was the central problem in the foundations of statistical mechanics. Weyl's was the first system for which a proof of the hypothesis was found.[62] Last in his expository paper, Weyl discusses the problem of mean motion. Having generalized the equidistribution theorem to several dimensions, he was able to go beyond the result of Bohl. As was noted, planetary motion can be approximated by a combination of uniform circular motions, with each such proper motion a term in the Fourier sum. Weyl makes the amusing remark that the planetary motions are related via a Ptolemian mechanism of epicycles in the approximation, rather than the astronomy of Kepler and Newton.[63] Fulfilment of the principle (*) above means that the representative point $z$ of Equation (2.14) visits the portions of the torus where it moves with the right average values. In 1938–1939 Weyl returned to the problem of mean motion. The existence of mean motion was defined as follows: Functions $\Phi$ of time have to approach a mean value in the sense of $(\Phi(t_2) - \Phi(t_1))/(t_2 - t_1)$ approaching a limit as $t_2 - t_1 \to \infty$ (1938, p. 891). This condition easily implies the one used by Gyldén and Bohl (Section 2.1(a)), in which one uses the sum of a multiple of time and a bounded function of time.

The foregoing discussion of Weyl's work drew connections into many directions. That fact lies in the nature of the matter, as is witnessed by Koksma's (1962) evaluation of the fields of research Weyl (1916) opened. I shall have occasion to return to it also in the sequel. Weyl's results also relate to what has been presented in Section 2.3 on laws of large numbers. The systematic reason for this is that the laws of large numbers can be seen as special discrete versions of ergodic theorems. Such theorems typically exclude a set of measure 0 as not obeying the general result. Above, we saw Hausdorff use a probabilistic language for the exceptional cases. Hardy and Littlewood said measure 1 results are 'almost always true,' while Felix Bernstein was ready to dismiss in advance any specific set of measure 0 as physically meaningless. One difficulty here is that the exceptional sets are usually left unspecified. For sure, in the case of Kronecker's theorem and the equidistribution theorem, one has a good idea of the set of exceptions: It is the set of rationals. But for more complicated equidistribution results and for ergodic theorems in general, no such simple characterization exists. In Weyl (1916, p. 345) the sceptical remark is made that 'one should not

[62] Note that König and Szücs proved an ergodic property in the sense of dense motion, but did not achieve an ergodic theorem since they did not consider the distribution. Ergodic theory will be the topic of Sec. 3.2.

[63] Cf. the Supplement for epicycles and related matters.

evaluate highly the value of theorems in which an unspecified set of exceptions of measure 0 appears.'

### *2.4(b) 'The rigid pressure of nature's causality'*

The shift from classical to modern science, in physics especially, was in one essential point a shift from a causal and deterministic to a statistical and probabilistic science. I shall later review some of the physical aspects of this development. Let us now see how it was reflected in the mind of Hermann Weyl. In an essay of the year 1920 he discusses the relation of causal to statistical approach in physics. As few people can be assumed having been following the *Schweizerische Medizinische Wochenschrift*, and as Weyl did not propagate his ideas elsewhere, at least not before his 1926 book, they must have gone mostly unnoticed at the time. But the paper is interesting as a rather early appeal for indeterminism, and certainly important for an appraisal of Weyl's philosophical thinking itself. Weyl was a follower of the philosophy of Edmund Husserl, and in the foundations of mathematics the most noted follower of the intuitionist Brouwer.[64] Here is a summary of Weyl's 1920 presentation of the phenomenological philosophy. The causal law gains content only through a decomposition of the unique world process into ever-recurring simple elements. This decomposition works through the taking apart of the spatiotemporal reality into bodies and events. An intuitively experienced process is seen as coming about through a spatiotemporal coming together of simple phenomena. In consciousness, phenomena are comprehended in the form of this-is-so, through their characteristic features. This kind of comprehension marks the origin of concepts and of the classification according to what is essential. On a scientific level, one is not satisfied with those elements that can be intuitively lifted in the decomposition process; instead signs of a hidden something are accepted, which leads to hypothetical elements (atoms, forces). Between the elements thus achieved, there are lawlike connections in time succession. In physics, one is led to the space-time coordinates and their unfolding. The causal law obtains the specific form: The succession of states follows a differential law (Weyl 1920, pp. 114–16).

But in addition to the causal approach, says Weyl, the statistical approach also plays a significant role since Clausius and Maxwell. Physical laws dealing with the atomistically structured matter have only a statistical meaning, they express statistical regularities. Is statistics 'a

[64] Cf. Weyl's Erkenntnis und Besinnung (1954), a sensitive look back at his life's work and convictions.

short-cut to some consequences of the causal laws, or does it show that there is no strict causal connection in the world, so that "chance" is to be granted as an independent potency limiting the validity of law?' (1920, p. 117). According to Weyl, physicists throughout are of the first opinion. Even if he does not share this opinion, he wants to show by an example how it is possible '*to derive statistical regularities from causal laws*' (p. 117). The example is that of a mass point moving linearly in a unit square (that is, billiards). The statistical behavior, with a continuous time, is contained in the time average $t_G/t$ for the various regions $G$. The limits of these averages should coincide with the measures of the regions. One says that the *probability* of finding the point at an arbitrary moment in a region $G$ is the measure of the region $G$. Hence probability is identified with the concept of time average (p. 118). Weyl argues that a system of $N$ points obeys independence of the component points. This allows him to apply the law of large numbers to the explanation of the equilibrium distribution of a closed gas system.

In discussing the relation of causality to statistics, Weyl sees as the main difficulty the time reversibility of causal processes as contrasted to the apparent time irreversibility of statistical, especially molecular, processes. The way out of the dilemma – and here Weyl says he sets himself against the dominant view of the day – is to grant statistics an independent role alongside that of 'law' (1920, p. 121).

Weyl's constructive ideas on the continuum of real numbers have a bearing on his philosophical thinking concerning the relation of causality to statistics. Following Brouwer, he sees the continuum as being in an infinite process of *becoming*, rather than as 'rigidly existing.' It follows that 'the quantitative relations in a part of the world $S$ intuitively given to me are not approximative simply as a consequence of the limited accuracy of my sense organs and measuring instruments, but *they are in themselves plagued by such a vagueness*' (1920, p. 121). The exact laws of nature give a strict lawlike connection of the present state of the world and the future ones. But from the vagueness of the world it follows that as the consequences of the present state keep unfolding in the future, the process of becoming of the present also continues forth (p. 122). Only 'in the end of all times' would the infinite process of becoming be completed. 'But in reality the future will again bring about the present over and over and make it ever more and more precise; the past is not closed and completed. Thereby the rigid pressure of nature's causality gives way...' (ibid.). Room is given for causally mutually absolutely independent 'decisions,' located according to Weyl in the elementary quanta of matter. These 'decisions' (put in quotes by Weyl) constitute what is the 'properly real' in the world. Such is Hermann Weyl's account of the basic structure of the world in 1920. Weyl had his own deep reasons for abandoning determinism, deriving from the foundations of mathematics. He felt he was putting himself against the mainstream of physical thought of the times. But it was more his reasons and arguments, rather than the indeterminist position itself, that were unique. Indeed, we shall see in the next chapter that already by 1920 probabilistic and indeterministic ideas had been penetrating physics for quite some time.

# 3

# *Probability in statistical physics*

## 3.1 CONCEPTS OF PROBABILITY IN CLASSICAL STATISTICAL PHYSICS

### *3.1(a) The definition of probability*

Around the middle of the last century, the *mechanical theory of heat* had won ground over caloric and other competing theories. The idea that heat consists of the motion of molecules was not new; it can be found already in Daniel Bernoulli in 1738.[1] But in the 1840s the principle of conservation of energy was finally established. It became the foundation of the science of thermodynamics and was referred to as its first law. The second law of thermodynamics is the famous *entropy principle*, a name given by Rudolph Clausius. In its strict form, it professes the necessary equalization of all temperature differences: The gloomy finale of the world stage is a heat death of the universe. Such a thermodynamical prediction loses its edge in statistical physics, which says that temperature equalization is not necessary but only extremely probable. In order to illustrate the statistical character of the second law, Maxwell imagined in the late 1860s a small demon, monitoring molecular motions and thereby being able to work against temperature equalization. These thermodynamic and probabilistic issues within physics also had a more general cultural significance. The prospect of heat death connected with a pessimistic sentiment seeing degeneration as the essential direction of things.[2] Its counterpoint was the emergence of Darwin's evolutionary theory with which some creators of the kinetic theory, Boltzmann especially, felt close sympathies.

The idea of heat as molecular motion, obeying mechanical laws, is the basis of the kinetic theory of gases. Kinetic theory, in contrast to statistical mechanics, proposes a molecular force law or a specific model

[1] See Brush (1976, 1983). Already Newton had proposed an atomistic theory. In it, however, gas pressure was due to a static repulsion, not the free motion of molecules.

[2] Brush (1978) offers good reading on these cultural implications of nineteenth-century science.

of molecular collision or interaction. The first use of probability here is by August Krönig (1856). His gas model consisted of a rectangular box containing perfectly elastic balls.[3] These are supposed to move with equal velocities, without collisions with each other, and parallel to the sides of the box. Krönig says that in comparison to an atom, the walls of the container must be considered very rough. Consequently the trajectories of the atoms are very irregular, so as to escape any computation. 'According to the laws of the calculus of probability, one can however assume a complete regularity instead of this complete irregularity' (p. 371). An argument, albeit not a strong one, was found for the use of average values.

The next contribution to kinetic theory is Clausius' 1857 paper 'On the kind of motion we call heat.' The billiard ball type of motion, in fact motion according to the laws of geometric optics, does not have to hold for the individual molecules. It holds on the average. Similarly, an average velocity can be ascribed to all the molecules, and no difference in the results of the theory is brought forth by such replacement (1857, p. 371). The law of the ideal gas of thermodynamics states that pressure times volume $PV$ equals $nRT$, where $n$ measures the quantity of gas (in gram-moles), $R$ is the gas constant and $T$ the temperature. In the kinetic theory, instead, one has for pressure times volume $PV$ the expression $1/3\ Nmv^2$, with $N$ the number of molecules of mass $m$ and velocity $v$, as in Krönig, or *average* velocity $v$ as in Clausius. It is clear that this first approach of Krönig and Clausius can only bring very crude results. For it completely eliminates the effect of variation in molecular motions. One cannot, for example, even formulate the idea of temperature equalization through molecular collisions. Clausius' next paper, of the year 1858, introduces the concept of average free path between molecular collisions. Its shortness explains why different gases take time to mix. After some criticisms, Clausius emphasized the statistical character of his conclusions, saying 'simplicity and regularity enter only when one considers... the average state of very many molecules.'[4] Clausius says he supports the view that in addition to translational motion, molecules have other motions. The shares the different motions take from the total kinetic energy, obey constant

[3] Contrary to what the word might suggest, perfect elasticity does not mean that the energy would be absorbed in a collision. On the contrary, it says no kinetic energy is lost, in this case in collisions with the walls of the container.

[4] Clausius (1858, p. 256). Schneider, in his excellent (1988) introduction to the original papers on kinetic theory, goes as far as interpreting Clausius saying that the laws of gas theory only hold with a high probability.

proportions. This constancy does not hold for the individual molecules, but only 'for the different motions in the gas as a whole' (1858, p. 256). There is no unambiguous statement that this constancy for the whole system would experience slight variation in time. Even if it is clear to us that no sharp boundary exists between few and many molecules, it took a long time before observable variation in a real gas was taken as a possibility. The first serious considerations about such variation or fluctuation are only found in the 1890s.

The calculus of probability had been used in error theory, in an astronomical connection, from the beginning of the nineteenth century. Probability played an auxiliary role there: It was thought that there exists a true if unknown value of the quantity of interest, say, the position of a 'fixed' star. Variation in observed results was fully accounted for through the assumption of errors external to the quantity itself. John Herschel published in 1850 a review of Adolphe Quetelet's work in which he considers true variation instead of error. Quetelet himself was an astronomer who advocated a 'social physics,' a statistics of humankind aiming at the definition of the 'average man.'[5] Herschel's publication of interest here contained a derivation of a two-dimensional normal distribution. He considered an iron ball that is repeatedly dropped on the floor. Under the assumption that the probability of a deviation from an absolutely vertical fall is a function of the distance only, not direction, so that deviations are symmetric in direction for the two orthogonal $x$–$y$ coordinates, and that repetitions are independent, he arrived at a two-dimensional normal distribution.[6] An aspect interesting for our purposes, also found in Quetelet, is the *reality* of the variation that is being observed in each repetition. There is no single 'true value' from which the measurements would 'deviate' because of external errors.

The argument of Herschel was taken up by James Clerk Maxwell in 1860. He substituted molecules of a gas for the iron ball, deriving the Maxwellian distribution law for the velocities of the molecules. He used the concept of mean free path introduced by Clausius (1858). Later he gave a new proof (1867) of the law, and introduced at about the same time his famous little demon. By the mid-1870s, he seems to have moved toward an indeterminist way of thinking.

Maxwell's derivation of his distribution law for molecular velocities in 1860 has been hailed as the first time a physical process itself is

[5] See several articles in Krüger et al. (1987).

[6] See Garber et al. (1986, pp. 9–10) for a reprint of the essential passage, and for references to the discussion on the role that Herschel's review had.

described by a statistical function (Garber et al., 1986, p. 8). Maxwell (1860) aims at founding investigations into the atomistic constitution of gases on 'strictly mechanical principles.' These mechanical principles govern the collision of perfectly elastic spheres. His proposition II (1860, p. 379) is 'to find the probability of the direction of the velocity after impact lying between given limits.'[7] An equiprobability assumption is made: Two spheres collide if the center of one is within the circle having as origin the center of the other sphere, and as radius the sum of the two radii of the spheres. 'Within this sphere every position is equally probable,' says Maxwell. Then he concludes to the spherical symmetry of his probability law: 'All directions of rebound are equally probable' (p. 379). His proposition IV determines what the distribution is 'after a great number of collisions among a great number of equal particles' (p. 380). A function $f$ has to be determined such that the number of particles in a spatial element of volume $dx\,dy\,dz$ is $Nf(x)f(y)f(z)\,dx\,dy\,dz$, where $N$ is the number of all particles. In forming a product of the $f$'s and $N$, Maxwell assumes here the components of motion $x$, $y$, $z$ and the particles to be independent. The latter cannot be strictly true of a closed gas system: If one particle steals a lot of kinetic energy, the rest have less to share. That effect, however, is negligible if there are many particles. Maxwell's argument is interesting as an example of physical reasoning leading to a definite probability law. It is the kind of argument that assumes physical (here: dynamical) independence and tries to infer uncorrelatedness. Maxwell goes on stating that since 'the directions of the coordinates are perfectly arbitrary,' $f$ must be a function of distance only so that it is symmetric in the sense of $f(x)=f(-x)$.[8] His conditions for $f$ lead to an exponential distribution law of molecular velocities, very similarly to Herschel's argument. In proposition VI Maxwell concludes that with two kinds of particles, the mean kinetic energies are the same (p. 383). In proposition X he derives the distribution law for Clausius' mean free path (p. 386). With its help, a connection between the micro- and macrophysical properties of gases can be established. One way for the experimental determination of molecular distances and dimensions, actually executed by Loschmidt in 1865, becomes possible.

In the 1870s, Maxwell seems to leave the doctrine of mechanical determinism, thus giving a more prominent place to chance and prob-

[7] Maxwell uses the word proposition here in the old Euclidean sense of a problem. All references to Maxwell are to his *Scientific Papers*.

[8] Since the distribution is a function of the absolute distances and the number of molecules, it does not really qualify as a statistical description of a physical process, contrary to Garber et al., for a process is described when *time* appears as an argument.

ability. I shall discuss these matters in connection with the dilemma of irreversibility below. In Maxwell's last paper (1879) he uses two concepts that are central to the foundations of statistical physics: ergodicity and ensembles. They are both inventions of Boltzmann, though he is not always credited for them, especially not the ensembles. Part of the reason may be that the word ensemble came to be used as an English expression only in the work of Gibbs. Boltzmann's term was 'Inbegriff von Systeme.' The concept of ergodicity will be the topic of Section 3.2. What is today called an ensemble appears in Maxwell as 'a large number of systems similar to each other in all respects except in the initial circumstances of the motion, which are supposed to vary from system to system, the total energy being the same in all' (1879, p. 715). One studies the time behavior of the distribution of the systems in the ensemble. Since Gibbs, Boltzmann's and Maxwell's 'large number of systems' have been known as the microcanonical ensemble. It is a continuous set of systems, with the characteristic property that each system is of the same total energy, and a probability distribution over the mechanical states. We could say that in an ensemble, each possible state is represented with a fixed density. Ensembles are a conceptual convenience, for in a real system, there is only a finite number of molecules so that at no single instant of time can a continuous distribution law have immediate physical reality. Ensembles therefore pose a problem of interpretation. Maxwell's talk of a large number of systems, instead of a continuous one, was a characteristic way of expression of a physicist who tends to see the infinite as an approximation to the sufficiently large finite. (A mathematician might think exactly the other way around).

The ensembles of Maxwell can be used in an epistemic interpretation of probabilities in statistical physics. It goes something like this: We really have only one system in front of us, and its exact mechanical state is only incompletely known. An ensemble is formed from all the systems compatible with that incomplete knowledge. Probability is an *a priori* measure of the degree of ignorance of the true physical state of the system.[9] The immediate objection to this approach is: Why would it work, as ensembles are not physically real objects? Why would the molecules care about the physicist's degree of ignorance? And how should one choose the ensemble distribution in case of different degrees of ignorance?

An interpretation of probability with a more palpable sense of physical reality is Boltzmann's. He compares Maxwell's and his own

[9] I shall have more to say on these matters in connection with Gibbs' statistical mechanics.

interpretation of the concept of probability as follows:[10] 'There is a difference in the conceptions of Maxwell and Boltzmann in that the latter characterizes the probability of a state by the average time under which the system is in this state, whereas the former assumes an infinity of equal systems with all possible initial states'. Boltzmann advocates a *time average* interpretation of probability, as also Maxwell notes (1879, p. 721). This interpretation is crucial for ergodic theory, to be discussed in Section 3.2, and it is also the notion of probability Einstein ended up with. Another of Boltzmann's inventions was the method of ensembles, which has led to claims that he was not always following the time average interpretation. As we shall see, Boltzmann thought the notion of an ensemble of systems is itself a 'Kunstgriff' in need of interpretation. The method of ensembles was later made the basis of statistical physics through the work of Gibbs. He properly gave to Boltzmann the credit for the very idea of representative ensembles, in the preface of his *Statistical Mechanics* of 1902. But he also says that one can avoid reference to ensembles. In that case the interpretation of probability would be based on its invariance under the time development of a single system (Gibbs 1902, p. 17).[11]

In his first paper on statistical physics, Boltzmann (1866) tries to give a mechanical proof of the second law of thermodynamics. He considers two kinds of molecules, with masses and velocities $m$ and $v$, and $M$ and $V$, respectively. What is the condition that they do not take, on the average, kinetic energy from each other? (1866, p. 10). If such is the case, after some time the sum of kinetic energies $mv^2 + MV^2$ obtains its previous value. Boltzmann uses a kind of periodicity assumption as a way for getting determinate average values. He then argues that the same values would be obtained in sufficiently long nonperiodic motions. In this particular situation he says that the integral (average) has to be taken between such time limits as 'give the average kinetic energy independently of any accidental feature.' His probabilistic assumption is the equidistribution of directions of motion (p. 13). Temperature is a function of average kinetic energy (p. 14). There we already see, in 1866, the origin of the time average concept of probability, which stems straight from the idea of temperature as defined by the average of kinetic

[10] See Boltzmann (1881, p. 582). The article is a short communication of Maxwell's work in the *Wiener Anzeiger*, which explains the third person used. All page references to Boltzmann's articles are to his *Wissenschaftliche Abhandlungen*.

[11] This kind of suggestion is made in Zermelo (1900), though without further hint of what it means. But it could be taken as invariance of time average probability in time. And in fact, von Neumann (1929, p. 30) presents Gibbs as an advocate of the Boltzmannian time average interpretation.

energy over time. In the 'analytical proof' of the second law, the above strategy is followed. First a closed molecular trajectory is assumed, from time $t_1$ to $t_2$ (1866, p. 23). Boltzmann says that the molecule repeats its motion in a time period of the same length, if not exactly similarly, in a way which gives the same average kinetic energy (ibid.). As a second step he claims to extend the conclusion to nonperiodic motions. Boltzmann's main result in his (1866) is 'a theorem of pure mechanics,' the principle of least action, representing the second law in the mechanical theory (p. 30). Entropy appears as a function of the logarithm of average kinetic energy (p. 28).

In a short paper (1868a) Boltzmann discusses the relation between the atomistic structure of matter and the description of physical processes through differential equations. Because of the discrete structure of matter, the coefficients in the differential equations describing motion keep changing. (This is due to collisions.) When can we ignore the irregularities of the internal structure of bodies in such a description, he asks. That is to say, is there a way of overcoming the effect of changes in the coefficients? Boltzmann suggests that if they change fast enough to their previous values, one can take the average values instead. In this way it becomes possible to use a continuous description of a process through differential equations, with no discontinuities. One objection to the feasibility of atomism, pertaining to its mathematical methods, is therefore overcome (p. 43). Here again, to justify the use of averages, Boltzmann imagines a periodic, even finite sequence of changes.

Boltzmann (1868b) is known as the paper which derives the Maxwellian distribution in the presence of external forces. He asks, how do material points behave when left alone for a long time, this being a type of question not studied in dynamics earlier (p. 49). The relative time of having a molecular velocity between $c$ and $c + dc$, 'during a very long time,' is its *probability* (p. 50, also p. 81). This passage in Boltzmann (1868b) marks the appearance of the time average notion of probability. As possible objections to the Maxwellian distribution, Boltzmann remarks that it is possible that the dynamical variables are dependent on each other and that this is maintained in the motion. Such a case would violate the distribution of states in time; however, the slightest disturbance would be enough to eliminate the dependence (p. 96). Another possibility is that all initial states lead to periodic motions not running through all possible states compatible with the total energy. There would in this case be an infinity of different possible temperature equilibria (p. 96). Boltzmann's intuitive thought seems to be that a unique equilibrium is brought forth by a motion which does not avoid any of the possible states.

The long memoir 'Further studies on the thermal equilibrium of gas molecules' (1872), contains several conceptual innovations. One is a new concept of physical state, as given by a probability distribution (p. 322). It comes together with the first time that a physical process obtains a truly probabilistic description, through a *time-dependent probabilistic law*. Probability gives the relative time one molecule is in a given state during a very long time. But Boltzmann says that this is the same as the relative number of all the molecules in that state at a single time (p. 317). With a great number of molecules, their relative numbers in different states would approximate a continuous probability distribution over the states. It is impossible to find out the probability law by an integration of the equations of motion; for this cannot be done for even three particles. (The latter is mentioned in 1871a, p. 240.) But the probability law obeys a certain form that can be determined from the equations without their integration (1872, pp. 317–18). Boltzmann aims at showing that the velocity distribution, a function of state $x$ and time $t$, approaches the Maxwellian one. The motion is mechanical: If the initial state and the force law are given at time 0, the future state is completely determined (p. 320). Still Boltzmann holds it obvious that after a long time all spatial directions of molecular motion have become equiprobable. Therefore, if the task is to find the velocity distribution after a long time, one can assume the equiprobability already from the beginning (p. 321). There will be a great number of molecules, with $f(x,t)$ the number of molecules in state $x$ at time $t$. It therefore makes sense to introduce the differential number of molecules $f(x,t)\,dx$. Boltzmann later divides the absolute number $f(x,t)$ by the total number of molecules $N$ to obtain a relative one (p. 400). Assume now an arbitrary (nonnormalized) distribution $f(x,0)$ given; the task is to determine $f(x,t)$. One checks how much $f$ changes in a small interval of time $\tau$, and this gives a partial differential equation that has to be solved for $f$, the fundamental *Boltzmann equation* (1872, eq. (16) on p. 334). Its solution gives the first example of a probabilistically described physical process. Boltzmann explicitly says that the state of the gas is completely determined by the distribution function $f(x,t)$ (p. 322). Here a new concept of state, different from the mechanical one, is born. The probabilistic law giving the state is a function of continuous time.[12] To prove the second law, Boltzmann considers the function that is the mechanical representation of *entropy* $E$, defined as $\int_R cf(x,t)\log f(x,t)\,dx$ with $c$ a positive constant

[12] A systematic theory of such functions, or stochastic processes as they are called, was developed only in the 1930s. It is well to remember the very first step taken toward such a theory.

and integration over the state space $R$. Normalization by replacement of $f$ with $f/N$ gives an expression in terms of probabilities. (This will make the logarithm negative, so to have a nonnegative entropy, one puts $c$ negative today.) Then with $x$ integrated out, $E$ is a function of time. For the time derivative of $E$, one has $dE/dt \geqslant 0$, with equality holding when $f$ is the Maxwellian normal distribution (p. 344). When entropy became to be understood as a measure of disorder, this result gave an explanation for why a normal distribution appears. Through later criticism, it became a central problem whether the above inequality for the time derivative of $E$ holds strictly. Boltzmann had mentioned that an exact proof requires a probabilistic consideration, already in his 1871 'analytic proof' of the second law (1871c, p. 295). In 1872 he says that the uniform distribution of directions of motion of molecules has exceptions: His example is molecules moving on straight lines and being reflected back along them from the walls. Boltzmann himself was the one to criticize Maxwell's 1867 attempt at an *a priori* proof that his distribution law is the only one invariant in time, that is, stationary. Instead, says Boltzmann, it is the only stationary one that gives positive probabilities for all values of kinetic energy (1872, pp. 360–61).[13] In more recent terms, granted that probability has to be continuous with respect to the geometric measure over state space, a unique distribution is sorted out by the requirement of invariance. Boltzmann's is a rather sophisticated view of the relation between the mechanical assumptions of the molecular theory and its probabilistic character. It is far removed from the popular image of Boltzmann as a mechanical determinist who gradually had to make concessions toward the introduction of probabilities, on the face of a series of 'objections' against his supposed determinism.[14] Instead, Boltzmann had by 1872 already a full hand against his future critics.

Next in the 'Weitere Studien' follows a section that is characteristic of Boltzmann's ideas on the role of mathematics in natural science. He replaces the integrals of the previous derivations with finite sums, by considering discrete kinetic energies $0, \varepsilon, 2\varepsilon, 3\varepsilon, \ldots, p\varepsilon$ for the molecules – for the sake of 'clarity and intuitiveness,' as he says (p. 346). This method of Boltzmann's is better known from his (1877b): To determine the distribution of energy, one has to find out how many ways there are to distribute $n$ molecules, with $k_0$ of energy $0$, $k_1$ of energy $\varepsilon$, $k_2$ of $2\varepsilon, \ldots, k_p$ of $p\varepsilon$, so that total energy is preserved. Each way of assigning to the molecules energies such that total energy is the same, is called a 'complexion.' The number of complexions satisfying the above energy distri-

[13]More on this is to follow in Sec. 3.2(a).

[14]A view to be found in Klein (1973), for example.

bution constraint is given by the combinatorial formula $n!/(k_0!k_1!\cdots k_p!)$. The probability $W$ of the above distribution is this number divided by the total number of complexions with the same total energy. The formula for entropy reads $S = k\log W$, or entropy is proportional to logarithm of probability. The above combinatorial scheme is known as Maxwell–Boltzmann statistics. It is a remarkable historical accident that Boltzmann's combinatorial way of treating problems of statistical physics proved important for Planck's derivation of his radiation law. In Boltzmann there was naturally no idea of the quantization of energy. For him, the discrete was easier to understand, though he also said that 'every infinity in nature never means anything else than a limiting process' (1877b, p. 167). In this particular case, for the limit of $\varepsilon \rightarrow 0$, the sequence of kinetic energies becomes continuous, 'and our mathematical fiction goes over to the previously treated physical problem' (1872, p. 348). Mathematically, the effect of Boltzmann's method is to reduce probabilistic problems to counting.[15] It gives a very high probability for having a distribution with entropy near maximum, and a correspondingly low probability for a deviant entropy. In the limit of $\varepsilon \rightarrow 0$ these probabilities become 0 and 1, their meaning certainly being more difficult to fathom than is the case for combinatorially defined low and high probabilities. Boltzmann's discretization of state space relates to a later idea in statistical mechanics, the 'coarse-graining' of the state space; specifically, attempts at explaining through coarse-graining macroscopic irreversibility, in contrast to microscopic reversibility, have been made. But Boltzmann's combinatorial approach originally did not seem to have any such goals.

At the end of his essay, after discussing the effect of invariants of motion for the values of time averages, Boltzmann introduces the concept of an *ensemble*. Apparently, few readers got to these last two pages of Boltzmann's hefty memoir. The principle on which the introduction of ensembles is based appeared already in Boltzmann (1871b): Just as one can consider each of the $N$ molecules of a gas as a separate system, one can also take a very great number of systems of particles, the difference being that these latter do not interact with each other (pp. 259–60). Ensembles are introduced in (1872) for the purpose of showing the proper relation between a mechanically defined entropy and the second law. One takes a body $A$, instead of particles, and then another one, $B$. 'Theoretically the effect of interaction depends... also on the states in which the bodies $A$ and $B$ are in the beginning' (p. 401). Nothing

[15] A reduction of continuous probabilities is made systematic in the nonstandard treatment of probability theory, cf. Nelson (1987).

of this is observed, which is due to the compensating effect of the great number of molecules in the interaction. 'To eliminate the effect of state, we take instead of one system ($A$) very many ($N$) equal ones, but in different states, and which do not interact.' That marks Boltzmann's step into treating an ensemble. Instead of the systems ($N$) interacting with each other, a ($B$) interacts on each of the ($A$)'s such that the beginning of the interaction occurs in the most varied phases. 'All effects that do not depend on phase, must come out as if only one system ($A$) interacted with one system ($B$) in an arbitrary phase, and we know that in fact thermal phenomena do not depend on phase' (p. 402). Having moved from a single system to an ensemble, Boltzmann studies which properties hold for the ensemble as a whole, and infers then back to an *arbitrary* single case: In an interaction with $B$'s going on for a long time, the systems have reached equilibrium and the formulas, previously derived for the molecules of a single system, hold for the ensemble. A division of the entropy by the number of systems gives the entropy of one system, whence the second law is derived, consequently a Maxwellian distribution also (p. 402). Why should an arbitrarily chosen single system have this distribution? The time average of each molecule is, according to Boltzmann, constant for sufficiently long times. Then, equally, the average over sufficiently many molecules at a single time is also constant. But Boltzmann repeatedly emphasized that there are exceptions to such conclusions of the kinetic theory. Contrary to a widely held opinion, Boltzmann is not in 1872 claiming that the second law and the Maxwellian distribution are *necessary* consequences of the kinetic theory. Even though theoretically the result of thermal interaction depends on the initial mechanical states of the bodies, Boltzmann's crucial step is based on admitting as an observed fact that thermal phenomena do not depend on phase (state). This, not any truth of mechanics, warrants the inference from the ensemble to an arbitrary single system.

Boltzmann's time average notion of probability is closely related to statistical or frequentist probability. If observations are made at discrete intervals of time, and with a finite number of discernible states instead of a continuous one, the integrals which express time averages in the dynamical systems of statistical mechanics become limits of relative frequencies. Thus Einstein, for example, uses at convenience terminologies from physics or statistics. Since the beginnings of the kinetic theory, intuitive probabilistic arguments had been taken to justify the use of an average quantity such as average velocity. The probabilistic counterpart on which the argument was based, was the law of large numbers. This is obvious in Boltzmann. His method of a finite number of energy levels makes probabilistic calculations take on the same combinatorial form as in gambling problems. And there the role of the law of large numbers was understood by everyone: It was taken to show that variation and irregularity in the small leads to regular behavior in the large. In the early 1930s, an abstract mathematical ergodic theory was created from the conceptual basis of ergodic theory in statistical mechanics. It brought into completion the analogy between time average probability and statistical probability by showing that the existence and uniqueness of time averages, or what is called the ergodic theorem, is from a probabilistic point of view nothing but the law of large numbers.

Gibbs' approach to statistical mechanics made popular the method of ensembles. Ensemble is just the French word for set or collection, possibly taken from Poincaré (1890, p. 56). There an 'ensemble of initial situations' is imagined. Somewhat later (pp. 68–70) Poincaré tries to show that quasi-periodic motions are the rule. The goal that Gibbs wants to realize in a general way with his ensembles is similar. As Boltzmann had stated when he introduced ensembles in 1871 and 1872, talking about a single system makes it difficult to express the idea that the macroscopic properties of gases do not depend on the initial state. Gibbs' ensembles systematize the formulation of such properties by expressing that they hold with a high probability. Another remarkable feature of the Gibbsian methods stems from the following: As he puts it, he found it best not to base his approach on any specific hypotheses concerning molecular structure or interaction. The main reason was the failure of the classical theory to explain the phenomenon of specific heats (1902, p. x). Therefore his statistical mechanics leads to such conclusions as one can draw from the general atomistic structure of matter and from the large number of molecules.

An ensemble is a set of a continuous number of systems of the same kind, together with a probability distribution telling what the density of different specific systems is in the set. The ensembles Boltzmann considered were called *microcanonical* by Gibbs, their defining property being that they all have the same total energy. Gibbs sees the use of ensembles as the third stage in statistical physics. The first one was the study of molecular distribution questions of a single system, and the second the study of time averages of a single system (1902, p. viii).

Ensembles represent an idea somewhat extraneous to physics. As Gibbs himself says, the systems in an ensemble do not have 'a simultaneous physical existence,' they are '...creations of the imagination' (p. 188). Their introduction also contains a marked epistemic component: For probability describes that which is imperfectly known as 'something taken at random from a great number of things which are completely described' (p. 17, and similarly, p. 163). This is the way the use of ensembles has been often construed subsequently, as a description of the degree of accuracy with which we know the state of the actual system in front of us. The ensemble distribution gives the a priori probabilities which are purely expressions of ignorance of the true state.[16] In a more extreme form, the whole method has been taken as a Bayesian-oriented method of statistical inference.[17] And it is true that there is an inferential component involved. In fact, such a component is noticeable already in Boltzmann's move from an ensemble to a single system. But Gibbs himself had no single preferred reading of ensembles. He had at least three alternatives, the first one of which we quoted above. It relates to some of the basic ideas of classical probability. There is a set of alternatives considered equally probable with the relative number of favorable cases giving the probability of success in one case drawn at random. Similarly, in Gibbs the probability is the relative '*extension in phase*,' with a random choice of the quantity represented by the independent variable. Gibbs thinks the single most important fact about statistical mechanics is that the measure ('extension') in phase space is conserved for isolated systems. It follows that the probability of the state being within given limits at one time is the same as the probability that it is within corresponding limits at another time. There is an argument for choosing a probability that is conserved in the motion: If the occurrence of an event at a certain time is equivalent to the occurrence of some other event at some other time, they should have the same probabilities. Gibbs says that one can in this way avoid reference to an ensemble if one prefers to do so (p. 17). His discussion here is very similar to Zermelo's two years earlier. Zermelo's basic idea was to reduce, via the conservation of volume in phase space, the probability at time $t$ to probability at time 0. He calls the argument for equiprobability at different times that Gibbs also uses, 'Laplace's probability law, according to which two events necessarily connected like cause and effect... have to be equally probable' (1900, p. 318).[18] This, he says, still leaves some arbitrariness in the definition of probability, but it is restricted to a single moment of time. Neither Zermelo nor Gibbs makes it very clear what kind of probability they have in mind here. Gibbs offers a further, third reading of probabilities toward the end of his book, telling that 'it is actually in large measure for the sake of representing more clearly the

[16] See, for example, Tolman (1938).

[17] This is the approach of Jaynes, see his (1983).

[18] If the cause were only a sufficient one, it could have a lower probability than the effect.

possible cases of the motion of a single system that we use the conception of an ensemble of systems' (p. 188). He also says that the microcanonical ensemble 'in many cases simply represents the *time-ensemble*' (p. 169). The latter is the same as the distribution of states in the time evolution of a single state, and it leads to more natural statistical mechanical 'analogies' of thermodynamical concepts than do ensembles proper. A single time-ensemble may depend on initial conditions, and in that case it is all the time-ensembles together (with a suitable weighting) that make up for a microcanonical ensemble (p. 180).

The real force of Gibbs' book does not seem to lie in ensembles. These were certainly helpful in making it easier to apply probabilistic arguments before measure theory was available for that purpose. For example, Gibbs discusses periodic motions, quasi-periodic motions, and motions that do not recur in either sense (pp. 139, 141). The question of the proportions of the different kinds of motions is formulated as a question of the distribution of systems in an ensemble. The probability of a certain kind of motion is zero if the corresponding number of systems is 'less than any assignable fraction.' While with a single system exceptions are possible, the use of ensembles results in such cases 'simply disappearing' (p. 141). The physical novelties that made Gibbs' book so important are the canonical and grand-canonical distributions. A *canonical ensemble* is obtained if the undetermined thing is the total energy. But here again, the significance of ensembles evaporates. What remains is a combination of a single system and an inference. The inference part consists in ascribing to the single system those properties which hold with high probability (or probability 1) for the ensemble. The single system has to be a sufficiently large system obeying a microcanonical distribution. Its small parts are canonically distributed according to an important result, meaning that the energy of the small component varies following a normal law (p. 183). This conclusion becomes exact in the limit of an infinite heat bath (ibid.). The grand-canonical ensemble is a further generalization into a situation where the number of particles is allowed to vary.

### *3.1(b) The dilemma of irreversibility*

The thermodynamic concept of entropy $S$ is defined by the equation $dS = dQ/T$, in which $Q$ is heat and $T$ temperature. Its value would be 0 in a reversible change of heat and work. But that would be an idealized limit in a frictionless world. Clausius' second law of thermodynamics says that the entropy grows in an irreversible change. It predicts the equalization of temperature differences (Clausius 1868, p. 419):

The more the universe approaches this limiting condition in which the entropy is a maximum, the more do the occasions of further change diminish; and supposing this condition to be at last completely obtained, no further change could evermore take place, and the universe would be in a state of unchanging death.

The second law finds its formal expression in $dS/dt \geqslant 0$, saying that entropy is a quantity S which increases with time. Thus entropy growth defines a preferred direction of time in the universe. Thermal processes proceed irreversibly in just that direction, as is also witnessed by every-day observation. That, however, is not the case with the laws of motion of classical mechanics. The equation $F = ma$, force equals mass times acceleration, will do here. The acceleration $a$ is the time derivative $dv/dt$ of velocity $v$; velocity is the time derivative $dr/dt$ of spatial position $r$, thus $a = d^2r/dt^2$. Position $r$ is a function of time and the substitution of $-t$ for $t$ results in exactly the same equation of motion, as differentiating twice cancels the negative sign. Classical mechanical motions are *time reversible*. The consequence would seem to be that the law $dS/dt \geqslant 0$ cannot be derived from mechanics, that is, there is no mechanical explanation of irreversibility. This is Thomson's (1874) and Loschmidt's (1876) *reversibility objection* to the (supposedly) Boltzmannian program of deriving the laws of statistical physics strictly from mechanics. The dilemma is: Whichever is wrong, classical mechanics or the entropy law?

Maxwell's *demon* stems from a letter of his of the late 1860s.[19] There are two adjacent containers of gas of different temperatures. There is a tiny opening in the wall separating them, with a gate operated by the demon. The demon lets fast molecules from the cold gas into the warm and slow ones in the other direction, thereby increasing the temperature difference instead of decreasing it, as would follow from the entropy law. The demon was meant to demonstrate that the atomistic constitution of matter turns the second law of thermodynamics into a statistical principle. This in itself stands in no contradiction with the assumption that the molecules should follow classical laws of motion. In fact, the remark has been made that the molecules cannot move randomly, or else the demon would be unable to predict their course and do its work properly (cf. Brush 1983, p. 89). It seems that Maxwell himself at a certain stage would have allowed the molecules to move according to mechanical principles, except for the collisions, which are described only probabilistically. The demon could therefore carry on its work even in

[19] See Brush (1976, p. 589).

an indeterministic world if not hit by a molecule. In 1875 Maxwell seems to be denying the validity of classical mechanics as applying to the invisible molecular motions: 'The peculiarity of the motion called heat is that it is perfectly irregular; that is to say, that the direction and magnitude of the velocity of a molecule at a given time cannot be expressed as depending on the present position of the molecule and the time' (1875, p. 436). It is impossible to follow the individual courses of molecules. Instead, what is possible is to make a (finite) grouping of molecules according to their velocities, and to register the change in the number of molecules in each group (p. 427). This marks the passage 'from the methods of strict dynamics to those of statistics and probability.' Maxwell at least considers the idea that molecular collisions are random events not governed by mechanical laws.[20] On a descriptive level, he had made an assumption of molecular randomness in the alternative (1867) derivation of his distribution law. Instead of assuming the components of one molecule to be independent, he assumes the twofold probability distribution $f(v_1, v_2)$ of the velocities $v_1$ and $v_2$ of two molecules to factorize as $f(v_1)f(v_2)$ (p. 43). This assumption was followed by Boltzmann.

There is a document which shows Maxwell's ideas about the future direction of physics, namely, his 1873 lecture, 'Does the progress of Physical Science tend to give any advantage to the opinion of Necessity (or Determinism) over that of the Contingency of Events and the Freedom of the Will?' It was published posthumously in Campbell and Garnett's biography of Maxwell in 1882. What happens to one's opinions concerning the question of the freedom of the will if one is educated with main attention given to physics, Maxwell asks (1873, p. 436). He says molecular science will have a powerful effect on philosophical thinking, for it leads to statistics, the only kind of knowledge available in molecular science (p. 438). Next Maxwell says the determinists base their arguments on the stability of evolution: A small variation in initial state is supposed to produce only a small variation in the future. From the finite accuracy of observations it follows that in the contrary case even approximately correct predictions are impossible (p. 441). The causal law, 'from the same antecedents follow the same consequents,' is a metaphysical doctrine 'of not much use in a world like this, in which the same antecedents never again concur, and nothing ever happens twice' (p. 442). Substituting 'like' for 'same' in the above will not do in case instabilities can occur (ibid.). At an instability, 'an

[20] See further Brush (1976, Sec. 14.4).

infinitely small variation may bring about a finite difference in the state of the system in a finite time' (p. 440). Instabilities are in other words points of discontinuity of the development of state in time. Maxwell calls them singularities (p. 442), and says that 'it is to be expected that in phenomena of higher complexity there will be a far greater number of singularities, near which the axiom about like causes producing like effects ceases to be true' (p. 443). He further contrasts 'singularities and instabilities' with 'continuities and stabilities,' saying attention to the former rather than latter has as consequence that 'the promotion of natural knowledge may tend to remove the prejudice in favor of determinism which seems to arise from assuming that the physical science of the future is a mere magnified image of that of the past' (p. 444). In his (1875, p. 419), Maxwell distinguishes between the visible macroscopic motion of a body, and the invisible microscopic ones of its molecules. While we are used to the Newtonian treatment of the former, especially with celestial bodies, the motion of the latter remains forever unobservable in detail (p. 420). Maxwell criticized the assumption that the future of physics would be a 'magnified image' of the past; he also thought that the magnified image of 'invisible motion' would not be the same as the Newtonian motion of the past.

Maxwell's demon tends to indicate that classical mechanics could be maintained, and the entropy law relegated to the status of a *statistical law*. If we look at Boltzmann's definition of entropy from the year 1872, we see that it is a probabilistic notion. Due to the incredible number of molecules, temperature equalization would always be observed. Similarly, for matter already in thermal equilibrium, spontaneous occurrence of noticeable temperature differences would not be observed, or even observable. The early champions of probabilistic physics admitted that the unobservable individual molecules vary in their dispositions, and that the equilibrium we observe is only a consequence of a very large number of molecules colliding in erratic ways. It follows at once that somewhere in between 1 and a very large number there is an intermediate region where *fluctuations* around an average value might perhaps be observable. But such considerations became important only in 1905. They played a certain role in Boltzmann, in his speculative explanations of the relative character of time asymmetry on a cosmological scale, for example.

Around 1875, Josef Loschmidt started criticizing some of the conclusions of the kinetic theory. He did not think correct the consequence of Boltzmann (1868b), that the molecules of air would have the same average kinetic energy independently of height. Instead of only the density being higher close to the earth, the molecules should move faster and the air thus be warmer. (As it is, but not for this reason.) Loschmidt was trying to argue for the existence of temperature equilibria, where not all parts of a system would have the same temperature. This possibility would mean an escape from the dreadful heat death of the universe. A particular remark of Loschmidt's came to be known as his reversibility objection. He said that if at some moment one reversed all the directions of motion of molecules, the universe would return to its previous state. This shows the impossibility of deriving the second law from mechanics alone (Loschmidt 1876). The same argument had been given by Thomson (Kelvin) two years earlier in a much more profound way. He suggested that Maxwell's demon reverse the velocities to get a flow of heat from cold to warm. Rather than being refuted by Loschmidt's objection, Boltzmann said it is 'very important for the right understanding of the second law' (1877a, p. 117). It provides 'an interesting sophism' where the faulty step has to be traced (p. 119). As we have seen, since 1868 Boltzmann had remarked many times on the existence of motions not leading to the Maxwellian distribution. But these distributions are exceptional, and the slightest disturbance would destroy them. In (1877a, pp. 120–21) Boltzmann says that 'there are infinitely many more initial states leading to uniform distribution than states leading to nonuniform.' Previous to this statement Boltzmann explained the matter by a comparison to a situation with a finite number of outcomes. Here a nonuniform distribution has a very small probability in comparison to that of a uniform one, and Boltzmann's 'infinitely many more' is a statement of what happens in the limit of a growing number of outcomes. A diminution of entropy is 'exceptionally improbable and can be taken as impossible in practice' (p. 121). This would seem to mean the denial of density fluctuations. But soon after, Boltzmann suggests the following. Suppose you reverse the velocities of all the molecules in the universe and follow its state into the infinite past. We are equally justified in assuming very probable a state of temperature equilibrium in the distant past as in the distant future. This is analogous to finding a gas far from equilibrium, after it has been left by itself for a long time. 'We would have to conclude that a very long time ago, the state distribution was uniform and the rare event of its becoming nonuniform occurred.' (p. 122). Obviously Boltzmann thought no one would ever observe anything of the kind, hence the conditional mode of the conclusion to be taken.

While the application of gas theory to the universe as a whole is 'highly suspicious' (1877a, p. 122), it points to the dependence between deviation from equilibrium and the number of atoms in the system. Some results in this direction are found in Boltzmann (1878). For a very great mass of gas in equilibrium, it is not excluded that a small part of it would possess a somewhat greater or smaller kinetic energy than the equilibrium average. The problem is to find an expression for the probability of deviation (p. 251). Let $n$ and $N$ be the numbers of molecules in the smaller part and the whole system. Boltzmann assumes $n$ is great, but to make for a small part, $n/N$ has to be small. With these assumptions, his calculations do not give appreciable probabilities for the deviations (pp. 259, 261). But this was not Boltzmann's final word on the matter, as we shall see.

In the 1890s Boltzmann was met with objections to his '$H$-theorem,' first around 1895 by various British physicists. In his short answers to these objections we find Boltzmann saying that the second law is a theorem of probability, 'the second law can never be proved mathematically from the equations of dynamics alone' (1895a, p. 539). If one describes the value of the quantity $H$ ($E$ in 1872) in a curve, this *$H$-curve* behaves in a peculiar way. If you take a value above the minimum, it is probable that the curve goes down.[21] It was very difficult for Boltzmann's adversaries to understand this, as the $H$-curve itself is symmetric in time. Boltzmann tried to explain the matter by comparison with dice. Another question is, why are those improbable $H$ values found at all? In the end of (1895a) he refers to an idea of his assistant Schuetz: If the universe is assumed large enough, any deviation from equilibrium of a part of it can have a great probability (pp. 543–44). In such parts visible motion and life can occur. In another reply of the same year (1895b), Boltzmann says the crucial assumption is that the laws of probability are applicable for finding the number of molecular collisions. External disturbances are not necessary for large systems with a mean free path greater than the molecular dimensions. These assumptions serve to justify the postulate of independence of molecular velocities as introduced by Maxwell (1867).

In 1896 the mathematician Ernst Zermelo published his criticism of the mechanical theory of heat, based on Poincaré's recurrence theorem. The long paper Poincaré (1890) on the three-body problem is the first one where a property is ascribed to mechanical systems with probability 1. Exceptions to the property are not impossible, but have probability 0. (In Boltzmann a similar probability is only meant as a limit.) The intuition, before measure theory was developed, came from the comparison of rational to irrational numbers. The theorem says that for all states $x$ but the exceptional ones of probability 0, of an isolated

[21] Or, equivalently; goes up from a value below the maximum of $-H$

mechanical system, the future state of the system will return arbitrarily close to the state $x$. It shows that the increase of entropy cannot be a necessary consequence of mechanics. While Zermelo thinks the recurrence theorem had not been earlier applied to gas theory, that was not the case. Poincaré himself makes in his (1893) remarks on the topic. His conclusion, not dissimilar from the one Boltzmann took, is that the second law will be 'reversed': The heat death of the Universe is 'a sort of slumber, from which it will awake after millions and millions of centuries' (1893, p. 206). He thinks that in the conflict between reversible mechanics and irreversible thermodynamics, the former might have to be abandoned. Now, as to Zermelo (1896a), irreversibility is there taken in a strict sense, as allowing no exceptions. This is a bit strange. Even Poincaré himself had allowed exceptions also in gas theory, on the basis of the existence of periodic motions (1894, p. 256). But Zermelo does not refer to that paper. He is led to the position that, as most states are recurrent after Poincaré's theorem, strict irreversibility can hold only for in some sense singular initial conditions (1896a, p. 485). Zermelo's central conclusion, even if taken with some reservation, is that either the second law of thermodynamics or the mechanistic conception has to be given away (p. 493). Obviously it is the second alternative that is in danger here.

Boltzmann's response (1896) to Zermelo's objection is significant in its discussion of possible deviations from a Maxwellian equilibrium distribution. He begins by saying that already Clausius and Maxwell had noticed that the results of gas theory are 'statistical truths.' Poincaré's recurrence theorem is correct; in fact, if a state has positive probability, it *has to* occur in a sufficiently long period of time (pp. 569, 571). But the successive occurrence times are enormously far apart. Boltzmann talks here of 'motions periodic in the mathematical sense,' intending with this what we would today call quasi-periodic motions. In an appendix, he gives what are obviously meant as generous bounds for with what precision the state has to recur: From these bounds it follows that a typical recurrence time for the state of a cubic centimeter of gas is on the order of $10^{19}$ seconds. Why one never observes a decrease of entropy, or increase of the neg-entropy for which the $H$-curve is drawn, is therefore explained as being only an enormous improbability (p. 572).

While Boltzmann repeated in so many words this explanation, we have here another occasion where he should not be taken entirely literally, in this case, about *never* observing entropy decrease. This is seen from when immediately after the above passage, he considers *pressure fluctuations* in a gas. He says that a sufficiently small valve in a gas container would experience measurable fluctuation in pressure in times much shorter than those of 'Zermelo's periods.' Still, they would be long enough to escape observation (p. 572). But he adds:

An argument against gas theory could be derived from such considerations only if phenomena of the said kind failed to occur in cases where, according to computation, they should occur in observable times. This does not seem to be the case.

This statement seems surprising, for the one who, by common opinion, was the first to be able to determine that fluctuations could be observable, was Einstein in his 1905 study of Brownian motion. Unfortunately Boltzmann does not explain what his computations are, but he does mention two cases that could be counterexamples of what he initially claimed to be unobservable fluctuations: 'In very small particles in a gas, motions were observed which can come from a sometimes greater, sometimes smaller pressure on a part of nonnegligible size of the particle's surface' (1896, p. 572). Boltzmann does not give any reference. He repeats his reasoning two pages later in the essay: With sufficiently few molecules, the second law ceases to be valid. In fact, he says, some experiments speak in favor of that, 'even if one is still far away from being able to speak of an experimental proof of its correctness' (p. 574). Fluctuations are an inevitable consequence of atomism. As Boltzmann says, 'with a finite number of molecules the Maxwellian distribution can never obtain exactly but only with a high degree of approximation' (p. 569). The variations predicted by the kinetic theory vanish only with an infinite number of molecules (p. 574). As to Zermelo's accusations of mechanism, Boltzmann says that gas theory claims to be only 'an approximate image of reality' (p. 574). Some of the reasons are the approximative character of any theory, as well as electrical phenomena the theory does not take into account.

In Zermelo's reply (1896b), emphasis is on properties of the $H$-curve. It is clear that Zermelo did not understand Boltzmann's argumentation. He contends that since the curve is time symmetric, a decrease and an increase of the quantity $H$ are equally probable (p. 795). Further, even if one assumed a high value of $H$, the value could 'equally well' be in an ascending as in a descending part of the curve (p. 797). Zermelo claims that probability cannot help in finding a preferred time direction of physical processes (p. 799). He rejects the time average notion of probability (p. 796) and says that the probabilities of states should be reduced back to the probabilities of the initial states (p. 800). Even though Zermelo's objection to gas theory has often been presented as an expression of antiatomism, there is no such unequivocal motive behind it. He is content with the statics of gases. The application of gas theory to time-dependent, irreversible processes, instead, is a failure Zermelo cannot see could be overcome (p. 801).

Boltzmann's second answer (1897a) to Zermelo again tries to explain the points of controversy. Why one finds systems out of equilibrium in the first place, is explained as follows. The part of the universe surrounding us is supposed to be in an improbable state, so that a small part suddenly isolated from it would also often be in such a state (p. 579). Boltzmann repeats many times that one assumes an improbable (non-equilibrium) state which then, with overwhelming probability, goes over to an equilibrium state. It is immaterial to this conditional probability how its condition became fulfilled – whether the system began from an improbable state, or achieved it during a very long time. Poincaré recurrence shows how enormously long the times would have to be. Instead of seeing an objection toward probability theory on the basis of recurrence, he says there is accordance (p. 582).[22] Somewhat later Boltzmann published a paper on Poincaré's theorem. There he criticized Zermelo's proof, saying the recurrence time in it could be infinite. His own proof concludes, similarly to Poincaré, an infinite number of recurrences (1897b, p. 593).

There was no immediate next item in the Zermelo–Boltzmann discussion. It seems that Zermelo was set back by the countercriticisms. Indeed, his 1899 Habilitationsvortrag for Göttingen tries to find a foundation, not a refutation, for the use of probability arguments in dynamics. Also, rather than denying mechanics, this work assumes it (Zermelo 1900, p. 317). The proper place of probability assumptions is with the initial conditions: These are supposed to be chosen independently (p. 318). Zermelo gives a proof that there cannot be a strictly increasing (or decreasing) function of state, thereby showing the impossibility of a strictly mechanical derivation of the second law (p. 319). But on the face of Boltzmann's writings, it is difficult to see what novelty there could be here.

Paul and Tatiana Ehrenfest published in 1906–1907 two expository papers which helped clear the air of paradoxicality that surrounded Boltzmann's $H$-theorem. They used a discrete probability model, the *Ehrenfest urn scheme*, for clarifying the overwhelming probability of entropy increase. As they note, they rely on Boltzmann in explaining the matter by a simple example from the calculus of probability (Ehrenfest & Ehrenfest 1907, p. 312). There are balls, numbered from 1 to 100, divided in two urns. There is a further urn of labels from 1 to 100. Each

[22] That of course was completely vindicated by later developments in probability theory.

time a label is drawn, with replacement, the corresponding ball is moved from the urn in which it is into the other one. In place of the continuous $H$-curve one plots the absolute difference $\Delta$ of the number of balls in the two urns at discrete intervals. This 'curve' is of course symmetric in time. Boltzmann's 'paradoxical' contention is that his $H$-curve is time symmetric, but that at the same time it is overwhelmingly probable that it goes down from a given high value. The problem can be put in the following terms: How is it possible that a value of $H$ away from the minimum could, with overwhelming probability, be a maximum? From the discrete model it is easy to see what is wrong with the thought, suggested by Zermelo (1896b, p. 795), that an increase and a decrease of $H$ should be equally probable, for *most* cases of $\Delta = 80$, say, are *points of local maximum* of the curve (Ehrenfest & Ehrenfest 1907, p. 313).[23] From the given value, $\Delta$ goes as a rule down, irrespective of whether we follow it from left to right or the other way around. Thus was brought home a point of Boltzmann's that had 'remained unintelligible' for Zermelo, even in 1906, as he admits.[24] The point is, in a more general formulation, that *irreversibility is a purely statistical effect.*

Boltzmann's discussion (1898) of the $H$-curve ends with the following amusing thought experiment. One can imagine a world where all natural processes run in the reverse time direction. But a human being who lived in this reversed world would by no means have an experience different from ours. What we call future, he would call past, and vice versa (1898, p. 637). Thus if the direction of heat flow determines time direction, it becomes a tautology that entropy increases with time.

## 3.2 ERGODIC THEORY

### *3.2(a) Ergodic theory in statistical mechanics*

Ergodic theory begins with Boltzmann's attempts at justifying the determination of average values in kinetic theory. In his (1868b) he had already interpreted the probability distribution in terms of time averages. In a modernized notation, the problem ergodic theory faces is the following: First, the average (expectation) of a function $f$ of exact microstate $x$, relative to the probability distribution (density) $p$, is given by the integral $\int_S f(x)p(x)\,dx$. This came to be called the phase average of the function $f$ over phase space $S$. Specifically, let $A$ be a region of

[23]The same point is made very clearly in the review Boltzmann and Nabl (1907, p. 521).
[24]See Zermelo (1906, p. 241).

the state space $S$ and $I_A$ the indicator function of $A$. Then one has for the probability $P(A)$ of the region $A$ the expression

$$P(A) = \int_S I_A(x) p(x)\, dx. \tag{3.1}$$

The situation is particularly simple if one uses a Hamiltonian formulation of mechanics, for with a system of constant total energy, the probability $p$ over state $x$ is uniform and remains so. With a finite number of atoms there is no immediate interpretation for this probability, or any other continuous one. Boltzmann's fundamental interpretive postulate says probabilities are time averages. These latter are defined, again in a somewhat modernized notation, as follows. Let $T(t, x)$ represent the dynamical law of evolution of the system under study: If $x$ is the state at time 0, the future state at time $t$ is $T(t, x)$. The *time average* $\hat{f}(x)$ of a function of state $f$, starting from state $x$ at zero time, is given by the limit, as $t \to \infty$, of an integral of $f(T(t, x))$,

$$\hat{f}(x) = \lim \frac{1}{t} \int_0^t f(T(t, x))\, dt. \tag{3.2}$$

Specifically, choosing the indicator $I_A$ for the function $f$, the time average becomes the average time the system spends in the region $A$, starting from state $x$. To justify the interpretation of the phase average probability as a time average, that is, to show that the number $P(A)$ is the average relative time the system spends in the region $A$, is the basic problem of ergodic theory. First of all, one has to show that the time average exists, and second, that it is unique, or independent of the particular mechanical state $x$ from which the evolution starts. That is, one requires $\hat{I}_A(x) \equiv \hat{I}_A = P(A)$. On this conceptual basis equilibrium statistical mechanics can be built.

Boltzmann, in 1871, was aware of a system of incommensurable harmonic oscillations which give rise to dense motions, known as Lissajous figures (1871b, p. 269–70). He assumes such a dense motion to be the case also for the atoms of a gas. He discusses the thermal equilibrium of a gas surrounding a warm body: 'The great irregularity of thermal motion and the manifold forces affecting bodies from the outside make it probable that the atoms of the warm body, through the motion we call heat, run through all the positions and velocities compatible with the equation of kinetic energy, so that we can use the equations developed above on the coordinates and component velocities of the atoms of warm bodies' (1871b, p. 284). At the end of the paper, Boltzmann stresses the hypothetical character of this way of justifying the equilibrium properties of gases. The statement I just quoted is the first general statement of Boltzmann's *ergodic hypothesis.* It is important to emphasize that Boltzmann in it gives two reasons for having ergodically behaving systems: 1. The motion of the molecules of the body is disturbed by external forces. 2. The motion of the molecules of the body is disturbed by the surrounding gas molecules bombarding it in an irregular fashion. The first of these reasons is the general one that bodies are affected from the outside by external forces, while the second reason is particular. If one assumes the external forces static, the assumption of constancy of total energy is not questioned. It would, however, exclude randomly varying external forces.

Boltzmann's ergodic hypothesis is invariably presented as the postulate that the trajectory of a system passes through all points of the phase space. Indeed, if this is the case, the average of a function $f$ over time, obtained by following the points of the trajectory, would be eventually taken over all the points and would equal the phase average. In symbols, assuming the convergence of the latter integral,

$$\int_S f(x)p(x)\,dx = \lim_{t\to\infty}\frac{1}{t}\int_0^t f(T(t,x))\,dt. \tag{3.3}$$

This simplified reading of Boltzmann is due to the 1911 review of foundations of statistical mechanics by Paul and Tatiana Ehrenfest. They called the above reason for assuming a single trajectory the 'Boltzmann–Maxwell justification' (p. 32, note 90), and since then it has been accepted as standard.[25] But that particular motivation for assuming ergodicity is not used by Boltzmann. We shall see that he had a different idea in mind. His statements, like anyone else's, if taken out of context and given a literal reading, can lead to a false interpretation. For example, he would here and there remark that there are periodic motions. Since the slightest disturbance would make them vanish, they are not important for him and need not be mentioned at every turn. A distinction between dense motions, as in the Lissajous case, and motions passing literally through all points later proved to be mathematically important.[26] For the points of a dense trajectory are a far cry from all the points of a phase space. (They have zero measure, for example.) The problem led to some of the first purely mathematical contributions to ergodic theory in 1913.

[25] The effect of the Ehrenfest review on twentieth-century understanding of Boltzmann and Maxwell is discussed in Brush (1971), p. 293 specifically.

[26] Boltzmann does mention dense motions explicitly in his (1892, p. 433).

Maxwell praises Boltzmann's approach through ergodicity because of its generality. There is no limitation on the intermolecular or external forces, except that they obey the constancy of total energy.[27] 'The only assumption which is necessary for the direct proof [of the distribution law of energy] is that the system, if left to itself in its actual state of motion, will, sooner or later, pass through every phase which is consistent with the equation of energy' (1879, p. 714). But as Maxwell at once adds, this kind of motion would not always be the case. One difficulty are strictly periodic motions. These would be special cases, however (ibid.). A second difficulty is:

The motion of a system not acted on by external forces satisfies six equations besides the equation of energy, so that the system cannot pass through those phases, which, though they satisfy the equation of energy, do not also satisfy these six equations. (*)

Maxwell does not elaborate more on this. But the matter is taken over from Boltzmann (1871a) and (1872). A short study of a special case is in Boltzmann (1868c), so that with the help of these works we can try to find out what is meant by the above quotation of Maxwell. The idea is to eliminate the effect of constants of motion through collisions. This was the second of Boltzmann's (and Maxwell's) reasons for assuming an ergodic path. In Boltzmann's 'Solution of a mechanical problem' the task is to determine the (time average) probabilities of position and velocity of a number of material points (1868c, p. 97). A special case of a point $M$ in a plane, with a given force pulling $M$ toward another fixed point $O$, is studied. The point $M$ is reflected elastically from a fixed straight line. The force law leads to two integrals of motion, the latter saying a certain function of the distance of $M$ from $O$ is a constant $a$. But this constant changes its value in collisions. The condition for the determination of the probability law is 'that every point of the plane will be run through with all possible directions of motion' (p. 103). This assumption from 1868 marks the birth of ergodic theory. Boltzmann describes the ergodic kind of motion as follows: 'If $M$ were a shining point and the motion extremely swift, the whole surface traversed by $M$ would appear uniformly illuminated' (ibid.).

In 1871 Boltzmann starts studying more general situations as encountered in kinetic theory. Molecular interaction takes place only at very short distances. Through this interaction, 'the various molecules

[27] This certainly sounds a bit strange as far as external forces are considered. One cannot say that a system with constant total energy experiences any random external influences, though it could be in a static gravitational field, say.

will run through all possible states of motion' (1871a, p. 236). Here again, one should not take Boltzmann to the accuracy of a mathematical point: In nature there are no strict bounds, but after a certain distance the interaction is so small that it can be set to null without error, as he says a little later (p. 248). The integrals of the equations of motion for $r$ particles give equations for positions $x$ and velocities $v, h_1(x_1, \ldots, v_r) = a_1, h_2(x_1, \ldots, v_r) = a_2, \ldots$, where the $a_i$ maintain their values in the evolution of state. The $a_i$ are the constants or invariants of motion of which there are $6r - 3$. But that a gas could have different (observable) states depending on initial conditions (and consequently on what values the invariants of motion obtain for these latter), is 'a priori improbable and not ever verified by experience' (p. 255). Between collisions there are only internal forces, so that the motion is linear and all the momentum components are preserved. The number of molecules in a given state can be an arbitrary function of the invariants as long as there are no collisions (1871b, p. 267). Here number of molecules in a given state means the same as distribution. When Boltzmann introduces the ergodic hypothesis later in this same paper, it is for the purpose of eliminating the division of all possible motions into separate kinds having different distributions according to the values of the invariants. The matter is further explained in (1872). After eliminating time from the equations, there are $6r - 4$ integration constants $a_1, a_2, \ldots$. They remain constant between collisions (p. 371). The velocity distribution is a function of time and the $a_i$, but it changes only at the collisions (p. 372). Boltzmann goes on to determine the distribution law for all possible evolutions, including the state variables and the values of the invariants. He obtains a general formula for the number of collisions per unit volume in time $\Delta t$, a partial differential equation whose solution is the sought for distribution law (eq. 55, p. 379). Why does Maxwell in his discussion of this matter have only six invariants of motion? One way to get that number would be to sum over translational and rotational momentum components of all the particles. Anyway, Maxwell follows Boltzmann in thinking that the values of the component momenta would decompose the phase space into different kinds of motions. According to the quotation of Maxwell, external forces could have the effect of making the momentum decomposition disappear. This is Boltzmann's first reason for ergodicity. And if there are no external forces, as Maxwell supposes in his quotation (*)? Maxwell relies in this case on Boltzmann's second reason for ergodicity (pp. 714–15).

But if we suppose that the material particles, or some of them, occasionally encounter a fixed obstacle such as the sides of a vessel containing the particles, then, except for special forms of the surface of this obstacle, each encounter will introduce a disturbance into the motion of the system, so that it will pass from one undisturbed path into another. The two paths must both satisfy the equation of energy, and they must intersect each other in the phase for which the conditions of encounter with the fixed obstacle are satisfied, but they are not subject to the equations of momentum.

By this argument, Maxwell feels he can 'with considerable confidence assert' the ergodicity of the ensuing motion (ibid.).[28] A further clue to Maxwell is found in Boltzmann's comments on him (1881). The discussion is in terms of a Hamiltonian formulation with generalized position and momentum coordinates $q_1, \ldots, q_n, p_1, \ldots, p_n$. The criterion of a *stationary* distribution $f$ is that $f \cdot dq/dt$ depends only on values of constants of motion for the initial state, but not on the time. Next ergodicity appears as the special possibility that the positions and momenta take on, in a sufficiently long time, all values compatible with the total energy. In this case $f \cdot dq/dt$ would have to take on the same value for all of them and would consequently have to be a constant (p. 590).[29] Stationarity is equivalent to the condition that the distribution $f$ equals $dt/dq$ multiplied by 'an arbitrary function of... the parameters' (1881, p. 590). (Parameter was Maxwell's term for a constant of motion.) Maxwell considers only the case that this function is a constant, which leads to the simplest stationary distribution. This special case is of course that of ergodicity, and it seems that nothing was done with the more general situation of stationarity in a long time. In 1932 von Neumann obtained a proof that a stationary system has a uniquely defined decomposition into ergodic parts. Statistical mechanics with an arbitrary number of constants of motion obtained its systematic treatment only in Grad (1952).

Boltzmann's last papers on ergodicity are (1884) and (1887). He takes into use his old 'Kunstgriff' of ensembles (1884, p. 123). A *Monode* is an ensemble of a special kind of stationary systems of the same energy. Individual systems are called elements of the Monode (p. 132). Assume that there are constants of motion $h_1, \ldots, h_k$. An ensemble may have the restriction that these constants have the values $a_1, \ldots, a_k$. If the only possible restriction is the equation given by the value of total energy, the ensemble is called an *Ergode*, if there are further restricting equa-

[28] If collisions with nonfixed obstacles, that is, the molecules, are allowed, the momentum components would be redistributed and only their six sums preserved. Then the remaining problem is, why does Maxwell not note that one of these sums is a function of the others and total energy.

[29] This is the way Einstein introduces the ergodic hypothesis.

tions, a *Subergode* (p. 134). This is the first time such words appear in Boltzmann.[30] First Boltzmann notes that the formulas for single systems hold also for ensembles; later he turns to the reverse inference: If every system of a Monode runs through all the states that the different systems take on simultaneously, one can substitute in place of the Monode a single system (p. 147, note 1). In Boltzmann (1887) it is explained for two examples, one being Lissajous figures, how collisions destroy all the other invariants of motion except total energy. But he says that it is difficult to express mathematically the ensuing result, namely, that the observable properties of a system depend in this case only on the internal and external forces and total energy, without using the concept of an ensemble (p. 265). That is what the ensembles were introduced for in Boltzmann (1872).

As can be seen, neither Boltzmann's nor Maxwell's earliest statement of the ergodic hypothesis exactly agrees with the ergodic hypothesis as it is usually understood: that a conservative system would visit every state of the phase space, or at least be dense in phase space, when left by itself. If one lifts Maxwell's statement out of context, it seems to make such a claim. A strict version of the ergodic hypothesis thus understood is that a single trajectory fills all of phase space. That would be very strange, for mechanical motion would not be deterministic only in the usual sense of one state determining the future (and past) states, but in an absolute sense; no 'Contingency of Events,' to use Maxwell's term of 1873, would be left in the world except the absolute time at which it was put running, while on the other hand it is clear that mechanical systems admit of varying initial conditions. In Boltzmann's example of ergodic motion, the Lissajous figures, there is an infinity of periodic motions, and Maxwell too says that there are special cases of periodic trajectories. These considerations contradict the possibility of ergodicity in the allegedly original sense of a unique trajectory. There are further equally simple reasons for not reading literally the assumption that one trajectory fills all of the phase space. As we have seen, the infinite was for Boltzmann only a limit. He was ready to substitute a finite number of energy levels such that in molecular collisions 'it is somehow brought about' that energy is exchanged by multiples of one level difference.

The common misconception that Maxwell and Boltzmann thought one trajectory fills all of phase space, seems to derive, as was said, from

[30] On p. 132 there is a slip, Ergode instead of Holode. Much earlier Boltzmann had called the potential function an *Ergal*; see (1875, p. 1). On p. 141 of the 1884 essay under discussion he speaks of the '*ergodisch*' property of a single system.

the picture given by the Ehrenfest review of the foundations of statistical mechanics. They present as motivation for the assumption under discussion the following argument: By the uniqueness of mechanical trajectories, there would be essentially only one trajectory, so that time and phase averages coincide. As proof of their claim they refer to Boltzmann (1887). But what they call the Boltzmann–Maxwell justification is nowhere to be found in Boltzmann. In the particular reference (1887), as we have seen, the concept of ergodicity is introduced through the existence of only one invariant of motion. Time averages are not equated with phase averages for the reason given by the Ehrenfests, but because the probability law is a function of exactly one invariant or parameter for ergodic systems (Boltzmann 1881, p. 590). It is the simplest stationary distribution (p. 591).

The Ehrenfests indicate that Boltzmann and others later forgot the ergodic hypothesis. According to them, he introduces in 1898, in his *Gastheorie*, vol. 2, p. 92, the probability distribution determined by total energy 'as the "simplest case" of a stationary distribution, and calls it "ergodic" without referring anywhere in his book to the ergodic hypothesis' (p. 33, note 94). That discrepancy did not lead them into asking what *Boltzmann's* ergodic hypothesis was and how it was motivated; instead they preferred to assume Boltzmann had forgotten it. The obvious alternative would have been that they, or someone else shortly before, had put up the single-trajectory hypothesis. In (1887) Boltzmann says that without the use of ensembles, it is difficult to express mathematically the idea that observable properties of a gas depend only on the forces and total energy. This latter, of course, is the macroscopic manifestation of, as well as the physical reason for, assuming the kind of invisible behavior of molecules Boltzmann in (1884) had called ergodic. In an ensemble there is a continuous number of systems forming the *Ergode*. Boltzmann is not saying they are all identical, as would follow from what the Ehrenfests suggest. They seem to think (1911, p. 31) that ergodic motions of the same energy would have the same values for all the invariants and differ only in absolute time.[31] But Boltzmann saw that the problem of invariants of motion is resolved differently (1887, pp. 260–61).

Let me sum up Boltzmann's position concerning ergodicity. The aim is to express that gases in equilibrium obey a single probability distribution from which all macroscopic properties are derivable. This is accomplished if the distribution is a function of external and internal

[31] These values would have to be constants over all states, hence they would all be multiples of the total energy.

forces and a single constant of motion only, the total energy. Theoretically one must admit an infinity of trajectories with restrictions that violate the distribution law. The simplest restrictions establish functional dependencies between states rendering the motion periodic. For such reasons one uses ensembles in order to express the independence of macroscopic properties of gases from their microscopic evolution. The advantage of ensembles is that they obey the probability distribution at a single moment, whereas in a single system one has to take time averages that may be sensitive to initial conditions. In an ensemble there are 'infinitely many more initial states leading to uniform distribution than states leading to nonuniform' (1877a, pp. 120–21). This argument was used for the probabilistic conclusion that an equilibrium distribution will actually be found. Ascribing the distribution to a single system also is a very similar kind of probabilistic inference: having a different distribution is 'a priori improbable and never verified by experience' (1871a, p. 255). The proper language for expressing these matters would be measure theory. But there was no such theory before the end of the nineteenth century, and its first application to statistical physics came only in 1913. Boltzmann's probabilistic analogies from finite situations such as found in lotteries, were more or less the best one could attain. The passage going farthest toward understanding the measure properties of sets of real numbers that I have been able to find in Boltzmann, is from his (1897b) paper on Poincaré's recurrence theorem. He assumes $N$ systems are given whose initial states are the points of a finite region (Gebiet) $g_0$ of extension $\gamma$. Infinitely many 'singular cases' $P$, that is, exceptional states, are possible. These are such as leave the region $g_0$ and never return. Recurrent points are denoted by $Q$ (1897b, p. 592). It follows from what Boltzmann proves that (p. 593):

> The number of points $P$ within the region $g_0$ must be infinitely small in comparison with the number of points $Q$. The points $P$ can never fill a finite connected region, and not even an infinity of unconnected regions whose total extension would equal that of a finite one.

Boltzmann also offered two physical reasons for assuming the ergodicity of single systems: external disturbances and collisions within the system. For the latter he gave examples of how an obstacle can disturb all the infinitely many possible trajectories (1887, p. 262, and already 1868c).

Boltzmann clearly accepts effects of external forces in his formulation of ergodicity in 1871, gravitation presumably being the most obvious candidate to deviate the courses of molecules. This is the case with Maxwell also as far as the first reason for ergodicity is concerned. Brush (1976, p. 367) says Maxwell postulated ergodicity only for systems interacting with their surroundings, and 'did not require this interaction to be of a random nature on the microscopic level; if the walls of the container are completely rigid, their action on the molecules is completely deterministic, having the effect of knocking the system from one orbit to another on the energy surface.' While this would hold for Boltzmann in 1871 and 1872, Maxwell also considers the case of no external forces, as the earlier quotation (*) shows. In a collision the trajectory or path jumps from one ergodic component to another, but energy is not exchanged in the interaction. How could the jump be caused by external forces in this case? In view of his earlier indeterminist ideas, it is interesting to speculate what his position here is. In the passage from one 'undisturbed path' to another, the paths must intersect at a common phase. If phase here is the same as state, the paths would have to coincide according to the basic property of mechanical trajectories. These paths are the 'collision' and 'rebound' trajectories before and after the collision. 'Intersection' in Maxwell occurs when the conditions for a collision are encountered. A simple way of making sense here is to note that the collisions are exactly the kinds of singularities Maxwell had described in his lecture on free will in 1873, that is, *discontinuities* in the mechanical evolution of state. What is continuous is the spatial position of the atom, but the velocity experiences a discontinuity in a collision. That is to say, the direction of the motion changes discontinuously, and therefore the three (or six) momentum components are discontinuously redistributed, only their sum being preserved. Through the instability introduced by the discontinuity, an infinitely small difference in the ingoing path could lead to a noticeable difference in the rebound path. Such motion would give a particular example of what Maxwell had proposed already four years earlier, namely, that the future mechanical state of molecules is not completely determined from their present state (Maxwell 1875).

Poincaré's paper on the kinetic theory of gases of 1894 contains the essential concepts that much later became the tools of the trade of ergodic theory: the requirement that the trajectories be dense, and that this hold, except for a set of initial conditions of probability 0. This is precisely what is meant by an ergodic hypothesis today. One reason for admitting exceptions to ergodic motions is given by Poincaré's recurrence theorem of 1890. Poincaré's paper on the kinetic theory (1894) gives, as he says, a general exposition of Maxwell's ideas. Those ideas in fact are all Boltzmann's as augmented by Poincaré's more strict mathematical requirements. Whatever the reason, Boltzmann is not mentioned at all in the paper.

'The postulate of Maxwell,' as Poincaré calls it, contains the following

(1894, p. 252): There are certain functions of state which remain constant under the motions. These constants or integrals of motion delimit the motions that are possible. 'Maxwell admits that, whatever the initial state of the system, it always passes an infinity of times, I won't say through every state *compatible with the existence of the integrals, but as close as desired* to any of the states.' What Poincaré calls 'Maxwell's theorem' (p. 253) shows, via equipartition, that time averages are proportional to volumes in state space.

'Maxwell's postulate,' however, rests on a fragile basis, says Poincaré. There are in every problem of mechanics periodic solutions, for which Maxwell's postulate is 'certainly false.' Therefore 'at least one has to adjoin to the enunciation of the postulate the restriction...: *but for certain exceptional initial states*' (p. 256). Since Maxwell's postulate would not hold for the solar system if, as it seems, it is stable, Poincaré thinks it likely that the postulate holds for some systems and fails for some others (p. 256).

As can be seen, Poincaré gives the ergodic hypothesis in the wording it receives in today's standard expositions. There is no idea of a single trajectory filling all of state space, for the possibility that all trajectories would be dense is at once denied on the basis of the existence of periodic motions; the latter would of course exclude the possibility of a single trajectory if it ever were suggested. What is more, the two features – *density*, or quasi-ergodicity as it was called for some time a bit later, that holds in a suitable sense *for most trajectories* – are presented as obvious augmentations of Maxwell's ideas, rather than devastating criticisms. Both can be found in Boltzmann already, though not in an equally striking form. As we shall see, not enough attention was paid to these two aspects in the next couple of decades, with some confusion and empty results as consequence. Probably Poincaré's paper, published in the *Revue générale des Sciences pures et appliquées*, escaped proper attention.

In Borel's application of probability to the kinetic theory (1906a) the unqualified statement of ergodicity as undisturbed motion filling all of the phase space is accepted as standard. But Borel would not be led into trying to give a strict derivation of time average probabilities from purely dynamical assumptions for the general reason that he thought all probabilities to be either assumed by convention or derived from other probabilities. His conventionalism with respect to probability is also the reason why he so easily introduces new ideas into the foundations of statistical mechanics. Thus talking about coin tossing, we can make the usual assumptions and derive the probability $1/2^n$ for a sequence of length $n$. But there could be other derivations of the same probability, based on different conventions. It is similar in all other cases. He first discusses the distribution of the minor planets, an issue raised by Poincaré (1896). What is the probability that they all are within the same semicircle? The answer is the same as for coin tossing, as follows if equal probabilities are given for the two halves and different planets taken as independent. But it can be justified differently. Likewise in the kinetic theory, where Maxwell's law was derived from the spherical symmetry of the directions of motion. If one starts with the deterministic assumptions of exact knowledge of state and law of motion, all probabilities would be 0 or 1, says Borel. But if we take a long time interval, we can ask for the probability that the system has a certain property at an *arbitrary* time of the interval. It is the same as the relative time the property holds in the interval. Being a mathematician, Borel does not take the existence of a limit of time average for granted. Using arbitrary times shows that time average probabilities make sense in a deterministic universe, a point made 12 years later by von Smoluchowski. However, the assumption of exact state already is mathematical idealization. Then Borel shows that an initial uncertainty would, because of the convex surfaces of molecules, be magnified by the dispersing effect of collisions. This would still leave the motion itself determined, only unknown and unknowable. But the indetermination of the given initial conditions is physical: one reason is the gravitational effect of distant objects such as stars. Borel assumes all the given quantities to be within small intervals, each value within such a small interval being equally probable. The motion of the system tends to disperse the initial indetermination over all states. By what is known as Poincaré's *method of arbitrary functions*, Borel is able to show that a particular probability law is determined independent of the specific form of the distributions over the initial intervals of indetermination (1906a, p. 1675).

Borel returns to his argument in 1913 and 1914, trying to show the origins of irreversibility and to replace the strict ergodic hypothesis by small random disturbances. Exact values of physical quantities are not physically real, specifically, there is a certain 'flottement' in defining the limits of a system (1913, p. 1698). Borel shows that molecular collisions lead to an *exponential separation* of trajectories or, as he puts it, of 'leaflets' of the set of states whose dispersion is studied (p. 1700).[32] The replacement by one centimeter of one gram in a star changes the

[32] Much later this nonlinearity in time development became the basis of what is today called chaotic behavior. It seems to be forgotten that gravitational instability as leading to exponential separation of molecular paths was invented by Borel if not earlier.

gravitational field by a fraction greater than $10^{-100}$. Since 'it is impossible to introduce the whole universe in our equations,' this much indetermination at least has to be admitted (p. 1701). In his 1914 supplement to the French edition of the Ehrenfest review, he emphasizes the difference between this indetermination and merely assuming the state of a system to be between given small bounds (1914a, p. 1722). The small indetermination is enough to allow external disturbances make molecular paths completely indeterminate in a fraction of a second.

The Ehrenfests in 1911 had made the distinction between ergodic and *quasi-ergodic* motions; the latter are the ones which come arbitrarily close to any point of phase space, that is, they are dense. They said that the existence of ergodic systems is to be doubted. One does not even have examples of the quasi-ergodic systems (1911, p. 31). They remark further that for the latter kind of systems, what they claimed to be the 'Boltzmann–Maxwell justification' of uniqueness of time averages, does not hold (p. 32, note 90). One year later the mathematician Michel Plancherel found a proof of the impossibility of ergodic systems, published in 1913. That same year Arthur Rosenthal published a different proof. It is based on Brouwer's fixed-point theorem. It is not difficult to show that the state spaces $R^n$ of dynamical systems all have the same continuous number of elements as the set of real numbers, independently of the dimension $n$. For example, there is a curve in the unit square filling it completely, known as the Peano curve. But the function defining the Peano curve does not give a correspondence between the two sets that would be continuous both ways. Brouwer's theorem shows that such cannot ever be the case: Dimension is a topological invariant, as one says. Thus as Rosenthal (1913, p. 797) points out, a curve filling the phase space cannot be a mechanical trajectory. He remarks several times that the number of trajectories on a surface of constant energy is nondenumerably infinite. Plancherel (1913, p. 1061) sets out to prove this claim by a measure theoretic argument. He shows, though only in outline, that a single trajectory has measure 0 whereas the set of all trajectories has positive measure.

After the impossibility results of Plancherel and Rosenthal, the obvious step was to try what the Ehrenfests had termed the quasi-ergodic hypothesis. However, a conceptual unclarity made the results in this direction empty. The notion of quasi-ergodicity first applied to a single trajectory. On the other hand, as the Ehrenfests note, there are continuously many different trajectories. It would have been important to keep apart the properties of one trajectory, of different sets of trajectories, and of the system as a whole. In the Ehrenfests it went without saying that all of the continuous number of trajectories had to be quasi-ergodic. This goes on the same page where they indicate that a point moving on a torus can trace a dense, aperiodic trajectory, or a periodic one (1914, p. 32, note 90). Not even Poincaré's recurrence theorem, allowing exceptions of probability 0, and the knowledge that there may be simple sets of periodic trajectories of measure 0, led to the reconsideration of the definition of ergodicity of a system as a whole. Thus Rosenthal's (1914) attempt at founding statistical mechanics on the notion of quasi-ergodicity assumes 'the single undisturbed motion of the system' to be dense in phase space (p. 894). That is to say, *every* trajectory is assumed dense. Rosenthal derives the result that the only stationary distribution must be constant over the different trajectories, or an ergodic distribution (p. 903). But its hypothesis is vacuously satisfied.

We saw in the beginning of this section that the proper definition of ergodicity as dense motion with exceptions of probability 0 was contained in Poincaré (1894). It went unnoticed by Rosenthal and others studying the ergodic problem. On the other hand, one can be sure that the matter was understood by Borel, the first one to propose a measure theory. His first paper on probability (1905b) deals with probabilistic problems for continuous sets in terms of measure theory. As he says, its scientific application is to the kinetic theory of gases, but in Borel (1906a) which deals with that application, measure theory does not appear explicitly. In his supplement to the Ehrenfest article he says of the strict ergodic hypothesis that 'it does not seem that this abstract hypothesis has been ever really meant by physicists' (1914, p. 1717).

The first ergodic theorems, or proofs of the existence and uniqueness of the time averages, appeared around 1910, as we saw in Section 2.4(a). Here the importance of a measure theoretic formulation allowing exceptions of measure 0 was not missed. The significance of these results, as connecting dynamical systems with pure mathematics and probability theory, was especially clear in the work of Weyl. An explicitly measure theoretic proof of Poincaré's recurrence theorem was given by Carathéodory in 1919. Increasing mathematical sophistication was, fortunately or not, countered by the ergodic theory's losing its role within physics, in favor of the more general methods of Gibbs.

### *3.2(b) The way to a probabilistic theory*

As we have seen, there were certain mathematical refinements in the conceptual basis of ergodic theory in the second decade of this century. These did not play any important role in physical applications. It is sometimes said that ergodic theory had no role at all in the later development of statistical physics, but this is not the case. In the statistical physics of Einstein ergodicity was postulated through its characteristic consequence, the uniqueness of limits of time averages. This assumption allows to determine them through the microcanonical measure. There are other ways of introducing ergodic behavior. Einstein says in 1902 that if there are further constants of motion besides total energy, the properties of the distribution of states are not determined by total energy solely, but depend on initial conditions as well. The avoidance of this situation is yet another way to introduce ergodicity, the one originally followed by Boltzmann.

The first ergodic theorem concerned the rotation of a circle with a given angle. As noted in Section 2.4(a), the result was formulated in different terms, but almost simultaneously, by the astronomer Bohl, the set theorist Sierpinski and that universal mathematician, Hermann Weyl, in 1909–1910. Weyl's paper 'Ueber die Gleichverteilung von Zahlen mod. Eins' of 1916 is the one widely known. However, a couple of years earlier Weyl had delivered a lecture on 'une application de la théorie des nombres à la méchanique statistique et la théorie des perturbations.' A formulation of the first ergodic theorem in number theoretic terms is as follows. If an irrational number is given, if it is multiplied by $2, 3, 4, \ldots$, and only the decimal part taken, the result will be a sequence of numbers between 0 and 1 that is uniformly distributed. The sequence is naturally only denumerable, so by equidistribution one means that the sequence visits each interval with the right limiting frequency. If a rational number is given, the above procedure results in a periodic sequence. The rationals have measure 0 in the set of all real numbers and are therefore exceptions. At that time, Lebesgue integration was naturally known to such a mathematician as Weyl. Earlier, in 1884, Leopold Kronecker had shown that the fractional multiples form a dense set on the unit interval. The result applies to any dimension. An example of density with a more physical outlook was published in 1913 by König and Szücs. They showed that, in the general case, the linear motion of a point inside a square or cube is dense. Weyl was able to generalize the equidistribution to any dimension, instead of only one. It followed that a case such as König and Szücs' dense two-dimensional motion also obeys equidistribution. Weyl was very well aware of the physical applications of his results, as the title of his 1914 talk indicates. He noticed that probabilities will coincide with the limits of time averages of statistical mechanics. Further, the astronomical application is noticeable. Since ergodicity leads to the convergence of time averages, in astronomical terms it leads to the existence of true mean motions, a great insight indeed.

The strong law of large numbers of Emile Borel is also related to the earliest ergodic theorems. This becomes clear from later developments where ergodic theory is made independent of the classical mechanical basis it had begun with. In some way the connection was already there in Bernstein (1912a), who related the ergodicity of rotations as studied by Bohl, Sierpinski, and Weyl, to the behavior of the successive integers in continued fractions. The probabilistic content of the ergodic theorem and related results was emphasized in Richard von Mises' views on the nature of the ergodic problem, starting in 1920. These are detailed in Section 6.3. Essential for him was to get rid of any mechanical concepts in the conclusion as well as the assumptions of the ergodic theorem. The underlying idea was that mechanical and random behavior are inherently incompatible. He represents a line of development that might seem not to have had any real effect. But that is not so, even if today only works of others are known, specifically those of von Neumann, Birkhoff, and others from the early 1930s. Von Mises was instrumental in formulating the *program* that led to a purely probabilistic ergodic theory, a program that did not go unattended.

Modern ergodic theory starts with the results of Koopman, von Neumann, and G. D. Birkhoff in the years 1931–1932. The development was very rapid, leading to a probabilistic ergodic theory and the theory of abstract dynamical systems. At the same time, measure theoretic probability obtained its definitive formulation in Kolmogorov and was at once put to use. Thus ergodic theory came to be seen as a special part of probability theory, or measure theory. I shall now give a rather detailed account of these crucial years.[33] In March 1931, Bernard Koopman, a 26 Ph.D. from Harvard, had a paper on 'Hamiltonian systems and transformations in Hilbert space' communicated in the National Academy of Sciences. The paper consisted essentially of the following remark. Take square integrable complex-valued functions $f, g, \ldots$ over the state space of a conservative Hamiltonian system. Define transformations $U_t$ over these functions by equations of the form $U_t(f(x)) = f(T(t, x))$. These transformations allow us to keep track of the values of $f$ under the dynamical motion $T$ of the state point $x$ without explicitly looking at the motion. The square integrable functions form a Hilbert space and the operators $U_t$ are unitary in this space, therefore linear and norm preserving. This observation marked the beginning of the operator treatment of classical dynamics. Note that

[33] Studies of the original unpublished sources were conducted at Harvard University archives during the fall of 1983.

indicator functions of subsets of the state space, through which probabilities are identified, belong to the Hilbert space.

Von Neumann was familiar with Hilbert spaces and their unitary operators, this having been his specialty in the formulation of the mathematical basis of quantum mechanics. He writes in his main paper (1932a, p. 71) that 'the possibility of applying Koopman's work to the proof of theorems like the ergodic theorem was suggested to me in a conversation with that author in the spring of 1930.' Only one year before the suggestion, the ergodic problem had seemed to him 'absolutely unsurmountable by the present standard of science' (1929, pp. 30–31). He also notes in 1932 that André Weil had in 1931 suggested a similar application. As a matter of fact, the collection of correspondence of George Birkhoff at Harvard archives contains a letter by Weil where he says that he had known all the results of Koopman's paper as well as many of von Neumann's already in 1928, but since he could not establish ergodicity in any specific cases, he did not publish this work.[34]

Von Neumann made an assumption of ergodic motions, formulated as the inexistence of other constants of motion than one which is constant over whole state space (1932a, pp. 78–79). He then showed what is known as the mean ergodic theorem. If one takes a part $A$ of the phase space $S$, the average (in other words, mean) of the time averages that the different motions spend in this region approaches the measure of it, as the time interval grows to infinity. If that measure is normalized into a probability $P$ and if $\hat{I}_A(x)$ stands for the limit of time average starting from state $x$, the result can be written as

$$P(A) = \int_S \hat{I}_A(x)\,dx. \tag{3.4}$$

Von Neumann's paper contains other remarkable insights, for example, the first sketch of the ergodic decomposition theorem. It states that the values of a complete set of invariants partition the state space into ergodic components. For example, in (3.4) there would be sets of states $x$ for which the time average $\hat{I}_A(x)$ has the same value.[35] His paper uses Lebesgue measure over the phase space of a dynamical system. Toward the end he discusses the empirical meaning of measure theory. Finite

[34] Prof. Weil tells that he became interested in the matter through Hadamard, who often spoke about the ergodic problem in his seminar in Paris 1928, and that he certainly did not have a proof of the von Neumann ergodic theorem for $L^2$ functions at the time, nor did he know about quantum theory or probability (personal communication).

[35] The probability $P$ is given as a mixture of the probabilities over the ergodic components. Of this latter, von Neumann seemed to have little idea.

accuracy of measurements corresponds to taking into account only sets of positive measure. His long paper 'Zur Operatorenmethode in der klassischen Mechanik' (1932b) is an extended version of the original article.

In October 1931 von Neumann told Koopman and Birkhoff about his result, the mean ergodic theorem. This led rapidly to Birkhoff's ergodic theorem. It says that under ergodicity, time averages approach unique limits not only on the average, but each single trajectory does this. There is a set of exceptions of measure 0. Measure theory was the natural way to formulate the result. Even the condition of ergodicity appears in measure theoretic terms. Let $A$ be a subset of state space and $T(t, A)$ the set of all states $T(t, x)$ where $x \in A$. Let $m$ be a measure over phase space. $T$ is a *measure-preserving transformation* if $m(T(t, A)) = m(A)$. $A$ is *invariant* if $T(t, A) = A$. A system is *metrically transitive* if all invariant sets have measure 0 or the measure of the whole state space. (In a probabilistic context that latter would be normalized to 1.) This definition of metrical transitivity, today usually called simply ergodicity, appears for the first time in Birkhoff and Smith (1928). Birkhoff published his result in the *Proceedings of the National Academy of Sciences*, as did von Neumann, but the paper came out in December 1931, whereas von Neumann's only came in 1932. In a note on 'Recent contributions to the ergodic theory' Birkhoff and Koopman (1932) go on to clarify the developments. There is no evidence in Birkhoff's correspondence of any reaction on the part of von Neumann, even though that has been sometimes suggested, or assumed. But von Neumann did publish a paper (1932c) on 'Physical applications of the ergodic theorem' in which he tried to argue that the mean ergodic theorem is all one needs in such applications.

In a review of Birkhoff's book *Dynamical Systems* of 1927, Koopman (1930) remarks that unstable solutions, if taken singly, are not physically meaningful. Birkhoff gave an informal talk in late December 1931 on 'Probability and physical systems,' where he said that, as Koopman suggested, single solutions are not meaningful since initial conditions cannot be determined precisely. Instead, one ought to get general characterizations of the class of possible motions. He refers to Poincaré as the first one to use (intuitively) considerations of a 'probability 1' kind: Poincaré's recurrence theorem is such a result. The systematic formulation of results of this kind is in terms of measure theory, which Birkhoff was using. He also says that physicists had 'on an intuitive basis formulated vaguely certain types of theorems, one of which in precise form is... the ergodic theorem' (1932, p. 366).

After Birkhoff's ergodic theorem, it took very little time to free it from the underlying classical dynamics. This was achieved simultaneously by Khintchine and Eberhard Hopf in 1932. Khintchine was one of the foremost experts in probability theory in his time. In the late 1920s he became interested in the mathematical foundations of statistical mechanics. His first technical paper on ergodicity is 'Zu Birkhoff's Lösung des Ergodenproblems,' where he says that 'Birkhoff's results are a really essential advance in the mathematical foundations of statistical mechanics' (1932d, p. 485). Because of the importance of the central result, he wants to give a simpler proof which generalizes it in two directions: He shows that the convergence of time averages holds for any stationary motions (that is, measure preserving transformations), and second, he considers any Lebesgue integrable functions of phase space instead of only indicators of regions. The first part is the essential one. With it, the ergodic theorem is shown as a result of abstract dynamical systems. These systems consist of a space with a measure and a measure preserving transformation, or several such transformations. Classical systems of course form a special case. The hard and difficult part, in this special case, is to show that the condition for the theorem holds, that is, that 'almost all trajectories go almost everywhere,' as one could say. In a footnote added after the paper was submitted, he notes that Hopf (1932b) had given a very similar result. The content of Birkhoff's theorem is that in the indecomposable case (just another expression for the measure theoretic criterion of ergodicity) limits of time averages are equal to phase averages.

In another paper of 1933, 'Zur mathematischen Begründung der statistischen Mechanik', Khintchine gives a more general philosophical discussion. It was published in the *Zeitschrift für angewandte Mathematik und Mechanik*, in a *Festschrift* for Richard von Mises. Khintchine compares the approaches of Birkhoff and von Mises in the following way: In the latter's theory, the succession in time of observations forms a Markov chain, so that the probability distribution of the events in the sequence depends only on the previous observation, but not on earlier ones. In cases in which this assumption is acceptable, it leads to great simplification. Birkhoff's approach is more general in that the effect of all of previous history can play a role. This statement naturally presupposes, even though it is not mentioned, that one refers by 'Birkhoff's approach' to a proper probabilistic generalization of the ergodic theorem of classical systems.

We see that for Khintchine, von Mises' theory was a special case of the purely probabilistic ergodic approach. In the former, a specific hypothesis is made about the probabilistic law for successive events. The latter only requires that we consider systems stationary in the probabilistic sense. Already in 1920 von Mises had published his ideas for an ergodic theory freed of mechanical assumptions. Khintchine in turn had published in 1929 a review on 'von Mises' theory of probability and the principles of physical statistics.' He there criticized von Mises' foundational system, and wanted to keep apart the ideas of statistical physics that he considered valuable. Khintchine's knowledge of von Mises' 1920 ideas on ergodic theory can be considered the route that led to the purely probabilistic formulation.

Next I shall discuss views on the relation between the probabilistic aspects of ergodic theory and quantum theory. The application of ergodic theory to quantum theory proper will not be addressed, even though there appeared works in this direction in the period under study.[36] It is remarkable how very little was published on the relation of ergodic theory and quantum theory. The probabilistic and indeterministic character of the new quantum mechanics became immediately well known. Soon indeterminism of the quantum mechanical kind became intimately connected with the very idea of physical probability. The development of ergodic theory within classical dynamics disturbs this ideal picture. But it did not seem to have disturbed some of the creators of these two theories. Von Neumann is a good example, for he contributed decisively to the proper mathematical formulation of both. He notes that there is a striking analogy between the operator treatment of classical and quantum mechanics, and that one can show a continuous change from quantum to classical mechanics by letting Planck's constant approach zero. But there is an essential difference, namely, that for a system with a finite-volume state space the spectrum is discrete in quantum mechanics, 'whereas it seems that in the classical problem, a pure line spectrum is the general case' (1932b, p. 595). In von Neumann's book on the mathematical foundations of quantum mechanics (1932), the classical and quantum theories are placed much more dramatically against each other. In that book the appearance of probability in the classical theory is only due to 'incomplete knowledge,' whereas in quantum mechanics probability appears because of the 'acausal' character of the theory (1932, pp. 2, 108).

Not every probability theorist was enthusiastic about quantum mechanical chance. Bohuslav Hostinsky was one of the first to treat games of chance as physical systems. Later he made important contributions to the theory of Markov chains and to continuous time random processes. He did not react favorably to the new mechanics. In the first volume of the famous journal of the Vienna Circle, *Erkenntnis*,

[36] Von Neumann (1929) is one example.

a discussion on foundations of probability is reported. Hostinsky says: 'There has been too much talk about indeterminism in the last years. In my view, there is determinism if in a given case a dependence between cause and effect has been established. If that does not succeed, it is better to say nothing than to talk about indeterminism' (1930, p. 285). A similar attitude was taken by Dirk Struik, possibly partly as a reflection of a materialist philosophy. He wrote in a 1934 paper on probability within ergodic theory that there is 'no probability without causality,' that is, probability is not physically well defined outside the dynamical framework of classical ergodic theory.

The idea of applying ergodic theory to the macroscopic domain of, for example, mechanically described games of chance, was continued by Eberhard Hopf, especially in his paper 'On causality, statistics and probability' (1934). This is one of the most important papers on physical probability, and will be discussed extensively in Chapter 5.

One more way was tried for claiming a relevance for classical ergodic theory in face of the new quantum mechanics. Koopman wrote in February 1932 to G. D. Birkhoff about his work on the spectrum of classical and quantal operators through which he tried to relate the two theories to each other. In his opinion, 'the insignificance of the individual streamline (trajectory) and the statistical interpretations suggest that there is a "principle of indetermination" in classical dynamics as well as in quantum theory.'

It appears that at least four different strategies were tried for claiming a role for ergodic theory within the development of physics. There were scholars such as Hostinsky and Struik who objected to the indeterminism of the new quantum theory. A second line was to try to let both theories live peacefully together by trying to push the analogies as far as possible.[37] Yet another approach was to apply the theory to macroscopic phenomena, as in the work of Hopf. He also hoped that new physical applications would prove the theory's relevance, specifically through the problem of turbulence in hydrodynamics (Hopf 1937, preface). A fourth way was that of Khintchine's. He had transformed the ergodic theorem into a result of probability theory, and was not therefore committed to classical dynamics. In the paper in honor of Richard von Mises (1933a), he commented on the changes in statistical physics in recent years: 'The continuous extension of state variables and the

[37] The analogy, as far as physical statistics is concerned, has not led to success. The ergodic theorems do not allow any obvious application to quantum systems, the reason being the feature that von Neumann had mentioned, namely, that of a pure point spectrum. The inapplicability became well understood only in the 1950s.

deterministic assumptions of the classical theory have gone more and more into the background. In addition to their historical importance, these two features of the classical theory seem to be applicable only as approximations in certain specific cases.' The aim of his paper was to give to Birkhoff's theorem 'a purely probabilistic formulation in which neither determinism nor the continuous character of states are mentioned.'

### 3.3 Einsteins views on probability

According to a widespread opinion, Einstein's probability concept was subjective; it related to knowledge. Everyone knows that he said, 'God does not play dice,' thereby denying randomness the ultimate role most quantum theorists wanted to give it. More generally, it is a very widespread idea that classical physics is incompatible with genuine chance. Probability and statistics result there from ignorance of the true determinist course of events, and not from any objective feature of nature itself. Einstein's views on probability will deliver a surprise in this respect. We shall see that the concept of probability had an important role in his early work, and that he ended up with very definite ideas about its meaning. This process took place between 1902 and 1904, in his papers on statistical mechanics.[38] One year later, at the height of his physical discoveries in 1905, he invented the theory of Brownian motion that was instrumental for the theoretical and experimental study of fluctuation phenomena. The particular way he thought probabilistic laws obtain their justification within classical statistical mechanics, was for him a model for the relation between the 'statistical quantum mechanics' and a further deeper theory he hoped would be found to explain it.

Einstein's background for his work on statistical mechanics was Boltzmann's gas theory. From here comes Einstein's conception of probability. We saw in Section 3.1(a) how Boltzmann had already in 1868 and 1871 identified probability as a limit of time average. First, though, Einstein started with something different. His paper of 1902 is similar to Gibbs' statistical mechanics, though he was unaware of Gibbs' work. In his next paper, of 1903, assumptions are made which are equivalent to the ergodicity of the system. First it is assumed that the system has a constant total energy, namely, that it is stationary or

[38] See Pais (1982, Ch. II) for Einstein's statistical mechanics proper. Since this and the next section were written, valuable comments on Einstein's early work have appeared in vol. 2 of his *Collected Papers.*

isolated from any surroundings. Second, the existence of other invariants of motion than energy is denied. As we have seen, this was precisely Boltzmann's ergodic hypothesis. There follows the assumption that there is for a stationary system a definite limit value for the time spent in a given region of state space. The consequence is that probability becomes identified as limit of time average.

The exact relation of Einstein to Boltzmann's several works is not known. In his early papers Einstein referred only to Boltzmann's *Lectures on Gas Theory*. Time averages do not appear explicitly there. However, in the paragraph on 'Ergoden,' Boltzmann refers to the time development of single systems. As to probability in general, Einstein had studied at least Poincaré's book *La Science et l'Hypothèse* (1902), where the latter discusses probability (ch. XI).[39]

In a short remark, Einstein (1911) says that had he known Gibbs' work at the time, he would have limited his own early papers to some remarks. On this basis it has been said that Einstein later adopted a Gibbsian standpoint. But even if we do not know what his remarks would have been, we shall see that he did not change his views on statistical probability in favor of postulated a priori ensembles. Einstein (1914a) leaves no unclarity in this respect. One should also remember that Gibbs was not so 'Gibbsian' about ensembles all the time, neither were ensembles foreign to Boltzmann.

The claim is sometimes made that Einstein's statistical mechanics abstracts from the classical mechanics of motion, and uses only one mechanical condition, the conservation of volumes of regions of the state space. This conservation property, corresponding to the conservation of total energy, holds for isolated mechanical systems which have a fixed total energy. We see that in fact Einstein assumed more: the nonexistence of invariants of motion other than total energy. He states this explicitly in his (1902, p. 419): The existence of an invariant of motion independent of total energy $E$ 'has the necessary consequence that the distribution of states is not determined by $E$ only, but must necessarily depend on the initial states of the systems.' The plural 'systems' here refers to his use of the microcanonical ensemble. If total energy is the only invariant of motion, 'the distribution of states of our systems... is produced by itself from any initial values of state variables which satisfy the condition for the value of energy' (pp. 418–19). His view is therefore that the distribution of states along individual trajectories (that is, the time averages) reproduces for any time evolution the microcanonical distribution. The validity of the assumption that total

[39] See Pais (1982, p. 133).

energy is the only invariant of motion depends on the dynamical law of motion of the system. Einstein's next assumption, the existence and uniqueness of limits of time averages, follows from the previous one. As a physicist he could allow the redundancy of stating his assumptions in various ways. The logical relations between these different formulations were settled with the birth of mathematical ergodic theory, as described in the previous section.

Einstein gives a more explicit explanation of the notion of probability in a somewhat surprising connection, in the famous 1905 article on the photoelectric effect, or light quantum hypothesis. In §5 of that work we read the following:

> In the calculation of entropy by the methods of molecular theory, the word 'probability' is often used in a meaning which is not in accordance with the definition of probability as it is given in the calculus of probability. Specifically, 'cases of equal probability' are often determined hypothetically, also when the theoretical pictures used are sufficiently fixed to allow of a deduction instead of a hypothetical determination. I shall show in a separate work [which did not appear] that one gets along completely, in considerations of thermal processes, with what are known as 'statistical probabilities.' By this, I hope to dispense with a logical difficulty which still hinders the execution of Boltzmann's principle. Here I shall give, however, only its general formulation, and its application for certain specific cases.

The entropy $S$ of a system is a function of the probability $W$ of its instantaneous state, and entropy growth is taken to be a passage to a more probable state. Then comes a derivation of the connection between entropy growth and probability. The result is: $S - S_0 = (R/N)\lg W$, where $S_0$ is the entropy of the initial state, $R$ the gas constant, and $N$ the number of molecules in a gram-mole. $W$ is a 'statistical probability': it is, as in 1903, the limit of time average.

The opening of the quotation, on the meaning of probability, is not unambiguous. The traditional definition of probability is that it is the number of favorable cases divided by the number of all possible cases which are here 'equiprobable.' In Boltzmann's formula for entropy, probabilities were determined combinatorially, as the relative numbers of 'complexions.' Why, contrary to appearance, is the meaning of probability in this 'equiprobability' derivation not in accordance with the one given in the calculus of probability? Because, it appears, Einstein meant that in the calculus of probability, probabilities are defined as limits of relative frequencies, that is, as 'statistical probabilities.' Further confirmation for this interpretation of the passage can be found by comparing it with a passage from Poincaré (1902, pp. 218–19). Poincaré there discusses a game of chance which, when observed for a long time, leads us to judge 'that the events are distributed in conformity with the calculus of probability, and here we have what I call objective probability.' The chapter on probability in Poincaré's book was one that Einstein had read, as was noted above.

The 'deduction' Einstein promises in the passage quoted, would be a replacement of combinatorial equiprobability hypotheses by a determination of limits of time averages. The existence of invariants of motion other than total energy makes these limits coincide with the distribution of states that is determined by the energy. He had in fact already once made such a replacement, in §7 of his paper of 1903. The state space is there divided into a finite number of regions of the same volume (microcanonical measure). To prove that the system spends asymptotically equal times in these regions, that is, to prove that they are 'equiprobable,' requires that there is only one independent invariant of motion. Einstein's promised deduction, therefore, would require a proof of ergodicity. The first such proofs for what were considered physically relevant systems appeared only in Sinai's work of the early 1960s. But Einstein's 'got along,' as he put it, without such a proof, with the considerable less difficult task of computing microcanonical probabilities, taking ergodicity for granted as he did.

The 'execution of Boltzmann's principle' is discussed twice later, first in a note (1907a) on the limits of validity of the second law of thermodynamics. According to the 'molecular theory of heat,' fluctuations of parameters are inevitable where thermodynamics says these are constants. The derivation of the fluctuation formula is treated in greater detail when, in 1910, Einstein returns to the foundations of thermodynamics and statistical mechanics. He formulates again what he calls Boltzmann's principle, as $S = (R/N) \lg W + \text{const.}$ He refers to the usual interpretation of the probability $W$, which is 'the [relative] number of different possible ways (complexions) in which the state, incompletely defined through the observable parameters of a system in the sense of molecular theory, can be thought realized' (1910, p. 1276). It can be questioned whether this interpretation of Boltzmann's principle can make any sense, 'without a complete molecular mechanical or other theory representing completely the elementary processes.' At about this time he had given up the hope of finding mechanical explanations to the quantized nature of radiation. His next step in the argument reflects this changing attitude. He says that Boltzmann's principle does have a content independent of any theory of elementary processes, provided that one views in a more general way the result of molecular kinetics according to which irreversibility in reality is only apparent.

In Einstein's review of 'the present state of the radiation problem,' of the year 1909, we find again an emphasis on statistical probability and Boltzmann's principle (1909, p. 187):

If one takes the point of view that the irreversibility of processes in nature is only an apparent one, and that an irreversible process consists of a transition to a more probable state, one must then give a definition of the probability $W$ of a state. The only definition that can be considered, according to my opinion, would be the following:

Let $A_1, A_2, \ldots, A_l$ be all the states which an isolated system of a given energy can take or, to be more precise, all states we are able to distinguish in such a system, with given aids. According to the classical theory, the system takes after a definite time a certain of these states (such as, $A_l$), and remains in it (thermodynamical equilibrium). But according to the statistical theory, the system takes, in an irregular sequence, all of the states $A_1, A_2, \ldots, A_l$ over and over again. [In a footnote, Einstein says that this is the only tenable view, as one can immediately see from the properties of Brownian motion.] If one observes the system for a very long time $\Theta$ there will be a certain part $\tau_\nu$ of this time such that the system is in state $A_\nu$ for precisely this time $t_\nu$. Then $\tau_\nu/\Theta$ will have a definite limiting value which we call the probability $W$ of the state $A_\nu$.

Einstein now defines entropy by Boltzmann's principle, and says that neither Boltzmann nor Planck have given a definition of $W$. They give a 'purely formal' definition of $W$ as a number of complexions. Complexions are logically unnecessary, since one can define their probabilities statistically. Einstein then makes the strong claim that the Boltzmannian relation between entropy and probability holds only if one follows his definition. Planck ought to have postulated the equations, $S = (R/N)\lg W$ and $W$ is the relative number of complexions, only 'under the additional condition that the complexions be chosen in a way in which they are found equally probable on the basis of statistical considerations, within the theoretical picture chosen by him.' Happily, Planck dispensed with this requirement, but 'it would not be appropriate to forget that Planck's radiation formula is incompatible with the theoretical basis Mr. Planck started with' (1909, pp. 187–88). In the ensuing discussion of the theoretical basis of Planck's formula, Einstein makes notice of the following possibility (p. 188, Sec. 6). He had applied Boltzmann's principle for the calculation of entropy from a 'more or less complete theory of the quantity $W$.' But the principle can be reversed, to yield statistical probabilities from observed values of entropy. A theory which gives values of probabilities differing from these, should be discarded. There is a clear turn in Einstein's thought at this point. His 'statistical probabilities' in the classical theory had been like Boltzmann's: The aim of ergodicity assumptions was to have physical conditions on classical systems such that limits of time averages could be identified with, as one says after Gibbs, microcanonical averages. Even if the conventional understanding of the program of ergodic theory is that it first aimed at deducing statistical laws from the underlying mechanical laws of motion, it was made clear above that such is not the case with Boltzmann. Einstein in turn hoped to derive his statistical probabilities from a 'complete theory,' though it would not necessarily have to be a mechanical one: Giving up the idea of such a derivation for quantized radiation, Einstein now suggests that we make a statistical inference of sorts, namely, one of discarding a theory which gives probabilities not consonant with observation. It should be emphasized, however, that he did not take empirically determined probabilities as a final state of matters, but insisted on a subsequent theoretical determination.

Einstein's last word on the concept of probability in statistical physics comes from two sources. The first is Einstein's discussion remarks (with Poincaré, Lorentz, and others) in the famous Conseil Solvay of the year 1911, appearing in print as Einstein (1914a). Einstein there insists on the necessity of a physical, time average concept of probability which renders Boltzmann's principle directly into a physical statement (pp. 356–57). Lorentz makes the remark that Einstein is not following the Gibbsian method, to which the latter answers: 'It is characteristic of this approach that one uses the (time) probability of a purely phenomenologically defined state. This has the advantage that one needs no specific elementary theory (for example, statistical mechanics) as a basis' (p. 357). A second work of interest here is a set of lecture notes of Einstein's course on statistical mechanics in the summer of 1913, taken by Walter Dällenbach.[40] Einstein considers there a point moving along a closed curve. The probability of finding the rotating point on a given part of the curve is defined as the limit of time average the point visits that part. Next Einstein considers another approach via ensembles: an infinite number of points is sent to rotate the curve. They form a stationary flow, which Einstein calls a 'Systemgesamtheit.' Probability is defined in the second approach as the relative number of points on the part of the curve under consideration. (Obviously relative number here reads the same as density.) Einstein shows that time and ensemble averages coincide in this case. Next he considers the combination of two uniform rotations. He notes, in more modern terminology, that

[40] I am most indebted to Prof. John Stachel, Editor, Papers of Albert Einstein, for a copy of these unpublished lecture notes, as well as for his bringing to my attention the paper Einstein (1914a). They will appear in vol. 3 of Einstein's collected papers.

irrational rotations lead to ergodic trajectories. Clearly Einstein had quite a good understanding of the nature of ergodic motions.[41] As we have seen, a mathematically satisfactory ergodic theory was at its beginnings at the time. The results it was striving at were obvious to Einstein from a physical and intuitive basis.

It is evident from the foregoing quotations that Einstein understood the relation between time averages and frequentist probabilities. Molecules are just like dice, only they do not have six faces and are not brought to rest at discrete intervals. They have a continuous number of orientations and an uninterrupted continuous motion. From the fundamental sameness of frequency and time average, it follows that the latter also must experience variation in time.

If a system is in thermal contact with a surroundings of constant temperature, any temperature differences should vanish. However, according to the 'molecular theory of heat,' this claim has, only approximate validity. The temperature of the system in fact fluctuates around that of the surroundings. These energy fluctuations are treated in Einstein's third paper (1904) on statistical mechanics. They follow a probability distribution of the familiar exponential form, and, as Einstein emphasizes, the probabilities are different from zero for all possible energy values, even if they are very small outside the value of highest probability. As we have seen, Einstein required of statistical physical systems that they have no other invariants of motion than total energy. For a single system, probability is then the same as the limit of time average. If an event has positive probability, it should eventually occur. Specifically, an energy fluctuation refers to the development in time of a single system. The interpretation of probability as time average leads to admitting the reality of such fluctuations. A second far-reaching consequence of the 'molecular-kinetic' theory of heat, in addition to fluctuations, is that there is no difference in principle between molecules and particles of greater size, possibly observable ones. It follows that one can consider such a single particle a 'system' whose kinetic energy fluctuates as a consequence of collisions with molecules. One can then calculate whether such thermal motion might be observable. These ideas initiate Einstein's third great achievement of the year 1905: the theory of Brownian motion. The existence of fluctuations is almost a consequence of the molecular structure of matter. Whether they would be observable was not much discussed

[41] Klein's biography of Ehrenfest tells of a discussion he had with Einstein on ergodicity in 1912. Ehrenfest had recently published his and Tatiana Ehrenfest's (1911) joint survey of the foundations of statistical mechanics.

before Einstein. Gibbs in his statistical mechanics (1902, pp. 74–75, note 4) was rather cautious about the matter, treating several cases where one could not observe fluctuations, without making a definitive claim that all cases are of that kind. And we have seen that Boltzmann in 1896 mentions some observations possibly indicative of Brownian motion, but does not specify what these observations are. He does say that the second law ceases to be valid with very few molecules. Einstein makes the same remark: The motion he predicts from the molecular-kinetic theory would, if observed, have as a consequence that 'classical thermodynamics could no longer be regarded as exactly valid even for microscopically detectable dimensions, and an exact determination of the true size of atoms is possible' (1905b, p. 549). Einstein himself made one such determination in 1905, and soon the matter was taken up with extensive experimental studies.

As I shall treat Einstein's works on Brownian motion proper in the next section, I will just make some remarks about his notion of probability in those studies. In Einstein's papers on Brownian motion, interpretative remarks are scarce. However, one can see that there he uses the time-average concept of probability (1905b, p. 553; 1906b, pp. 372, 373). At times he refers to the 'irregular' and 'fortuitous' character of molecular motion, and he introduces probabilistic concepts and conditions on this basis. One example is when he assumes the probabilistic independence of the motion of a particle from that of other particles (1905b, pp. 556, 558), and another, the assumption of the independence of the motions of one particle at sufficiently separated intervals of time (p. 556). The emergence of a uniform distribution of particles suspended in a solution is a consequence of the motion of molecules which 'change their position in the most irregular manner thinkable' (1908, p. 237). For the present we can sum up the most essential features of Einstein's study of fluctuation phenomena: He accepted the molecular theory and its inherently statistical character, with probabilities referring to the behavior of a single system in time. This interpretation of probability gives immediate reality to fluctuations as physical phenomena occurring in time whose conditions of observability can be determined. By these achievements Einstein laid the theoretical basis for the experimental study of Brownian motion and related consequences of the atomistic constitution of matter. Within a few years a new science of molecular statistics was established, silencing the last antiatomists.

It is clear that Einstein saw the time averages based on continuous time, and the relative frequencies based on discrete observations, as two aspects of a general concept of statistical probability. There is in fact a short paper (1914b) of his, completely forgotten until recently, which shows a deeper involvement with general statistical problems. Einstein invented in the paper some of the basic concepts of time series analysis, calling it 'a method for the statistical evaluation of observations from apparently irregular quasi-periodic processes.' The problem is to characterize the statistical properties of an empirical quantity $y$ over time $t$, the number of sunspots being mentioned as an example. To this task, Einstein introduced the concepts of spectral density and of the autocorrelation function of a stationary time series.[42] The former, called 'intensity' by Einstein, was meant to describe the statistical character of the observational series. He showed how it can be determined from the autocorrelation function, and discussed methods for its mechanical computation.[43]

Finally a few words about quantum mechanics. Einstein contributed greatly to the introduction of probability in the old quantum theory, as we shall see in Chapter 4. His well-known 'uneasiness' with the 'statistical quantum theory' is closely related to his views on the relation between statistical and ordinary mechanics of the classical sort. In fact, one sees that the uneasiness was not about theories being statistical, but rather about the kind of justification we are able to give to statistical theories and laws. In his comments on the essays in the volume *Albert Einstein: Philosopher-Scientist* (1949) he formulates his view as follows: The quantum theoretical description ought to be interpreted as a description of an ensemble of systems. The statistical predictions of quantum theory are unacceptable only if the claim is that they are a description of individual systems. Instead, Einstein hopes that a future theory could be found that would give a complete description of the latter. The relation of the present statistical quantum theory to that future theory would 'take an approximately analogous position to the statistical mechanics within the framework of classical mechanics' (1949, p. 672). Further, 'if it should be possible to move forward to a complete description, it is likely that the laws would represent relations among all the conceptual elements of this description which, per se, have nothing to do with statistics' (p. 673). Einstein believed that the probabilistic laws of statistical mechanics arise from the properties of an underlying mechanical motion. The analogy that he suggests is, that

[42] Both terms are of more recent origin. The spectral theory of stationary processes was developed in Khintchine (1934).

[43] See Yaglom (1987) for a specialist's commentary on Einstein's achievement. That paper is preceded by an English translation of Einstein's note. A longer, so far unpublished German version of Einstein's paper will appear in vol. 4 of the *Collected Papers*.

a future complete theory of quantum phenomena lead to statistical laws in a similar way. Probability and statistics are in no way unacceptable or undesirable concepts in physics, provided that their use is justified from a basic complete theory. Any other way of introducing probability into physics has to postulate it as a basic undefined notion. It forms therefore a hypothetical element of physical theory, which Einstein did not accept as a final state of matters.

## 3.4 Brownian motion and random processes

The kinetic theory had transformed the second law of thermodynamics into a statistical principle. A state of thermal equilibrium is only the overwhelmingly probable one. But then, since deviations from equilibrium have positive probability, they should be observed if we wait long enough. That 'long enough' was, as the Zermelo–Boltzmann debate showed (Section 3.1(b)) absolutely enormous, even for a cubic centimeter of plain ordinary gas. The reason is that the probability is extremely close to zero. Nevertheless, we can at least conceive in our minds of some very small number of molecules, for which the probability of deviation can be large. Boltzmann, who contributed more than anyone else to gas theory, said that with a few molecules only, the second law ceases to be valid (1896, p. 574). That is, behind the irreversible macroscopic world, there exists an unobservable, reversible microworld. Boltzmann had suggested in 1896 that certain observations of tiny particles in gases and liquids might offer evidence in favor of the fluctuation of pressure on different parts of a particle, but let the matter rest at that (1896, pp. 572, 574; 1897a, p. 584).[44]

Brownian motion, the erratic movement of microscopic particles suspended in a liquid, was discovered in 1827. Toward the end of the last century its cause came to be seen in the atomistic structure of matter. Nothing more than this qualitative connection was made with the kinetic theory, and the first theories of Brownian motion had to wait until Einstein in 1905 and Marian von Smoluchowski in 1906. These theories were confirmed by experimental studies of the most varied fluctuation phenomena. Foremost among the experimenters was Jean Perrin. Von Smoluchowski's theory of Brownian motion was directly based on probabilistic hypotheses concerning collisions between the Brownian particle and the molecules of the medium it is in. Einstein's theory, instead, started from a more traditional physical basis and had

[44] See also Sec. 3.1(b) for a more detailed account of Boltzmann's suggestions in this direction.

a specific continuous time random process as its end result.[45] The classical statistical physics he used is built on mechanical concepts, which easily suggests the use of continuous time. A physical literature evolved on Brownian motion and related phenomena. But a mathematical theory was missing. There were only two isolated attempts to develop such a mathematical theory with continuous time, Louis Bachelier's in 1900 and Filip Lundberg's in 1903, neither of which treated physics, but applications in economics and insurance. The former survived in Bachelier's book on probability in 1912, while the latter became generally known outside Scandinavia only after 1930.

The mathematical theory of continuous time random processes started taking shape as late as 1929, first with no connection to the physical literature that had appeared since Einstein's pioneering work. In the 1930s the physical and mathematical research traditions were brought together in Kolmogorov's general measure theoretic treatment of stochastic processes. I shall here follow the earliest theories of Brownian motion and some of the conceptual problems involved, and then discuss briefly the earliest mathematical developments in random process.

### *3.4(a) Brownian motion*

In 1827, the botanist Robert Brown was observing, under the microscope, a liquid containing pollen particles. They were in a state of seemingly incessant motion. The subsequent nineteenth-century history of Brownian motion is a model case of qualitative hypothesis testing.[46] Brown first thought that the motion of the microscopic particles suggests life, but abandoned this hypothesis as inorganic matter behaved in the same way. After Brown had recognized the motion, it was found that others had seen it before, but not realized its importance. One alternative hypothesis after another was eliminated: The motion is not caused by currents in the liquid, or any mutual effect of the Brownian particles, and so on. But the phenomenon remained a curiosity in the last century, with occasional experimental studies that kept adding to the list of what did *not* cause the phenomenon: The cause was not to be found in possible small temperature differences, or vaporization, or

[45] Recall that the first one to describe the state of a physical system by a time-dependent probability distribution had been Boltzmann in 1872.

[46] A historical overview of Brownian motion studies until 1906 can be found in de Haas-Lorentz (1913). See also Brush (1976) and the very useful book on von Smoluchowski by Armin Teske (1977).

dissolution, the last because the motion seemed to continue forever. It was observed that temperature has an effect on the speed of motion and that it depends inversely on the size of the particle. In the 1870s electric charges as causes were studied and dismissed. Some years later, Carbonelle suggested that the motion is caused by impacts from the surrounding molecules. He thought that in the Brownian motion of a small bubble of gas, there are random changes in the number of molecules leaving or entering the bubble. In 1888 Louis-Georges Gouy published the qualitative explanation that has since been accepted as correct. 'Thus Brownian motion, as the only one of physical phenomena, makes visible for us a state of constant internal agitation of bodies which has no external cause. One cannot avoid relating this fact to modern kinetic hypotheses and seeing in it a weakened result of thermal molecular motion'. (Gouy 1888).[47] The constancy of the motion had been the ground for eliminating many of the suggested explanations. It had to be concluded that the motion took place without any cause that would have created a thermal disequilibrium. Hence it seemed that Carnot's principle could not be valid at microscopic dimensions, for there heat seemed to become spontaneously converted into visible motion.

In 1900 Felix Exner studied the speed of motion of the Brownian particles. It was known to go down with size and up with temperature. According to the kinetic theory's equipartition principle, a particle has the same average kinetic energy as a molecule of the surrounding liquid. But from his observations, Exner arrived at velocities that would give a kinetic energy much too small for the Brownian particle. For equal kinetic energies, molecules should move only about 30 cm a second instead of a few hundred meters, or else the Brownian particle should move much faster.

Einstein's theory of Brownian motion is firmly based on the statistical mechanics he had developed in 1902–1904. Since it is based on statistical mechanics instead of kinetic theory, it is rather generally applicable. No specific kind of molecular interaction need be assumed. I have reviewed Einstein's conceptual background in the previous section. Let us now see in some detail how he derives the continuous time probability law for a Brownian particle. Einstein's argument combines two aspects he has been admired for: simplicity and a methodical following of ideas he thought right. Einstein's starting point in his theory of Brownian

[47]Teske (1977, Ch. 12) offers an evaluation of early observations and hypotheses on Brownian motion. He says (p. 175) that only two of (the?) seven works on Brownian motion between Gouy and Einstein follow Gouy's view.

motion is his doctoral dissertation, published as Einstein (1906a), but finished in April 1905 right before his first paper on Brownian motion (1905b). In the thesis he determines Avogadro's number $N$ by relating it to the diffusion coefficient $D$ by the formula

$$D = \frac{RT}{6\pi kr} \tag{3.5}$$

where $R$ is the gas constant, $T$ absolute temperature, $k$ the coefficient of viscosity, and $r$ the radius of the diffusing molecules. The formula results from a combination of van't Hoff's law and Stokes' law. In his elementary exposition of Brownian motion, Einstein (1908) explains his use of the quantitative van't Hoff law for osmotic pressure as follows. There is a container of liquid with a movable porous wall in the middle such that the molecules of the liquid pass the wall. Bigger molecules that do not pass through are placed on one side, which according to the kinetic theory results in an osmotic pressure on the wall. Thus, with or without wall, a concentration of dissolving molecules creates a pressure in a liquid, of a magnitude determined by van't Hoff's law. Stokes' law, on the other hand, gives the force needed to move a spherical object through a liquid. The diffusion coefficient formula results from requiring the force of osmotic pressure with which the particles diffuse in the liquid to be in equilibrium with the force (friction) with which the liquid resists the motion.[48]

From the point of view of the kinetic theory, there is no difference in principle between molecules of the dissolving substance and particles suspended in the liquid; the only difference is size (Einstein 1905b, p. 550). Therefore the particles perform the same kind of irregular motions as the molecules themselves. Einstein derives the probabilistic law of Brownian motion for a symmetric one-dimensional diffusion. The movements of $n$ different particles are assumed independent, as are the movements of the same particle for different time intervals sufficiently far apart. The displacement $\Delta$ obeys a symmetric probability law $\Phi$. Particle density at time $t$ is denoted by $f(x,t)$. The diffusion coefficient $D$ is identified as the average speed of mean-square displacement in time $\tau$ (p. 558),

$$\frac{1}{\tau}\int_{-\infty}^{\infty} \frac{\Delta^2}{2}\Phi(\Delta)\,d\Delta = D. \tag{3.6}$$

It follows rather simply that the time development of the distribution

[48] See Brush (1976) and Pais (1982, pp. 88–92) for details.

function $f$ is governed by the equation (1905b, p. 558)

$$\frac{\partial f}{\partial t} = D\frac{\partial^2 f}{\partial x^2}. \tag{3.7}$$

The right probability law is the solution of this diffusion equation. Its explicit expression is

$$f(x,t) = \frac{n}{\sqrt{4D\pi}} \frac{e^{-x^2/4Dt}}{\sqrt{t}}. \tag{3.8}$$

This distribution law is, as Einstein at once notes, the same as the distribution of 'probable error.' It follows easily from (3.6) that the *mean displacement* in a given direction is proportional to the square root of $2Dt$ (p. 559). Note that particle size but not mass figures in the displacement formula, through the diffusion constant $D$ as given by (3.5). Einstein ends his paper by considering particles of diameter 0.001 mm in water at room temperature. The mean displacement in a second would be $8 \times 10^{-5}$ cm, and in a minute, $6 \times 10^{-4}$ cm. A new way of connecting Avogadro's number to observable phenomena is found as a by-product (p. 560).

In his 1905 paper Einstein wrote that 'it is possible that the motions to be treated here are identical with what is known as "Brownian molecular motion,"' which was at once confirmed. Einstein's theory pointed at the right observable quantity to focus upon, namely, the mean-square displacement. In a short note (1907b) he calls the attention of experimentalists to the impossibility of observing the velocity of a Brownian particle, a point made also by von Smoluchowski earlier, in 1906. It follows from the equipartition of energy that a particle of mass $m$ has a definite average velocity $v$. Einstein considers one case investigated by Svedberg. There the velocity as determined from average kinetic energy would be far too high to be observable in a microscope. The actual path of a Brownian particle is not observable; what one sees is only a mean velocity over the interval of observation. Einstein says that this mean velocity will appear to the observer as an instantaneous velocity. 'But it is clear that the velocities so obtained do not correspond to any objective property of the motion studied – at least if the theory corresponds to fact,' he concludes (1907b, p. 42). When the right observable quantities had been pointed out, a breakthrough in the experimental science of molecular statistics soon occurred, with Jean Perrin as the leading figure.[49]

[49] See Pais (1982, Ch. 5), and the monograph Nye (1972).

Let us make a digression of a philosophical character before continuing with theories of Brownian motion. Do we really *see* the motion of Brownian particles? This is not so, for observed speed proved to be a mere chimera. That should explain, at least in part, the scarcity of quantitative data on the phenomenon. The observable magnitude that can be brought into relation with the kinetic theory is the mean displacement of the particle as Einstein's theory suggests.[50] A theory of the motion had to be invented before the proper magnitude to observe was located. But even if that event was a great success for theoretical science, let us not think that experimentalists were completely as loss without a good theory.[51] In a way experimenters were describing what they observed – the motion of the Brownian particle. But maybe they did not reflect on what they saw, and under what conditions. As to what they saw, the motion observable through a microscope is not the same as the true motion of the particle. For it experiences in normal circumstances up to some $10^{20}$ collisions a second, so that one sees average effects of these collisions. Second, there are all sorts of conditions relating to seeing the particle's motion, physiological and technical such as the resolution of the eye and the aperture of the microscope, and so on.[52] Anyway, that one does not see the true motion, is also the explanation of the deviant figures concerning the particle's supposed velocity. The kinetic theory's equipartition principle gives a velocity that is much too high, as noted by Exner in 1900 and by others even earlier. In fact, if the particle moved at that speed, it would have to vanish at once from the field of view of the microscope. The discrepancy was taken to be a decisive fault of the kinetic theory. The evidence of the senses weighed more in this case. To that evidence, an *interpretation* of the data was silently added. An observational report would say that the particle is moving in all directions, being in such-and-such locations at such-and-such times. It was possible to measure the time and distance with a certain precision; so far all is well. An error was committed when, starting from these figures, a *velocity* was ascribed to the particle. Therefore what the theory of Brownian motion gave to the experimentalists was a new interpretation of their data: the figures told about

[50] Here one can follow one particle in nonoverlapping intervals, or several particles at the same time.

[51] Maybe they were at a loss anyway: von Smoluchowski says that the phenomenon of Brownian motion was never studied in detail, even if it was observed a thousand times. Cf. Teske (1977, p. 177).

[52] The role of the observer in Brownian motion is studied in the last chapter of de Haas–Lorentz (1913).

mean displacement, not velocity.[53] The theory also led to the result that the displacement is proportional to the square root of time, and *that* was crucial knowledge for experimentalists.

There were a couple of other problems of a philosophical character about Brownian motion that puzzled physicists, in addition to the question whether we see it. These problems were: Can Brownian motion be taken as a violation of the second law of thermodynamics? And last, does classical physics permit the increase of observational accuracy beyond any limit, or is there a definite lower bound caused by thermal motion? Von Smoluchowski (1912) argued that one would not be able to gain work from the Brownian motions. As to the last question, Brownian motion sets a lower limit to an observational situation. Sufficiently fine measuring instruments will register their own thermal noise, as first made clear by Ising (1926). Barnes and Silverman (1934) give a review of these problems. Kappler in 1938 placed a torsion balance with a mirror under low pressure, and registered photographically a (very slow) movement of light reflected from the mirror. He was able to show that he had captured 'the true motion of the torsion balance' (1938, p. 389), that is, traces of individual molecular impacts on the mirror, acting as a Brownian particle, had been magnified on the film. One final remark relating to the observation of Brownian motion: The idea is sometimes expressed that one could improve observational accuracy by averaging. The dangers invited by such averaging were well known to statisticians. One has to keep separate the error term (variation) in observations and the numerical accuracy of the readings. The last significant digit of a single observation is the last significant digit of an average. Hence this way, too, is barred from leading to unlimited precision.

Von Smoluchowski had studied density fluctuations in a gas in his article in the *Boltzmann Festschrift* published in 1904. These fluctuations are violations of a strict reading of the second law. Brownian motion is a particular result of a fluctuation. The pressure that the surrounding medium exerts on a particle from all sides varies. Von Smoluchowski had been postponing the publication of his theory of Brownian motion for several years. In his paper (1906) he tells that he had hoped to conduct an experimental study also. But when Einstein published his theory in 1905, von Smoluchowski felt that he, too, had to give out his version. It is built up differently from the one of Einstein's, being more directly probabilistic in character: The individual motions are described as consisting of a succession of nearly linear parts with

[53] If there were that many data; cf. footnote 49.

randomly changing directions at points of collision. The random changes occur at discrete intervals. Being based on assumptions about collisions, it is an approach in the style of the kinetic theory, whereas Einstein's theory has the abstract character of statistical mechanics.

The single impact of a molecule on a Brownian particle is small. Previously it had been thought that the great number of impacts from all sides would average out any possible net effect on the particle. Von Smoluchowski illustrates the 'Denkfehler' committed here by a comparison with gambling (1906, p. 762). The error would be the same as in thinking that one would not be bound to lose in a casino a sum greater than the bet of a single trial. But the mean deviation of gain or loss is on the order of the square root of the number of trials. Von Smoluchowski moves at once to the consideration of Brownian motion. It differs from the gambling situation in two ways: First, the average change of velocity in a collision is not constant but depends on the velocity $C$ of the particle. Second, it is not symmetric either, for an increase is the less probable the greater $C$ is (p. 763). The condition of a stationary, equilibrium state is that the average kinetic energy of the Brownian particle equals the average kinetic energy of the molecules of the surrounding medium. But this would give particular velocities something like a thousand times too high. Von Smoluchowski offers as an explanation of the discrepancy, the unobservability of the true motion through a microscope. That motion is an 'extraordinarily complicated zigzag line, whose linear parts are much smaller than the dimensions of the particle itself' (p. 764). Those linear pieces number $10^{16}$ to $10^{20}$ a second.

Von Smoluchowski gives a probabilistic description of the Brownian motion process, based on the following simplifying assumptions: The velocity of the particle is assumed constant. Each collision causes a very small change in direction. The angles of these changes are assumed equal, of size $\varepsilon$, taken in a random direction in space. The lengths of the individual linear pieces are likewise assumed all the same, the 'true mean free path of the particle,' denoted by $l$ (pp. 765–66). There are two cases to be investigated, according to whether the mean free path is small or great in comparison to the size of the particle. If the particle size is small in relation to the mean free path, successive impacts can be assumed independent. That would be the case for Brownian motion in a sufficiently rare gas, whereas in a liquid it would be the opposite. The independence assumption is the justification for assuming a random direction of change of motion. Von Smoluchowski expresses this by requiring all the planes, as determined by the given direction of motion and a new one, to be equally probable. We can think of the choice also in this way. There is a cone with its axis aligned from the sharp end to the end of a linear part. The next linear part goes from that end point to the circumference of the base of the cone. Therefore the choice is one of a random direction in the circle forming the base of the cone. With all the above assumptions put together, von Smoluchowski is studying what could be described (for the modern reader) as a rather simple *random walk process*. His probabilistic problem is to determine the *mean-square displacement* traversed, as a function of $l$, $\varepsilon$, and the number $n$ of linear parts (p. 766). The problem for a liquid is more complicated.[54]

Von Smoluchowski's formula for mean-square displacement shows proportionality with the number of steps $n$. Since he assumes a constant velocity for the particle and an equal length between collisions, the steps of his process take place at equal time intervals. The proportionality of mean displacement and number of steps is therefore a discrete version of the proportionality for continuous time that we found in Einstein's theory.

Von Smoluchowski remarks twice on the 'unexpected consequence,' or the 'surprising phenomenon' of the independence of the mean displacement from the mass of the particle (pp. 769, 776). It does not seem that special attention would have been paid to that aspect of Brownian motion before. It had been clear that size and temperature are the only relevant factors; why that was so, was not declared a problem. Another matter on which von Smoluchowski remarks concerns the second law of thermodynamics. Brownian particles are interesting in this respect, for, as he says, a way is conceivable that could turn heat into mechanical energy. 'It is one of the many ways of gaining work from heat, were only our experimental tools fine enough. But it is of greater interest, since it does not seem as completely unrealizable as the catching of individual molecules by a Maxwellian demon' (p. 778).

In the 1910s, the study of fluctuation phenomena became well established. In a typical situation one would follow the change in the number of Brownian particles in a certain region at regular intervals. If the motions of the particles are assumed independent of each other, with all positions within the region equally probable, the number of particles in the region follows a Poisson distribution. Since that distribution is determined by one parameter, the average number of particles, the frequencies of the different numbers of particles are independent of what the particles and the liquid are like. The numbers

[54] Einstein's statistical mechanical approach applies equally well to gases and liquids since it is not based on assumptions about collisions.

observed at different times provide an example of correlated random events. The probabilistic dependencies were called 'probability after-effects.'[55] These dependencies can be expressed in the form of transition probabilities from a number $m$ to a number $n$ of particles in the region, in a given period of time. Through the consideration of recurrence times, that is, average times for when the particle number first returns to a previous value, the statistical character of irreversibility could be illustrated. Experimental work verified the existence and magnitude of von Smoluchowski's theoretically predicted density fluctuations. Svedberg and others used for their studies liquids with colloidal particles. Radioactive decay was also taken to belong to this science of molecular statistics, partly because the same Poisson formula as one finds for density fluctuations, was found to apply there too.[56]

### *3.4(b) Continuous time random processes*

Von Smoluchowski's probabilistic theory of Brownian motion gives an answer to a problem posed by Karl Pearson in 1905, known as the problem of *random flights*: to determine the probability that the distance traversed in a plane in $n$ steps of equal length is within given bounds when the angle of turn after each step is arbitrary. Von Smoluchowski's process turns by a fixed angle, but is three-dimensional. Since the spatial orientation of the angle is arbitrary, Pearson's case is covered by von Smoluchowski's theory. Pearson published his problem in *Nature* in 1905, but it is unknown whether von Smoluchowski had noticed it. He had worked in England in the late 1890s and had even himself published a short paper in the same journal. He refers to Pearson's problem in his (1916, p. 558) together with Rayleigh's random flight problem of the year 1880.[57]

From a more general point of view, von Smoluchowski's Brownian motion process, Pearson's random walk and Rayleigh's random flight are discrete time continuous state space *Markov processes.* That nomenclature, of course, is of much later date, from the early 1930s. The most obvious way of arriving at a Markov process is to consider sums (or averages) of independent random variables.[58] That is also what von Smoluchowski in effect does.

[55] Jaw-breaking 'Wahrscheinlichkeitsnachwirkungen' in the original.

[56] These developments are reviewed in Fürth (1920) and Chandrasekhar (1943). For molecular statistics see also Nye (1972).

[57] See Dutka (1985) for the history of the theory of random flights.

[58] This is how Markov arrived at the idea in 1908.

An even more elementary example of a random process is met with von Smoluchowski's theory of fluctuations in molecular statistics. In applications of the theory, one counts at regular discrete intervals the number of Brownian particles in a given space. The particle count process gives the earliest physical example of what since the early 1930s have been called *Markov chains.* Whereas in von Smoluchowski's theory of Brownian motion there is a continuous state space, it is finite here. The characteristic Markov property – the probability of the next event of interest depends on the present event but not on previous ones – says here that the probabilities for the number of particles in the next observation depend only on the number at present.

There are two main lines of development in the theory of Markov chains and processes. One leads from an originally dynamically formulated ergodic hypothesis to a purely probabilistic formulation. This is the work of von Mises from 1920 and later. The random processes connected with Brownian motion were one of the principal examples von Mises had in mind, and I shall illustrate his approach somewhat in Chapter 6. Von Mises assumed the macrostates of a statistical mechanical system to form a Markov chain. He developed the matrix formalism for such chains, summing up the theory in his *Wahrscheinlichkeitsrechnung* of 1931. The second main line in the development of Markov chains comes from ideas of mixing of liquids, as introduced by Gibbs. Poincaré reformulated the mixing problem in discrete terms in (1912, p. 301) as one involving the mixing of a deck of cards. After a long delay, the matter was taken over again by Hadamard in 1927 and Hostinsky in 1928. Von Mises did not come to grips with the continuous time theory, whereas Hadamard makes notice of that possibility in his (1928), and Hostinsky publishes his first papers on it in 1929.[59]

Classical statistical physics inherited a continuous time parameter from its mechanical basis. Probabilistic laws depending on continuous time had been used by Boltzmann and others, and even by Einstein in his theory of Brownian motion. Von Smoluchowski's theory of Brownian motions had a discrete time and a continuous state space, and his theory of density fluctuations finally had both a discrete time and a finite state space. Its predictions were verified by Svedberg and Westgren around 1915.[60] The observation of molecular fluctuation phenomena produced discrete time series. The problem of formulating a theory with continuous time is suggested from the general conceptual

[59] These developments are studied in great detail in the monograph Antretter (1989).

[60] Cf. Chandrasekhar (1943, Ch. 3) for these matters.

basis, from Einstein's use of continuous time, and from the inherent theoretical interest. Since a Brownian particle has no definite velocity, its behavior in short times presents a challenging problem. Physicists responded to that challenge: Langevin (1908), von Smoluchowski in several works, Fokker (1914), Planck (1917), and Ornstein as foremost.[61] But these physical works remained unknown to the mathematicians who started shaping a more mathematical theory of continuous time random processes in the late 1920s. Up to that time they had almost completely neglected it. The exceptions were Louis Bachelier, who in 1900 dealt with problems arising from the change of stock exchange rates, and Filip Lundberg in 1903. Lundberg's work, done in Swedish, considered the risk probabilities an insurance company faces, these probabilities being functions of continuous time. The form of such probability laws stems from the assumption of 'rare events' occurring independently of each other. It is the same Poisson law that Rutherford had found for radioactive decay in 1900 and 1903.[62] Norbert Wiener's 1921 theory of Brownian motion introduced probability in function space, the space of realizations of a random process, but did not succeed in having influence before the 1930s.

Bachelier's first work was a dissertation, *Théorie de la spéculation* (1900), with Henri Poincaré as one of the two examiners. Its topic was a probabilistic treatment of the changes in stock exchange rates. The workings of the Paris stock market are explained rather in detail. That is not of specific interest here, but the mathematics Bachelier tries to create. His later book *Calcul des probabilités* of 1912 is devoted to continuous probability, and I shall follow that presentation. It is, unlike Bachelier (1900), an accessible source, namely, one of the few books on probability of the time. The book also displays the connections to the theory of Brownian motion. Bachelier's basic idea is very clear: A discrete number $n$ of observations leads to complicated expressions, he says (1912, p. 152). The limit $n \to 0$ was earlier taken as an approximation. 'We suppose a very great number ($\mu$) of experiments in sequence, such that the succession can be considered continuous' (p. 153). The experiments follow each other in 'infinitesimally small equal intervals,' so $\mu$ gives total time. 'This consideration furnishes a valuable image that makes us conceive the transformation of probabilities in a sequence as a continuous phenomenon,' says Bachelier (p. 153). The change of exchange rate obeys a law of chance subject to

[61] See Chandrasekhar (1943) for a review of these.

[62] To be discussed in the next section. The intuitive ground of the Poisson law of rare events is shown in Feller (1968, pp. 156, 446).

an *infinity* of varying influences. The problem is: with the rate set at 0 now, to determine the probability after time $t$ that it is between the values $x$ and $x + dx$. In 1900 Bachelier's notation had been $p_{x,t}\,dx$, in 1912, $\tilde{\omega}_{0,t,x}\,dx$. Bachelier assumes independence from previous variation. Independence is 'due to the complexity of causes' that 'makes all things happen as if they were independent' (p. 279). The Markov property follows:

$$\tilde{\omega}_{0,t,x}\,dx = \int_{-\infty}^{\infty} \tilde{\omega}_{0,t_1,z}\tilde{\omega}_{t_1,t,x-z}\,dz\,dx \tag{3.9}$$

for any $t_1, t$. Bachelier introduces an idea he calls 'radiation of probability,' which is the analogy of diffusion as applied to probabilities (p. 323). Let $\tilde{\omega}_{t,x,\ldots}\,dx$ denote the probability that the exchange rate is between $x$ and $dx$ at time $t$. Let $\zeta(u, x, t, \ldots)$ be the probability that the rate increases by $u$ in the time interval from $t$ to $t + dt$, assuming it was $x$ at time $t$. One says that $x$ has 'emitted' toward $x + u$ the 'quantity of probability $\tilde{\omega}_{t,x,\ldots}\zeta(u, x, t, \ldots)\,du\,dx$' (p. 323). Special cases studied by Bachelier include:

> 'Case of independence': $\zeta(u, x, t, \ldots)$ depends only on $u$ and $t$.
> 'Uniform case': $\zeta(u, x, t, \ldots)$ does not depend on $t$.

Bachelier's work did not have much influence. When Kolmogorov presented his general theory of Markov processes in 1931, he made rather acerbic comments about the lack of rigor in Bachelier's mathematics. In his (1935) review of the state of research in probability theory, he went as far as saying that Bachelier's work is incomprehensible. However that may be, Bachelier's book with its main theme, the study of probability laws depending on continuous time, was there for anyone to look at. It made the connection between the theory of Brownian motion and diffusion, and it contained the basic property of Markov processes as well as definitions of special kinds of processes such as the stationary ones.

The mathematical study of continuous time random processes emerged suddenly in 1929. Several probabilists were doing it that year, de Finetti, Hostinsky, and Kolmogorov.[63] Why does this come so late? One aspect at least is that it was difficult to imagine a continuously acting chance. Continuous time formed part of determinist classical physics. In Bachelier that determinism prevails, and only the 'infinity of influences' is responsible for things proceeding *as if* guided by chance

[63] Harald Cramér also published an exposition of Lundberg's theory of risk in 1930.

(1912, p. 277). In others, the change from a deterministic to a probabilistic science brought along a discretization of time.

After having put down these somewhat daring suggestions about continuous time as leading to a deterministic description, and a probabilistic one reciprocally requiring a discretization of time, I found Reichenbach's 1929 essay 'Stetige Wahrscheinlichkeitsfolgen' (continuous probabilistic sequences). The paper was received by the *Zeitschrift für Physik* on November 30, 1928, and it is fairly long as well as technical in character. Reichenbach clearly must have been working on it for quite some time before the above date. He says that it was in the nature of the usual concept of probability that it could only be applied to discrete events (p. 275). Reichenbach's reason for this view is that he thinks probabilities are to be interpreted frequentistically, and frequencies apply only to discrete events. He also finds (although we may know better) that continuous time was previously unknown to probability theory (p. 278). Still, he says, continuous processes occur everywhere in physics, as with the trajectory of a Brownian particle.

> That process has so far been always cut artificially into a discrete one and the old concepts of probability applied. The method gives naturally useful results, but it does not lead us closer to understanding such processes. It leads to the well known paradox that the individual states of the particle are treated as independent... even though one knows that there obtains a continuous causal chaining of these states, which excludes probability. The paradox is resolved if one succeeds in transforming the strict causal determination of the continuous evolution into a probabilistic one (p. 276).

That is what Reichenbach set out to do. The traditional idea is that knowledge of nature can always be improved. Thus, ignorance of initial conditions is represented by a probability that can be made approach 1 (pp. 274, 305). Motivated by his indeterminist philosophy, Reichenbach instead postulates *another*, separate probability over the possible trajectories also.[64] A consequence of the second probability is that the old kind of causality holds only 'infinitesimally' (p. 277). He develops the mathematical theory at length, but I have not been able to find any other references to it, except for a couple in Reichenbach's own work. Still, written before any of the other mathematical papers on continuous time processes started appearing in 1929, it is one more witness to a change in views about probability.[65]

[64]The suggestion brings to mind the path integrals of quantum mechanics; cf. the end of Ch. 4.

[65]For a continuation of these developments, see Secs. 7.3, 7.4(e), and 8.3.

Was the simultaneous invention of theories of continuous time random processes based on unrecorded influences, or was it an expression of *Zeitgeist*? Maybe someone will find out some day. A possible source of influence could be the international congress of mathematicians held in Bologna in 1928, where many of the people involved were present. Hadamard spoke there about the mixing of cards and its relation to statistical mechanics, and it does not take much imagination to see the possibility of a continuous time theory on the basis of his paper (1928). On the other hand, *Zeitgeist* may have made its presence felt through quantum mechanical probability, two years old at the time, and particularly through the quantum indeterminacy that had made its entrance the previous year.

## 3.5 RADIOACTIVITY BEFORE ITS EXPLANATION IN QUANTUM THEORY

Radioactive decay is a phenomenon discovered in 1896 but only explained by quantum mechanics. Today one can say that radioactivity provides the primeval absolutely indeterministic event.[66] But that knowledge was slow in coming, from the discovery made by Becquerel in 1896 to Einstein's realization in 1916–17 that it had to do with quantum theory to Gamow and Gurney and Condon's explanation in 1928 of alpha decay in terms of quantum mechanics. Radioactivity is sometimes seen historically as one of the factors that led into the acceptance of indeterminism. Statements to that effect may be due, at least to a first approximation, to the rationalizing tendency of the brain, to its remarkable ability to see things as they ought to be. Certainly such statements are typically written *after* 1928, with two or three exceptions.

The role of radioactivity in probabilistic thinking has been discussed by Amaldi (1979), Pais (1986) and van Brakel (1985), among others. Amaldi is enthusiastic about radioactivity as 'a pragmatic pillar of probabilistic conceptions' and laments that historians of quantum theory are not paying attention to it. Pais sees in radioactivity two puzzles that remained unresolved before modern quantum theory: What is the source of energy of the radiation, and why do some atoms decay and others not?[67] The decay follows the exponential probabilistic

[66] It can even be heard as a tick of a counter, so you take a break or something if there is one in the next five seconds. That is entirely decided by quantum mechanical chance (and you!).

[67] From Pais (1977). He adds a third problem – do all atoms eventually decay – in his book *Inward Bound*.

'half-life' law, so that the latter question can be put as: What is the mechanism producing the half-life law? Van Brakel (1985) finally discusses in great detail 'the possible influence of the discovery of radio-active decay on the concept of physical probability.' His conclusion concerning that role is by and large in the negative.

The early study of radioactivity can be fruitfully compared to that of Brownian motion. Both were seen as statistical phenomena where fluctuations from an average are observable.[68] After its accidental discovery, the first problem with radioactivity was to find at least a qualitative explanation. In Brownian motion such an explanation was found in the 1880s. But before Einstein's theory of 1905, attempts at quantitative description failed because the Brownian particle's supposed velocity did not exist. With radioactivity it was the other way around: Search for a qualitative explanation came to an end only through quantum theory. The radioactive atom's sought-for cause of decay did not exist. The quantitative law, on the other hand, was uncovered by Rutherford already in 1900. It is difficult to say when the law was understood as statistical in nature.

Rutherford (1900) was studying the radioactivity 'emanated' from thorium oxide. His experimental arrangement, and the apparent ease of the conclusion, are as if from a textbook. He placed the thorium in one end of a long tube; the other end had a larger space with an arrangement for measuring electric conductivity. Air was blown from the thorium end of the tube for a while, and the air in the space became thereby ionized. Then the airflow was cut, and the drop in current registered at regular intervals. Arguing that the intensity of radioactivity at time $t$ is directly proportional to the current, that is, the number of ions produced at $t$, a mere eight readings were enough for Rutherford to fit a continuous curve to his observations. He concluded that 'the intensity of the radiation given out by the radio-active particles falls off in a geometrical progression with the time' (1900, p. 7).

For the rate of decay of the number $n$ of ions produced by radioactivity, Rutherford gives the equation

$$\frac{dn}{dt} = -\lambda n. \tag{3.10}$$

If at $t = 0$ there are $N$ ions produced, their number $n$ at time $t$ decreases

[68] Kohlrausch (1926) reviews the study of the statistical aspect of radioactivity up to a time when quantum mechanics was developed.

by the law (p. 11)[69]

$$\frac{n}{N} = e^{-\lambda t}. \tag{3.11}$$

Putting $e^{-\lambda t} = 1/2$ and solving for $t$, one gets the half-life of the radioactive substance studied. This is the time it takes for half of the remaining radioactive atoms to decay. The parameter $\lambda$ has a different value for each of the radioactive substances.

Rutherford's decay law is continuous in time. A hint at the discontinuous character of the decay process is found in Rutherford and Soddy (1903), who identify it with 'the expulsion of a charged particle' (p. 580). There is no explicit discussion of the decay law's holding in the average.

Rutherford was first searching for an external source of energy. One naturally tried to alter the external circumstances in order to find out what effect it has on the radiation. Radioactivity proved to be resilient, the rate of decay remaining unaffected by any manipulation. The view emerged according to which the unknown cause must lie in the atom itself.

Probabilistic concepts were explicit in the study of radioactivity at least by von Schweidler (1905), who considered *fluctuations* in the decay rate. On general probabilistic grounds, a small number of atoms must lead to a deviation of the factual process of diminishing of radioactive atoms from the ideal probabilistic law (p. 1). Von Schweidler's object was to study whether the dispersion term given by the calculus of probability can reach a level that could be empirically shown, and in fact, radiative fluctuations were observed one year later by Kohlrausch (1906). Fürth in his review of fluctuation phenomena (1920) places radiative fluctuations as one chapter among others. Thus radioactivity was taken as a statistical phenomenon quite comparable to, say, density fluctuations of Brownian particles in a liquid. Fürth does not question classical physics: Even if the mechanism of radioactive decay is unknown, he thinks the atom is like 'a tiny planetary system' (p. 76). Then, with a great number of component parts, its 'conditions of equilibrium become so complicated that they are not predictable in practice.' It follows that 'the configuration of component parts becomes so unfavorable that one of the particles has to go from a closed trajectory into an open one' (ibid.). The event can be taken as a random phenomenon; the statistical law is a consequence of a great number of

[69]Obtained by integrating (3.10), as in Rutherford and Soddy (1903).

individual cases. Classical models of radioactive decay were proposed and studied rather extensively several years before Fürth's book.[70] These models contained a complicated nucleus with a high degree of freedom, which would account for the statistical law. The types of explanations suggested were similar to the ones we shall meet in Chapter 5.

Not everyone was satisfied with the kind of classical explanation Fürth envisaged. It was thought that classical models could not display the exterme variation of lifetime one observes in radioactive decay. Maybe it was also thought that the 'tiny planetary system' of Fürth and others would have to be stable and regular in its behavior like its big brother, the solar system, as some statements by Rutherford indicate (cf. Pais 1986, p. 122). That was precisely the view of stable classical mechanics von Smoluchowski had challenged in his paper (1918) on the origins of chance and probability in physics. As an explanation of the random variation observed in classical systems, he suggested what is known as sensitive dependence on initial conditions. In order to show that there is no contradiction between 'lawlike' causes and 'random' effects, von Smoluchowski constructs a mechanical model reproducing precisely the exponential law of radioactive decay. To complicate matters further, he says that he 'of course does not believe radium atoms to really possess such a structure' (p. 262). Instead, radioactivity can be taken 'as the most complete type of "randomness"' (p. 261).

Even if purported classical explanations were sometimes met with disbelief, the role of the 'acausal' character of radioactive disintegration has been mainly seen after the fact. Van Brakel (1985) has surveyed the literature and come to the conclusion that 'before 1925 there is no publication in which the "indeterministic" nature of radioactive decay is considered to be a remarkable aspect of the phenomenon' (p. 379). On the other hand, he finds many publications claiming such a role, all written *after* 1928. If there was an early acausal view of radioactive disintegration, as has been claimed, it was held by a few physicists only, and not properly published. Einstein, in his paper on quantized radiation (1917), concludes that spontaneous photon emission follows the same Poisson probability law as radioactivity. This remark gives a connection between radioactivity and the old quantum theory. As Pais (1982, pp. 410–12) stresses, Einstein expressed in this connection his 'earliest Unbehagen [discomfort] about chance.' That feeling does not make him a very good candidate for a protagonist of acausality.

Specialized articles in professional journals are often reserved about

[70] These have been discussed in some detail in Amaldi (1979, pp. 15–18).

the motivations of the work. Speculation does not easily fit into the impersonal ‘objective’ style of communication. Hendrik Kramers was Niels Bohr’s closest associate in the 1920s. He wrote together with Helge Holst a small popular book on Bohr’s theory of the atom, whose German edition of 1925, written just before the invention of quantum mechanics, contains interesting speculations about radioactivity and related topics, such as atomic energy. The chapter of interest was written by Kramers. Speaking about ‘quantum jumps’ between the stationary energy states of an atom, he asks: ‘Is there an unknown micro-mechanism regulating the jumps in a uniquely determined way, that is, does every jump have its cause, or do probabilistic laws perhaps belong to the kind of *fundamental* physical laws which cannot be reduced to anything deeper?’ He says that exactly the same question was posed already when the laws of radioactive decay of atomic nuclei became known (p. 138). Usually there have been causal singular processes behind probabilistic laws – lotteries and the kinetic theory are mentioned as examples – but with atomic processes it is possible that the law of causality fails (p. 139). That possibility became true sooner than Kramers could ever have expected, for less than half a year later, his young colleague Werner Heisenberg invented quantum mechanics.

# 4

# *Quantum mechanical probability and indeterminism*

## 4.1 PROBABILITY IN THE OLD QUANTUM THEORY

Probability was connected to quantum theory right from the start, in 1900, through the derivation of Planck's radiation law. But not much attention has been paid to the concept of probability in quantized radiation, or in the 'quantum jumps' from one energy level to another in an atom, and related problems. A whole chapter or book, instead of a section, could be written on the background of quantum theory in statistical mechanics and spectral analysis with this aspect in mind.[1] Probabilistic properties were included in very many of the most important papers dealing with radiation between 1900 and 1925.[2] It became also clear in time that Planck's law could not be derived from classical physics. Einstein admits this around 1908. Later he said, in a work of 1917, that it is a weakness of the theory of photon emission that 'it leaves to "chance" the time and direction of the elementary processes,' thereby, according to Pais (1982, p. 412), making explicit 'that something was amiss with classical causality.' It remains somewhat open how the origins and shifting interpretations of probability in the old quantum theory affected the acceptance of the probabilistic interpretation of the new quantum mechanics of 1925–1926, and of the indeterminism that found its confirmation in Heisenberg's uncertainty relation in 1927. For example, statements can be found which indicate that Born's probabilistic interpretation of Schrödinger's wave mechanics had already become obvious after the latter's proof of the mathematical equivalence of matrix and wave mechanics (cf. Pais 1986, p. 260).

The old quantum theory starts with Planck's radiation law in 1900. The 'blackbody' law assumes radiation to be emitted in discrete quanta

[1] The physical developments have been studied by Klein, Kangro, Kuhn, and Hoyer, among others. For references to this literature, see Ch. VI of Pais' Einstein biography, which also provides a good starting point for these matters.

[2] See, for example, the collection of papers on the 'old' quantum theory in van der Waerden (1967) or ter Haar (1967).

$h\nu$, where $h$ is of course Planck's constant, the quantum of action, and $\nu$ the frequency of the radiation. In his derivation of the radiation law, Planck used Boltzmann's method of discrete energy levels.[3] Boltzmann's reasons for dividing the phase space into a finite number of cells with side length $\varepsilon$ were connected to probability: There was no proper probability mathematics for showing that a property of states has probability 0 (or 1) for a continuous set of states. The meaning of probability 0 was illustrated by a combinatorially calculated *low* probability that goes to zero in the limit. Planck, in order to derive his radiation law, was forced to use Boltzmann's combinatorial scheme with $\varepsilon = h\nu$. One can speculate whether the development of modern physics would have been delayed had Boltzmann had measure theory at his disposal.

Bohr (1913) combined Rutherford's (1911) model of the atom with Planck's quantum of action. There is a sequence of stationary electron orbits around the nucleus, and each of these corresponds to a definite energy level $\varepsilon_i$ of the atom, starting from the lowest $\varepsilon_1$. *Transitions* between stationary states are connected by Bohr's quantization rule:

$$\varepsilon_m - \varepsilon_n = h\nu. \tag{4.1}$$

This law, connecting energy difference with frequency of light emitted, forms the basis of the derivation of molecular spectra.

Two kinds of probabilities can be found in the old quantum theory, namely, probabilities of energy states and *transition probabilities* between energy states. Einstein's crucial 1917 paper, 'On the quantum theory of radiation,' introduces both.[4] In this work, Einstein sees as the aim of quantum theory the determination of the energy states $\varepsilon_n$ and the quantum theoretical probabilities (1917, pp. 121–22). A gas of temperature $T$ is assumed, the probabilities of the different energy states of its molecules, or the relative frequencies as Einstein says, being given by the formula

$$W_n = p_n e(-\varepsilon_n/kT) \tag{4.2}$$

where $k$ is Boltzmann's constant, $T$ the temperature, and $p_n$ a 'combinatorial weight' of the $n$th energy state. The motivation for the distribution comes from statistical mechanics. The second kind of probability in quantum theory concerns the transitions between energy levels, as denoted by pairs of indices ($n$, $m$). First, a (differential) probability law is assumed

$$dW = A^n_m \, dt \tag{4.3}$$

[3] Explained in Sec. 3.1(a).

[4] The paper was also published as Einstein (1916).

for the transition to a lower energy level 'not affected by external causes.' In this connection Einstein remarks that 'the statistical law assumed corresponds to that of radioactive reaction' (p. 123). The energy $\varepsilon_m - \varepsilon_n$ of a definite frequency $\lambda$ is emitted in the transition. Next the gas molecules are assumed to be under directed radiation, which leads to absorptions of radiation energy by the molecules, that is, jumps to higher energy levels. These obey the probabilistic law

$$dW = B_n^m \rho \, dt \tag{4.4}$$

where $\rho$ is the intensity of the radiation. But, and here comes the essential step in Einstein, also a transition to a lower level is possible as an effect of radiation, with the probability

$$dW = B_m^n \rho \, dt. \tag{4.5}$$

Einstein then argues that emission has to be 'directed,' with all directions being equally probable. The molecules exchange energy with the radiation under which they are placed. An equilibrium of that exchange and the statistical mechanical energy distribution as given by (4.2) requires that the number of transitions to higher energies be on the average the same as the sum of the two kinds of transitions to lower levels (p. 124). In more recent terminology, the number of spontaneous and induced emissions has to match the number of absorptions. This condition leads easily to Planck's radiation law and to Bohr's quantization condition (4.1).

It is clear from Einstein's paper that he considers the quantized nature of radiation as very definitively established. Both absorption and emission of energy are quantized and directed. A weakness of the theory is, as mentioned, 'that it leaves to "chance" the time and direction of the elementary processes' (1917, p. 128). To this self-criticism of Einstein's it can be further added that the theory is not able to determine the probabilistic laws, for the coefficients $A_m^n$ and the two kinds of $B$'s are only described as 'constants characteristic of the index combinations in question' (p. 123). At best they could be determined empirically from the observation of radiation intensities.

In view of his later opposition to an indeterminist quantum mechanics, it might seem surprising that it was just Einstein who introduced probability into quantum theory. But as we have seen from our discussion of his views on probability, he had no prejudices against the use of probabilistic concepts. On the contrary, some of his most important achievements were based on probabilistic methods. In Brownian motion, the essential step is based on the determination of the character and magnitude of statistical fluctuations. Similarly in the theory of quantum gases, to be considered below. But just as the probabilities of Brownian motions came out of a classical mechanics of the microscopic level, so the statistical laws of quantum behavior should come out from some fundamental theory. When Heisenberg found such a theory in 1925, Einstein did not find that it fulfilled the strict constraints he had set for the derivation of probabilistic laws. Heisenberg's quantum mechanics owed to Einstein (1917) the essential idea of working with quantities that depend on *two* stationary states.

The next important step in the introduction of probability to quantum theory, after the transition probabilities of 1917, is also due to Einstein. Inspired by Einstein's probabilistic treatment of radiation in 1917, S. N. Bose had in 1924 given a derivation of Planck's radiation law. It was based on a new way of counting the number of states and was purely probabilistic in character. In Boltzmann the molecules of a gas are treated as probabilistically independent. This leads to the classical Maxwell–Boltzmann formula for the number of 'complexions' of a given energy $E$ (as in Section 3.1(a)). In Bose the molecules are instead positively correlated. In 1924 and 1925 Einstein developed a quantum theory of the ideal gas that is based on the new 'Bose–Einstein' statistics. I shall only be concerned with one specific aspect of this theory, namely, the derivation of Einstein's fluctuation formula for the number of molecules. A similar consideration had led him already in (1909) to the idea of a dual wave–particle nature of radiation.

In his second communication on 'The quantum theory of the ideal monoatomic gas' (1925), Einstein assumes given a volume $V$ of gas. It is separated from an unbounded system of the same nature by a wall that lets through only molecules of energy 'infinitesimally close' to $E$, that is, molecules with energy between $E$ and $E + \Delta E$. Let $n_\nu$ be the average number of molecules within this energy range. The problem is to determine the fluctuation $\Delta_\nu$ in the number of molecules. Einstein obtains (p. 9) for the mean-square fluctuation $\overline{\Delta_\nu^2}$, relative to the number of molecules $n_\nu$, the formula

$$\overline{\left(\frac{\Delta_\nu}{n_\nu}\right)^2} = \frac{1}{n_\nu} + \frac{1}{z_\nu}. \tag{4.6}$$

If one assumes the molecules independent, the mean-square deviation $\overline{\Delta_\nu^2}$ is proportional to $\sqrt{n_\nu}$. Normalizing with division by $n_\nu^2$ gives the first term in Einstein's formula. The second term, $1/z_\nu$, appears in radiation as an 'interference fluctuation,' as Einstein says. It relates to the number $V/h^3$ of cells of size $h^3$ in the volume $V$ by the formula (p. 5)

$$z_\nu = 2\pi\left(\frac{V}{h^3}\right)(2m)^{3/2}E^{1/2}\Delta E. \tag{4.7}$$

Here $m$ is the mass of a molecule. For a gas $z_\nu$ can be interpreted 'by adjoining to the gas in a suitable way a radiation process' (p. 9). The way to do this had been laid down in de Broglie's 1924 theory of matter waves: To a free material particle of mass of $m$ there corresponds the frequency $\nu_0$, as determined by the equation relating the energies,

$$mc^2 = h\nu_0. \tag{4.8}$$

If the particle is at rest, a standing wave of such frequency is adjoined to it. If it moves with velocity $v$ relative to a coordinate system, there exists a wavelike process with the modified frequency

$$\nu = \frac{\nu_0}{\sqrt{1 - v^2/c^2}}. \tag{4.9}$$

The meaning of the second term in the fluctuation formula (4.6) becomes now clear: '...a scalar wave field can be adjoined to a gas, and I have assured myself through computation that $1/Z_\nu$ is the average square of fluctuation of this wave field,' Einstein writes (p. 10).[5]

Einstein's fluctuation formula (4.6) proved important for the development of quantum mechanics in at least two respects. First Erwin Schrödinger also was working on the statistics of quantum gases. He learned of de Broglie's matter waves from Einstein[6] and was led through matter waves to his wave mechanics. Second, by adjoining a scalar wave field to matter, Einstein arrived at a computation of his 'interference fluctuation' $1/z_\nu$, writing presumably with interference in mind that 'here more than a mere analogy is involved' (1925, p. 9). What did he use matter waves for? He used them for determining the mean-square fluctuation of $n_\nu$. But this can only be if the wave field determines a statistical law that the number of particles in the volume $V$ obeys. As we shall see, in some way this connection between matter wave and probability was used in Born's probabilistic interpretation of Schrödinger's wave function. At the same time it pointed at a characteristic feature of quantum probability: the appearance of interference terms.[7]

Our discussion of the old quantum theory has led us to transitions between energy states, to matter waves, and to interference phenomena.

[5] There is a misprint, uppercase $Z_\nu$ instead of the $z_\nu$ otherwise used in the paper.

[6] This is seen from a correspondence reprinted in *Briefe zur Wellenmechanik* (ed. Przibram).

[7] See further Pais (1982), Ch. 23 for quantum statistics and Ch. 24 for Einstein's role in the birth of wave mechanics.

Einstein implanted the concept of probability into each of these three themes which, as pursued later by Heisenberg, Schrödinger, and Born, respectively, bring us to the creation of quantum mechanics proper and to the realization of its true probabilistic character.

By the early 1920s the view became accepted, at least by the leading researchers, that classical mechanics is incompatible with discrete quantum phenomena. Specifically, the old theory could offer no reason for the discrete energy states, absorptions, and emissions in the Rutherford–Bohr 'planetary atom.' With Einstein (1917) these were treated by probabilistic methods, even though there was no way of determining the probabilistic laws of energy states and transitions from theory. The status of probability therefore remained open: What was thought of the appearance of probabilistic notions in the most fundamental part of physics?[8] Some discussion of this matter can be found in the German 1925 edition of Kramers and Holst's popular book on Bohr's quantum theory we met already in Section 3.5. It contains a review of the interaction between light and matter, written afresh for that edition by Kramers. He was there asking whether the probabilistic laws for the quantum jumps between stationary states could be the most fundamental ones, that is, irreducible to deterministic laws (p. 138). Were this the case, processes in nature would not be uniquely determined from the state of things at a given time. 'One must beware of labeling such a view an epistemological impossibility.' For the law of causality is not a logical necessity, and it is conceivable that it 'breaks down in atomic processes' (ibid.). Probabilistic laws are ones that 'leave a free choice within a definite region, a kind of nonuniqueness.' Kramers concludes the matter by saying that the choice between causality and probabilistic laws is 'at present more a matter of taste, and possibly remains so for ever' (p. 139). These lines must have been written at almost exactly the same time as their author was preparing his joint paper (1925) with Heisenberg, on the dispersion of radiation by atoms. In the spring and summer the latter pursued, in Born's words, 'some work of his own, keeping its idea and purpose somewhat mysterious.'[9] A fundamental change in quantum theory was about to take place.

## 4.2 THE PROBABILISTIC INTERPRETATION OF QUANTUM MECHANICS

After the above preparations, we come to the invention of matrix mechanics and wave mechanics in 1925–1926, and to the recognition of

[8] We know already that Einstein was not satisfied with the way this happened.

[9] Born in van der Waerden 1967, p. 36.

their probabilistic and indeterministic character. I shall not try to outline any of the developments of the physics proper here. An admirable presentation of the development of the old quantum theory and of matrix mechanics is found in van der Waerden's *Sources of Quantum Mechanics*. Another useful reference is Tomonaga's *Quantum Mechanics* that presents the physical theory, both old and new, following in broad terms the historical development. I shall only discuss matters as far as probability and indeterminism are concerned. Here is a brief time chart that should help steer through the discussion that follows[10]:

*July 1925*: Heisenberg invents quantum mechanics. *September 1925*: Matrix mechanics of Born and Jordan. *January 1926*: First physical success of the new theory, the derivations of the hydrogen spectrum by Pauli, and by Dirac. Schrödinger invents wave mechanics. *March 1926*: Schrödinger shows equivalence of wave and matrix mechanics. *By June 1926*: Schrödinger denies 'quantum jumps' in favor of a 'classicist continuum theory.' *July 1926*: Born's probabilistic interpretation of wave mechanics. *December 1926*: Schrödinger gives up the idea of a pure continuum theory. *March 1927*: Heisenberg's uncertainty relation, the definite step to indeterminism.

By the 1920s, Rutherford's classical model, as supplemented by Bohr's quantization rules, had become more and more doubtful in the study of atomic structure. Heisenberg (1925) attacked the problem with the conviction that a dead end had been reached in the old quantum theory as based on classical electron orbits and Bohr quantization. He wanted to have a completely new kind of physical theory that, as he wrote in the abstract, 'is founded exclusively upon relationships between quantities that in principle are observable,' a theory that would determine the energy levels and transition probabilities of atoms (p. 268). The latter come out as absolute squares of the theory's *transition amplitudes* $a(n, m)$, that is, the complex numbers representing quantum mechanical states. An amplitude $a(n, m)$ can be given as the sum over all intermediate transition amplitudes $a(n, k)$, $a(k, m)$, where the product of the latter two gets summed over all possible values of $k$. In the definition of a product of transition amplitudes one can get a different result if the product order is reversed, as Heisenberg notes.[11]

[10] Based on the date of submission of an article. One can only envy those olden days for the speed of dissemination of new results. Publication usually followed submission in less than two months. Both dates can be found in the bibliographies of Mehra and Rechenberg's book series.

[11] I shall add some more remarks on Heisenberg's paper below, in discussing the origins of Born's probabilistic interpretation of wave mechanics.

The proper mathematical formalism of Heisenberg's quantum mechanics was matrix algebra, as Born very soon realized. Each transition amplitude $a(n, m)$ is a function of two integers. If one lists each value, running through $n$ and $m$, one is easily led to a system of rows and columns, or a matrix. The resulting matrix mechanics was first presented in Born and Jordan (1925b). Their mathematical technique was new to physicists. Shortly afterward Dirac (1925) worked out his own algebraic theory, on the basis of Heisenberg's original article.

With Schrödinger's wave mechanics, in contrast to the abstract and formal algebra of matrix mechanics, physicists could get along with their tried methods of differential equations. As mentioned, Schrödinger was fast in pointing out a certain mathematical equivalence between the two approaches. He also suggested the first interpretation of the formalism. It is a bit strange that both matrix and wave mechanics were first formulated in symbolic terms, then used in the deduction of physically remarkable consequences, and only later directly interpreted.[12]

From de Broglie's doctoral thesis of 1924, Schrödinger found the matter waves of frequency $\nu_0 = mc^2/h$. Further $\nu = \nu_0/\sqrt{1 - v^2/c^2}$ is the frequency of the wave associated with a particle moving uniformly with velocity $v$, as in formulas (4.8) and (4.9). Then the wavelength $\lambda$ is $u/\nu$, where $u$ is the phase velocity $c^2/v$ of the wave. De Broglie was able to explain somehow the stationary electron orbits on this basis: The electron has to resonate an integral number of times in one full orbit,

$$2\pi r = n\nu. \tag{4.10}$$

Here $2\pi r$ is the length of the electronic orbit. I shall not go into the details of how wave mechanics was born from this rudimentary basis. One story is that Schrödinger was giving a report on de Broglie's thesis at a seminar in Zurich upon which Debye at once asked for a wave equation. Some weeks later Schrödinger came back and had it worked out: namely, the equation for the hydrogen atom in Schrödinger's first paper (1926a, p. 362) of the four-part 'Quantisierung als Eigenwertproblem.'[13] The general form of the famous Schrödinger equation

$$\operatorname{div}\operatorname{grad}\psi - \frac{1}{u^2}\psi = 0 \tag{4.11}$$

[12] The same is true of Dirac's transformation theory. But he is well known for his 'playing with equations,' in order to see what they give.

[13] The writing of Schrödinger's series, and related papers such as (1926c,e) spans half a year, from January to June 1926. The most detailed account of the history of wave mechanics is Mehra and Rechenberg, vol. V. See specifically their Ch. 3 for the rise of wave mechanics, and p. 419 for the story about Debye.

with $u$ the phase velocity, appears in Schrödinger's second paper on wave mechanics (1926b).

Schrödinger's first interpretation of his state function $\psi$ was as follows (1926f): A particle is represented as a superposition of waves, a 'wave packet,' and $\psi$ gives the intensity of electric charge. It would thus have immediate physical reality, and would display a kind of continuous mechanism for the way in which particles repel each other at short distances. The problem, fully realized by Schrödinger, is of course that it carries no energy or momentum, neither can it be conceived as a wave motion in three-dimensional space except for a free particle; so where's the reality? Still, wave mechanics was greeted by most physicists, who found it more familiar to work with than the abstract matrix mechanics of Heisenberg, Born, and Jordan.

It is a bit strange, as Pais (1986, p. 258) says, that Schrödinger 'never liked quantum probability.' He certainly was against quantum jumps, even in the 1950s (cf. his 1952). But Schrödinger had been an *indeterminist* already in 1922, a philosophy that has clear similarities with the views one finds in part IV of Franz Exner's huge book (1919), and that was in general motivated from the Viennese tradition in statistical mechanics since Boltzmann. Schrödinger had considerable expertise in statistical physics, his publications including, for example, probabilistic studies of radioactive disintegration, as well as the quantum statistics of ideal gases right before the wave mechanics.[14] Schrödinger, in his inaugural lecture of 1922, explains his philosophical position as follows[15]: Causality is a habit of thought. Physical processes instead are statistical in an immediate way, and that habit is only brought about through the extraordinary probability of our usual observations of the macroscopic world. Schrödinger does not want to deny the possibility of postulating a causal structure for the elementary atomic processes, but it would be 'circular,' 'a duplication of natural laws resembling the doubling of objects in nature through animism' (1922, p. 11). Admitting such a possibility, he describes his position as *acausal* rather than indeterministic. Even planetary motions are acausal. The extraordinary precision in their behavior is only due to the enormous number of atoms. Further views of this kind in Schrödinger can be found in a letter of 1924 he wrote to Hans Reichenbach[16]: It is probable that a statistical view of all of nature will be adopted 'in a few decades' (p. 69). On the atomistic level 'laws will be assumed of a kind where for *sharply* determined

[14] His work on quantum statistics of gases is described in Hanle (1977).

[15] The lecture was only published in 1929.

[16] Published by Reichenbach in 1932, in the Vienna Circle journal *Erkenntnis*.

conditions...a whole continuum of possible results corresponds.' We see that by 1924 Schrödinger professed exactly the kind of indeterministic laws on the atomic level that are characteristic of quantum mechanics.[17] The 'riddle' of induction, that is, why we have those causal habits, reduces according to Schrödinger to the following: In a great number of repetitions of the 'sharply defined initial conditions,' the distribution of results is determinate, say, a uniform one (p. 70). The riddle becomes the question, why is there a determinate distribution. Such a determinateness, too, proves to be a characteristic feature of quantum mechanics. Schrödinger's estimate for the time it would take for a change to a statistical view of nature was off by one order of magnitude: It took only three years from 1924.

It is indeed peculiar that the development of wave mechanics made Schrödinger the chief challenger of an indeterminist view of quantum phenomena. His opposition possibly started after the equivalence proof of wave and matrix mechanics. In wave mechanics he saw, at least by the summer of 1926, a way back to a visualizable continuum theory. The German favorite of Schrödinger was of course *anschaulich*. This turn probably started taking place after the equivalence proof of wave and quantum mechanics in March 1926. In another letter to Reichenbach, right after the first paper on wave mechanics was finished, Schrödinger still writes that he stands on Reichenbach's side 'in the fight against the prejudice of absolute determinism.'[18] By June 1926, transition probabilities were to be replaced by intensities derived from a space–time description. Schrödinger scored great successes with his retreat to a continuum theory.[19] Heisenberg was horrified at what he considered a dismissal of all the central insights achieved in quantum theory since Planck's discovery in 1900. Of Schrödinger's lecture in Munich in June 1926 he wrote that '...one becomes 26 years younger listening to it. Schrödinger throws all things "quantum mechanical" over board: photoelectric effect, Franck's collisions, Stern–Gerlach effect, etc.'[20] In the discussion of Schrödinger's lecture, Heisenberg was under heavy attack.[21] But the defense was on its way already, in the form of Born's probabilistic interpretation of the wave function $\psi$.

[17] One might want to add that a discrete spectrum of possible results, instead of a continuum, would be typical of quantum mechanics. The essential point is that there is a one-to-many relation between initial state and the end result.

[18] Written February 1, 1926. From the Hans Reichenbach Collection, University of Pittsburgh Libraries.

[19] See Mehra and Rechenberg, vol. V, Sec. IV.4 for documentation.

[20] From a letter to Pauli, in Pauli (1979, pp. 337–38).

[21] See Mehra and Rechenberg, vol. V, pp. 802–804.

Heisenberg's quantum mechanics had been used successfully for the determination of the stationary states of the hydrogen atom in January 1926. In an attempt to study quantum mechanically the collision of electrons with atoms, Born (1926a, p. 864) found only wave mechanics proper to the task. The reason is that matrix mechanics originally only described the energy states and transitions between them, whereas in wave mechanics the notion of trajectory of a particle still could be the object of physical statements in a modified sense. On the basis of the equivalence of wave and matrix mechanics, one could try to interpret the wave function $\psi$ in corpuscular terms. The probabilistic interpretation that Born gives for $\psi$ is 'the only possible one' (p. 865).[22] Schrödinger's wave mechanics is able to tell what the effect of collisions is, but in that effect 'it is not a question of a causal relation' (1926a, p. 866). One can make statements about particle trajectories in the sense of giving the probabilities of the different effects, here, the probability distribution over the possible directions of rebound after a collision. Then Born continues:

Here all the problems of determinism emerge. From the point of view of quantum mechanics there exists no quantity which in an individual case causally determines the effect of a collision.

Lack of causal determination could have been taken as a matter of fact about the formalism. But subsequent remarks make it clear that Born was making a decided step into quantum mechanical indeterminism.

In the more elaborate version (1926b) on the quantum mechanics of collisions the absolute square $|\psi|^2$ is (correctly) identified as the quantum mechanical probability density. That probability is given as a *third* way of interpreting the formalism: The first one is Heisenberg's original (1925) in which 'an exact representation of processes in time and space is impossible,' and the second Schrödinger's in which waves are given a physical reality (1926b, p. 803). The contrast to Heisenberg is interesting, for Heisenberg had attempted to eliminate completely the notion of path of a particle in his first paper on quantum mechanics. Born was returning some way back from that point of view, admitting trajectories of particles, however, with the proviso that 'for the occurrence of a definite trajectory, only a probability is determined' (p. 804). His 'third interpretation' with $|\psi|^2$ as a probability is based on a remark of Einstein's: Waves relate to quanta of light like a guiding 'phantom field' (Gespensterfeld) that determines the probabilities of the different trajectories (p. 804). No reference is given, and it has been said that the

[22] In a footnote Born changes this to the square of $\psi$ being proportional to probability.

probabilistic interpretation of quantum mechanics has its origins in an unpublished speculation of Einstein's from the earlier 1920s (cf. Pais 1986, p. 259). My own view on this matter has been substantially presented in the comments to Einstein's fluctuation formula (4.6). I shall add some historical evidence to this systematic point of view at the end of this section.

Physicists often speak about 'the probability interpretation' of quantum mechanics. In the foundational discussions on probability, the problem lies one step further: Probability in itself is a mathematical notion in need of interpretation. For Born it goes without question what the interpretation of the numbers is that taking absolute squares of grinds out of the quantum mechanical formalism. These numbers are the *statistical frequencies* of the different quantum mechanical states (1926b, p. 805). For the individual paths of particles only probabilities can be given. The probability law, on the other hand, 'evolves according to the causal law', that is, the Schrödinger equation for $\psi$ (p. 804). In his final remarks Born says that the singular events are not 'causally determinate' (p. 826). The problem of hidden variables arises: If one is not satisfied with indeterminism, 'one is not forbidden to assume further parameters not included in the theory that determine the singular events' (ibid.). As a comparable example of such parameters in physics Born gives the notion of exact state in classical mechanics. He tends to think that hidden parameters could not be found in quantum mechanics. Whether or not such a possibility exists, 'it will not change the practical indeterminism of collision processes' (ibid.).

Now we come to the reception of the interpretation of the absolute square of the quantum mechanical state function as a probability. After the first half-year of quantum mechanics, the second half brought Schrödinger's wave mechanics. That latter was received with enthusiasm, as it seemed to be offering a way back to more classical images. It was clear that the function $\psi$ of Schrödinger's wave mechanics gave some kind of a density, but a density in a continuous wave theory receiving an interpretation in electromagnetic terms.[23] The equivalence proof of wave and matrix mechanics connected $\psi$ to a discrete treatment in terms of particles—or at least that was the way Born used it. Schrödinger proceeded in the opposite direction, trying to get rid of the discontinuous transitions of electron levels. A complete turn had taken place in his thinking, as compared to only a few months earlier.

Schrödinger became acquainted with Born's probabilistic interpretation of $\psi$ and of the indeterminist view of collisions in August 1926.

[23] In Schrödinger's fourth paper on wave mechanics (1926f), finished in June.

This is shown from his correspondence with Wilhelm Wien.[24] He referred back to the acausal position, as in his inaugural lecture of 1922, saying: 'Today, however, I no longer wish to assume, with Born, that such a single event is "absolutely accidental", that is, completely indeterminate.' Then, in October, Schrödinger payed a visit to Bohr in Copenhagen where Heisenberg and Dirac were also present.[25] Schrödinger's attempt at replacing Born's probability distribution of the position of one electron by an electron density points at the essential debate. What he suggested was a move very familiar from foundational discussion on probability. One has to admit that the statistical law enjoys an undeniable reality. Yet, its application can be restricted to a great number of individual cases. Such would be the situation in a quantum gas where the net statistical effect would be observable as a radiation intensity. Bohr and Heisenberg tried to convince Schrödinger that a proper treatment of quantum phenomena requires discontinuous single events escaping classical visualizability. This would mean that probability refers to the indeterministic single events.

In light of today's wisdom, Schrödinger would have done better had he changed his view in the direction of Bohr and Heisenberg (as, in fact, he did a few months later). It must be admitted, though, that Heisenberg, too, had to step back a bit from his initial position, for the microworld does not escape visualizability completely. The concept of position retains a certain validity, and it is only the combination of position and momentum (or velocity) that cannot be applied in full, as is dictated by Heisenberg's uncertainty relation. As to Schrödinger's theory: In a letter of Wien to Schrödinger of August 20, 1926, we find some of the first doubts by an advocate of that 'continuum theory': 'It seems to me that one cannot get at the photoelectric effect without involving statistics.'[26] Schrödinger replied that 'the photoelectric effect offers (along with the experimentally ascertained quantization of direction) the greatest conceptual difficulty in the execution of a classicist theory.'[27] By December 1926 he must have had severe doubts concerning the feasibility of an 'anschauliche' interpretation of his theory. 'The exchange of energy and momentum between the electromagnetic field and "matter" does *not* in reality take place continuously,' he had to admit (1927, p. 261).

It is interesting to contrast Schrödinger's epistemic standpoint in 1922

[24] Partly quoted in Mehra and Rechenberg, vol. V, p. 827.

[25] Mehra and Rechenberg (vol. V, pp. 820–28) present, in order, all that is known about this visit.

[26] Partly quoted in Mehra and Rechenberg, vol. V, p. 830.

[27] Cf. Mehra and Rechenberg, vol. V, p. 830.

against his harking back in physical theory to more classical views in the spring of 1926. In 1922 he saw there to be 'clear comprehension in the world of appearances,' the essential features of which are 'determined from the pure number concept, the clearest and simplest one created by the human spirit' (1922, p. 11).[28] The assumption 'of "proper," true, and absolute laws in the infinitely small' would be an improbable '*doubling of natural law*' (ibid.). In October 1926 he wrote to Wien about his talks with Bohr in Copenhagen. Bohr had said, writes Schrödinger, that 'visualizable wave pictures work as little as the visualizable point models, there being something in *the results of observation* which cannot be grasped by our erstwhile way of thinking.' For Schrödinger instead, 'the comprehensibility of the external processes in nature is an axiom.'[29] To that extent at least, there is continuity from 1922 to 1926 in his basic epistemology. Schrödinger's 'axiom' would be fulfilled by 'the best possible organization among the different facts of experience.' To extend the discrete character of observations to the atomic world would have been precisely the 'best possible organization' he talks of in the letter to Wien, had he followed his philosophy of 1922. Why, then, the insistence on a pure continuum theory of the atom in 1926? That is one of the more mysterious aspects of the quantum drama whose chief players Schrödinger and Heisenberg were. Even if Schrödinger was defeated, he did not return to his previous position. Whereas in 1922 he had definitely thought improbable and circular the 'doubling' provided by a continuum theory, he later saw the essential epistemological choice as a matter of taste.[30]

Born had made the new quantum mechanics *probabilistic*. As we noted, he already saw this notion as something entirely different from the concept of probability in classical statistical physics: It meant a renunciation of determinism. That, in spite of the alternative delivered by Schrödinger's continuous waves, became the accepted view. Often the remark of Einstein that Born credits, is given as the only background for the decisive developments that led to the new quantum probability. Born himself has recollected the events in 1954 and 1961, one notable aspect of the latter being the explicit pronouncement that even classical mechanics is indeterministic. The reason is that there are unstable evolutions where a finite initial accuracy becomes worthless (as a basis

[28] This I think refers to the discreteness of our immediate experience, its division into some kind of units.

[29] Letter to Wien, October 21, 1926, from Mehra and Rechenberg, vol. V, p. 825.

[30] See Schrödinger's talk (1929) on the occasion of his becoming member of the Prussian academy of sciences.

of prediction), and only probabilistic statements can be made. Born had published several papers on this argument in the 1950s, and it is clear that his later view of probability in quantum mechanics was affected by the admission of probability as an essential component also in classical mechanics. The question then is whether that was his view already in 1926, that is, whether he admitted any such role for probability in the classical case in his original work. As we have seen, he speaks there of the 'practical indeterminism' of collision processes, an aspect that would remain even if hidden variables were found. But as the above quotations show, in 1926 he thought it was the new quantum mechanics that meant a farewell to determinism.

To end this section, I shall back up the emphasis I have laid on Einstein's 1925 quantum gas paper: The outputs of a xerox machine, that miraculous aid of historical studies, replace recollections and notes that are of necessity incomplete. It lets you have all the papers spread around, to be filled with remarks in order to work out connections otherwise difficult to remember. My copy of Heisenberg's 1927 paper on the uncertainty relation has the long footnote on p. 176 all encircled in black ink, with the marginal comment: *H's account of the birth of q.m. prob.* The footnote reads[31]:

> The statistical meaning of *de Broglie* waves was first formulated by A. Einstein (1925). This statistical element in quantum mechanics plays subsequently an essential role in M. Born, W. Heisenberg, and P. Jordan (1926), esp. chap. 4, §3, and in P. Jordan (1926a); it is analyzed mathematically in a fundamental work of M. Born (1926b) and used for the explanation of collision processes. The justification of the probability-*ansatz* from the transformation theory is also found in the following works: W. Heisenberg (1926), P. Jordan (1926b), W. Pauli (1927, note [p. 83]), P. Dirac (1926b), P. Jordan (1926c). The statistical aspect of quantum mechanics is discussed generally in P. Jordan (1927) and M. Born (1927).

That was written before March 23, 1927. Heisenberg obviously had the papers and the story sorted out quite well. First of all, he locates the influence of Einstein to the paper on quantum gases (1925). Born and Heisenberg (1927, p. 165), a paper presented by Born to the fifth Solvay conference in the fall of 1927, also acknowledges that same paper as a source of the probabilistic interpretation: 'Einstein (1925), in deducing from de Broglie's bold theory the possibility of a "diffraction" of material particles, had already tacitly admitted that it is the number of particles that is determined by the intensity of the waves.' Second, returning to Heisenberg's footnote, he sees no abrupt change in the appearance of

[31] I have changed the style of references.

Born's interpretation, but presents it instead as a direct continuation of earlier work. Something can be added to this development, though, for Heisenberg seems far too modest here. We saw that Born in his (1926b) spoke of 'the first interpretation' of quantum mechanics in Heisenberg (1925). That work, as well as the two papers Born and Jordan (1925a,b), can be added to Heisenberg's above list. They are the papers, together with the Born–Heisenberg–Jordan paper, that precede Born's interpretation. The earliest of these (Born and Jordan 1925a) was written at the same time as Heisenberg was developing his quantum mechanics. All these papers use the (if I may say) pre-born transition probabilities between states. Thus Heisenberg's first interpretation of his formalism was that it gives the transition probabilities $|a(n,m)|^2$, in terms of the 'transition quantities' $a(n,m)$. The transition probabilities were 'observable' through their connection to the intensity of light emitted. Indeed, these probabilities were not only an interpretation of Heisenberg's formalism: Heisenberg's aim was precisely to create a theory that would determine the energy levels and transition probabilities.[32] The novelty of Born (1926a, b) was that he gave quite a general rule for computing quantum mechanical probabilities for any variables the wave function is defined over.

## 4.3 THE UNCERTAINTY RELATION

Heisenberg's 1927 paper is of course remembered for another reason than a footnote-long history of probability in quantum mechanics. It renders precise the idea of indetermination that was about to surface in Born's papers on the probability interpretation and in a letter Pauli wrote to Heisenberg in October 1926. The latter had a job in Copenhagen at that time. The long letter by Pauli contains discussion of the probability interpretation, and some of the essential ideas that in 1927 led to Heisenberg's uncertainty relation. Pauli asks, 'why is it that one is allowed to prescribe with arbitrary precision only the $p$'s, and not the $p$'s *as well as* the $q$'s? (Here $p$ denotes momentum and $q$ position.) And later, 'one can look at the world with the $p$-eye and with the $q$-eye, but if one wants to open both eyes simultaneously, one is led astray' (Pauli 1979, pp. 346–47). Both eyes open would give $c$-eyes, for classical. Bohr, Dirac, and others were informed of these ideas, for the letter circled around in Copenhagen. Heisenberg's reply to Pauli's letter is evidence of the radicalness of the ideas he was entertaining at the time, for

[32] See further the discussion in van der Waerden (1967, pp. 28–35). A full derivation of 'Einstein's $A$'s and $B$'s' is given in Dirac (1927).

there he speculates whether space and time could be only 'statistical concepts.'[33]

The indetermination, or *Unbestimmtheit*, in Heisenberg's terminology in his 1927 paper 'On the visualizable content of quantum theoretical kinematics and mechanics,' is 'the real reason for the appearance of statistical connections in quantum mechanics' (1927, abstract). His many examples show concretely what the physical implications of quantum mechanics are. In his very first paper on the new quantum mechanics, he wanted to base the theory 'exclusively upon relationships between quantities which in principle are observable' (1925, abstract). Until the uncertainty paper it remained a formalism, the workings of which were not properly understood.[34] The paper is not technical, but tries instead to explain the uncertainty relation in simple terms, which has sometimes led to a somewhat confused impression. For example, Heisenberg gives the uncertainty relation as saying: 'The more accurately the position is determined, the less accurately the impulse is known and the other way around' (p. 175). Determination and knowledge are used as synonyms so that we could add: determined *by us*, not in itself. On the other hand, position and momentum at a given time 'can be thought determined' in the classical theory, whereas in quantum theory they are 'undetermined in principle' (p. 177). Passages such as these point at an interesting sequence of more and more general distinctions between what is given more or less directly and what is theoretical construction: 1. The uncertainty relation as applying to an individual case, versus an uncertainty relation as derived from the theory of quantum mechanics. The latter is a probabilistic result applying to a great number of cases. 2. The requirement of an operational definition of concepts, versus the principle that 'only the theory decides what can be observed.'[35] 3. The distinction into epistemic versus semantic. These themes recur in Heisenberg's well-known Chicago lectures of 1929 on the physical principles of quantum theory to which I shall also refer in the following. Let me proceed to substantiate the above points:

1. Heisenberg (1927, p. 174ff.) first derives the uncertainty relation for an individual electron's position and momentum. The most accurate measurement of the former is based on the scattering of light. In the interaction of a quantum of light and the electron, the momentum of the

[33] Heisenberg's reply is printed in Pauli (1979, pp. 349, 350).

[34] Born (1927) states this, as well as Pauli in the letter referred to above. But the interpretation in terms of transition probabilities was there, of course.

[35] Quoted from Heisenberg's recollections of the discovery of the uncertainty relation (1969, p. 92).

latter changes discontinuously, the more so, the shorter the wavelength of the light. The wavelength gives the best possible accuracy of the position measurement. The uncertainty formula, written as

$$p_1 q_1 \sim h \qquad (4.12)$$

follows by elementary arguments, essentially the law $\lambda = h/p$ for the wavelength (p. 175). That law shows the inverse proportionality between the wavelength $\lambda$ and momentum $p$. It is considered as 'empirically established' in 1929 (cf. Heisenberg 1930, pp. 14–15). The uncertainty relation gives a lower limit to any possible position measurement (1927, p. 175). Heisenberg analyzes several possible (real and Gedanken-) experiments. In 1927 he is pretty carefully talking about lack of determination, whereas two years later he says things, such as 'every experiment destroys some of the knowledge of the system which was obtained by previous experiments' (1930, p. 20).

The elementary derivation of the uncertainty relation is paralleled by another derivation within quantum mechanics. This is not fully worked out in 1927, but the idea of deriving uncertainty from quantum mechanics as a statistical dispersion relation is there already (1927, p. 180). An exact derivation was given by Kennard (1927, pp. 337–39). In Heisenberg (1930, pp. 15–19) there is a full derivation, in which the relation reads

$$\Delta p \Delta q \geqslant \frac{h}{2\pi}. \qquad (4.13)$$

Here $\Delta p$ and $\Delta q$ are dispersions calculated using $\psi$, that is, they are probabilistic concepts stemming from the quantum mechanical probability itself.

2. Heisenberg compares quantum theory to relativity: The requirement of defining concepts through measurement prescriptions led in relativity to a critique of the notion of absolute simultaneity. Likewise for the notions of position and momentum in quantum mechanics (1927, p. 179). On this basis one would of course classify Heisenberg's early philosophy as operationalism. But I shall argue that this would be too one-sided a view, and that there is an essential theoretical component that cannot be eliminated in favor of a pure operationalism. Heisenberg himself compares relativity and the uncertainty principle, saying that once relativity theory was formulated, the requirement of relativistic invariance became a condition for any suggested new physical theory. Similarly, the uncertainty relation is 'a requirement that can be useful for the finding of new laws' (pp. 179–80). The maxim that Heisenberg recollected in (1969, p. 32) as a clue to the uncertainty principle, 'only

the theory decides what can be observed,' was Einstein's. A lot has been written about Einstein's operationalism of 1905. But here, too, qualifications are needed. For example, his statement about the role of theory fits perfectly the 'advice' he gave only two years after special relativity to experimental researchers of Brownian motion. As we saw in Section 3.4(a), they had been chasing an inexistent concept, the velocity of a Brownian particle. Einstein's theory of Brownian motion of 1905, instead, told what the observable magnitude is that one should be looking for, namely, the mean displacement. Heisenberg's philosophy has a similar double aspect: On the one hand we have real and Gedanken-experiments, that is, more or less concrete individual situations, where we try to fix what can be done. An example is Heisenberg's discussion of position measurement with light of wavelength $\lambda$. The wavelength used gives a lower bound for the accuracy of the measurement procedure. If light of shorter wavelength is applied, momentum can be measured less and less accurately, as Heisenberg's examples demonstrate. The other aspect is that we have the theory of quantum mechanics in which the meaningfulness of exact simultaneous values of position and momentum is denied.[36]

3. From one point of view, the uncertainty relation is a limitation of knowledge, an epistemic principle: In 1929 Heisenberg says that 'the uncontrollable perturbation of the observed system alters the values of previously determined quantities.' It gives a lower limit 'to the accuracy with which they can be known' (1930, p. 3). Form another point of view, it is a *semantic* principle: 'An analysis of the words "position of an electron," "velocity,"...,' has been given (1927, p. 184). A long footnote in the 1929 lectures (1930, p. 15) reads a bit like another contemporary antimetaphysical program[37]:

> ...the human language permits the construction of sentences which do not involve any consequences and which therefore have no content at all – in spite of the fact that these sentences produce some kind of picture in our imagination; e.g., the statement that besides our world there exists another world... One should be especially careful in using the words 'reality,' 'actually,' etc., since these words very often lead to statements of the type just mentioned.

Heisenberg's quantum mechanics had started with the bold idea of abandoning the notion of trajectory of particles. Space–time having thus gone lost, *meaning* had to replace the given reality of classical or even

[36]Compare also Weyl (1919, p. 8): 'Essential for measurement is the difference between "giving" an object through individual ostension on the one side, and giving it in a conceptual way on the other.'

[37]That of the Vienna Circle, and Carnap especially.

relativistic physics, its 'ideal world' as Heisenberg sometimes said. The uncertainty relations shortly afterwards became the backbone of Bohr's complementarity principle of 1928. Bohr, too, was eager to dispense with reality, a notion that did not fit the Copenhagen philosophy of quantum mechanics.

Heisenberg's uncertainty relation of 1927 very definitely made modern physics indeterministic. Different aspects of the introduction of probability in physics have led us to ask for the roots of the idea. How much indeterminism was there in physics before quantum mechanics? One's general questions tend somehow to evaporate once one gets immersed in the details of actual developments. But let's try: We have seen at least one indeterminist within the classical theory already, namely, Maxwell in 1873. We have seen many accept the objectivity of statistical law within that theory: Boltzmann, Einstein, von Smoluchowski. Already in 1872 Boltzmann had boldly declared that a probability distribution gives the complete description of physical state in the kinetic theory. Einstein gave up the hope of finding a mechanical explanation of quantized radiation in 1908 or 1909. In 1917 he made his statement about 'chance' in the emission of photons (a regrettable feature of the theory for him). The last of the above three expressed his disbelief in a mechanical explanation of radioactivity in 1918, in his analysis of the concept of probability in physics. In a different line, von Mises in 1919 built up a theory of probability where randomness was a basic notion. In 1920 he judged classical mechanics to be 'idle' in problems with a great degree of freedom, and set himself the task of constructing a purely probabilistic statistical physics. Franz Exner, in his 1919 book, declared himself an indeterminist, and in 1922 was eagerly followed by Schrödinger (until the establishment of quantum mechanics, at least!). In 1920 Hermann Weyl also expressed his indeterminism, against the accepted view of physicists as he said. He even endowed atoms and quanta with a 'free will' to make 'decisions.'[38] – So it seems that there is some ground for a (less serious) remark: Around 1920 everyone started thinking he was the only indeterminist.[39]

The beginnings of the mathematics of quantum mechanics had been groping. Born realized that Heisenberg in 1925 did matrix multiplication for combining the transition from one state to a next to a third. Then

[38] He had a very special kind of reason for this, for he saw the determinism of classical mechanics as arising from the nonconstructive character of the theory of real numbers; cf. above, Sec. 2.4(b).

[39] Forman's (1971) discussion of von Mises, Schrödinger, and Weyl is an attempt at locating the cultural influences promoting indeterminist ideas.

he noticed the connection to Hilbert's theory of integral equations. Dirac, upon reading Heisenberg's original paper, worked out a formalism of his own (1925). These approaches, together with Schrödinger's wave mechanics, were united in the operator theory of Hilbert spaces. The person behind this piece of mathematics was Johann von Neumann, whose book 'The Mathematical Foundations of Quantum Mechanics' became the standard reference for decades.

From the point of view of probability theory, the most interesting aspects of quantum mechanics are its indeterminism and its phenomenon of interference.[40] The latter would be classified as correlation in ordinary probabilistic terminology. But the quantum mechanical concept of state is such that correlations cannot be explained in the classical way, as one event being partly 'responsible' for the occurence of another, or through two events sharing some common background cause in the past, which makes them occur correlatedly. The effect of interference is that probabilities are computed somewhat differently from ordinary probability theory in quantum mechanics. The best illustration of the way to compute probabilities quantum mechanically comes from the famous double-slit Gedanken experiment. Note first that the probability amplitude in quantum mechanics is a complex number, and only its absolute square is a probability. In the double-slit situation, there is a screen with two narrow openings, and behind the screen, at some distance, there is a film which registers hits by particles, photons say, which are sent toward the screen. If only one slit is kept open, there will be a dispersion pattern, typically a normal distribution of points of hit. The same if the other slit is kept open. If both slits are kept open, the statistical distribution of hits, that is, the intensity of photons of the film, should, according to classical particle conceptions, be the sum of the two individual patterns. In quantum mechanical experiments, instead, there is an *interference pattern* such as one would expect from a wave motion going simultaneously through both slits. A classical wave motion, however, would be in difficulty explaining the phenomenon, for it is possible to lower the intensity of photon radiation so that only one photon passes in the apparatus at a time; still, the interference remains there. Interference leads to the peculiar way of computing quantum mechanical probabilities. Let us consider diffraction on the film in one direction $x$ only. Then there is a wave function $\psi_1$ for the first slit, which gives the probability amplitude $\psi_1(x)$ from which one gets the probability distribution itself, for the location of the photon in the $x$ direction,

[40] Interference of probabilities was first discussed by Jordan (1926c). He even tried to put up an axiomatic postulate system for quantum probabilities (pp. 813–14).

$P_1(x) = |\psi_1(x)|^2$. Similarly, $\psi_2$ is the wave function for the second slit, with a probability distribution $P_2(x) = |\psi_2(x)|^2$. If there were no interference, the distribution resulting from having both slits open would be given by the sum $P_1(x) + P_2(x)$. In probabilistic terms, the particle's going through the first and second slits are 'mutually exclusive events,' hence one can sum the respective probabilities. But the quantum mechanical state of the composite system is described by $\psi = \psi_1 + \psi_2$, and it follows that the probability distribution for the composite system is

$$\begin{aligned} P(x) &= |\psi_1(x) + \psi_2(x)|^2 \\ &= |\psi_1(x)|^2 + |\psi_2(x)|^2 + 2|\psi_1(x)||\psi_2(x)|\cos\delta \end{aligned} \tag{4.14}$$

with $\delta$ the phase difference between the two amplitudes.

Further topics of great interest for probability and indeterminism in quantum mechanics can only be mentioned. First, in 1936 Garrett Birkhoff and von Neumann initiated the study of quantum logic. An extensive literature has grown on the topic since.[41] Second, the probabilistic character of quantum mechanics was brought closer to ordinary probability theory in Feynman's theory of path integrals (1948).[42] Third, questions about the completeness of quantum mechanical description, and about the local character of physical effect, were initiated by the well-known 'EPR paper' of Einstein, Podolsky, and Rosen (1935). Since the results of John Bell in the mid-1960s, this topic has been very much in the center of foundational studies on quantum mechanics.[43]

[41] A useful guide is Hughes (1989).

[42] See Feynman and Hibbs (1965) for an extensive treatment of quantum mechanics along these lines.

[43] A good starting point is Bell's book *Speakable and Unspeakable in Quantum Mechanics*.

# 5

# *Classical embeddings of probability and chance*

## 5.1 SUBJECTIVE OR OBJECTIVE PROBABILITY: A PHILOSOPHICAL DEBATE

In the mechanical world view of last century's physics, the future course of events was thought determined from the present according to the mechanical principles governing all change. If there was any ignorance, it was completely located in the ignorant person's mind. It follows that probability stands only as a kind of index of the degree of ignorance. Laplace is, more than anyone else, responsible for this classical concept of probability. It can be found in his *Essai philosophique sur les probabilités*, written as a popular preface to the second (1814) edition of his extensive *Théorie analytique des probabilités*. There, in the classic passage on Laplacian determinism, we find him imagining 'an intelligence which could comprehend all the forces by which nature is animated... for it, nothing would be uncertain and the future, as the past, would be present to its eyes' (p. 4). The exactness of planetary motions was of course the practical reason for such confidence in determinism. But Laplace made a giant extrapolation from astronomy to the smallest parts of nature: 'The curve described by a simple molecule of air or vapor is regulated in a manner just as certain as the planetary orbits; the only difference between them is that which comes from our ignorance' (p. 6). In a deterministic universe there are no true probabilities, for complete knowledge would make all probabilities be 0 or 1. But since we do not quite acquire the status of Laplace's 'vast intelligence,' there remains a task for probability: to determine the cases of equal probability relative to our ignorance, and to compute the probabilities of composite events on this basis.

The role of probabilistic knowledge in the Laplacian scheme obtains an extreme illustration in his famous calculation of the probability that the Sun will rise tomorrow. His calculation can be explained by assuming an urn of labels 'sunrise tomorrow' and 'no sunrise tomorrow' in an unknown composition. If all compositions are assumed initially equally probable, the probability of drawing the label 'sunrise tomorrow,' conditional on it having been drawn on the previous $n$ trials, is $(n+1)/(n+2)$. In Laplace one finds the probability computed under the assumption that the Sun has risen on every morning for the last five thousand years. Feller (1968, p. 124) comments that 'a historical study would be necessary to appreciate what Laplace had in mind and to understand his intentions.' That is easy, for Laplace explains the matter in his *Essai.* The assumption of 5000 years and a determination of the probability of sunrise come from Buffon. Laplace has some criticisms to make concerning Buffon's calculations. His own determination of the probability of the Sun's rising tomorrow is intended to illustrate the dependence of probability on the hypotheses made, and on the evidence assumed available. We are free in the choice of such hypotheses: Laplace says that 'when the probability of a single event is unknown, we may suppose it equal to any value from zero to unity' (p. 18). Immediately after giving the probability stemming from the assumption of 5000 years of sunrises, he adds that 'this number is incomparably greater for him who, recognizing in the totality of phenomena the principal regulator of days and seasons, sees that nothing at the present can arrest the course of it' (p. 19). That is, all probabilities are relative to knowledge and ignorance, and the completion of knowledge in the limit simply eliminates them.

The preceding story appears in a strange contrast with the emergence of the objectivity of *statistical law.* It is a phenomenon discovered by the new science of statistics in the last century. The stability of frequencies had been known long since from gambling. In social statistics, perfect regularity seemed to follow from the sheer number of acts; each of them, taken singly, is subject to the free will of the acting individual or to some other fortuitous circumstance. But the sum total follows a rigid law, predictable with remarkable certainty, as was found to be the case for the most varied kinds of phenomena.[1] From statistical data, statistical probabilities can be derived which are not much altered by the addition of future data. Statistical or frequentist probability therefore has the essential sign of objectivity. Let us ask next, what is knowledge of probabilities good for? At one end of the spectrum of possibilities stands the Laplacian idea: Such knowledge is good only for the ignorant individual in guiding his judgments. But if probabilistic knowledge is based on statistics from a large number of events, it can be used for predicting the statistics of future events, predictions which often are astonishingly accurate. This is the other end of the spectrum.

The founding fathers of statistical physics, Krönig, Clausius, Maxwell,

[1] See various articles on the rise of statistical thinking in Krüger et al. (1987).

Boltzmann, all started from the assumption that the observed perfect regularity in the behavior of gases is a statistical effect from a large number of molecules. The precision by which this regularity was predicted from a probabilistic statistical physics, was sometimes called 'statistical determinism,' a phenomenon likened with the determinism of ordinary mechanical motions. The position of the founders of statistical physics concerning the relation between the mechanical world view and the statistical behavior of aggregates of molecules has been reviewed in Chapter 3. As we saw, Maxwell and Boltzmann were both quick to recognize the insufficiency of the mechanical view. The supposedly compact mechanical doctrine of the last century was relegated to the status of a useful analogy in Boltzmann, and rejected outright in Maxwell. Probabilities were not determined from the degree of ignorance. Instead, in the statistical physics of gases, the probability numbers are determined from the total energy of the system. An estimate is obtained by a thermometer reading, and here no one thinks the value has anything to do with ignorance.

In the twentieth century, too, the idea of objective probability has been closely connected to the development of physics. Before the recent popularity of chaos theory, it was common to make a twofold classification of physics into deterministic classical physics and indeterministic quantum physics. Similarly, probabilities were classified into epistemic and objective. Of the four combinations prima facie possible, only classical-deterministic together with epistemic probability, and modern-indeterministic together with objective probability were thought feasible. We have witnessed this trend in connection with von Neumann's discussion of his proof of the impossibility of hidden variables in quantum mechanics. But it is not difficult to show a certain conceptual independence between probability and determinism. The first case to this effect is offered by the philosophy of probability of Bruno de Finetti, to be discussed more deeply in the last chapter of this book. He found that the usual conception of epistemic probability, as based on a deterministic view of science, was outmoded by modern physics. Therefore around 1930 he sought new foundations for subjective probability that would better agree with what he thought was 'the new spirit of indeterministic science.' He formulated a view which remains totally noncommittal with regard to the ultimate reason of ignorance: whether it lies in the nature of things themselves, or is imposed upon the external world by us. Thus subjectivism in probability will not decide on the determinism–indeterminism issue. In a line complementary to de Finetti's, it can be argued that objective probability can make sense also within classical physics.

The determinism of the mechanical world view continues to have its ghostlike effect on the philosophy of probability. The successes of classical statistical mechanics and the explanations of random behavior in dynamical systems provided by ergodic theory have not been sufficient to oust it. Instead, since the late 1920s, quantum mechanical phenomena have usually been viewed as the only true and irreducible source of randomness in nature. A whole philosophy of probability has been built along these lines, namely, Karl Popper's *propensity interpretation* of probability. The view of the place of chance in nature, which lies at the basis of this interpretation, has been taken for so obvious that contrary statements, even very explicit ones, remain misunderstood. The English edition of Popper's *Logic of Scientific Discovery* contains as one appendix a letter by Einstein from the year 1935.[2] Referring to earlier correspondence, he writes:

> I wish to say again that I do not believe that you are right in your thesis that it is impossible to derive statistical conclusions from a deterministic theory. Only think of classical statistical mechanics (gas theory, or the theory of Brownian movement). Example: a material point moves with constant velocity in a closed circle; I can calculate the probability of finding it at a given time within a given part of the periphery. What is essential is merely this: that I do not know the initial state, or that I do not know it precisely!

Popper suggests in 1959 that Einstein was here thinking of a subjective interpretation of probability (p. 208, note). But that seems a philosophical prejudice: The determinist Einstein must have thought that probability relates to ignorance. As we have seen in Section 3.3, Einstein instead shared the Boltzmannian notion of probability as a time average. And with a finite number of outcomes at discrete times, time average probability becomes a statistical probability.

Einstein's choice of his example was not a haphazard one. As it happens, we know quite precisely what he thought about it, for it is discussed extensively in the Dällenbach lecture notes of 1913.[3] First he considers a closed trajectory, drawn in a plane, with length along this curve as the only variable. The probability of finding a point rotating on the closed curve in a differential element of length $ds$ is determined as follows: We consider the point during a very long time $T$. Let the point be in $ds$ altogether for the time $t$ during $T$. The differential probability $dW$ is $\lim(t/T)$ as $T \to \infty$. We see that there is nothing subjective in Einstein's view of probability in this case. It is the limit of time average. One can also interpret the time average

[2] Popper (1959, pp. 459–460). Popper's book appeared in German original in 1934.
[3] See above, p. 119.

probability as the probability of finding the point in *ds* at a randomly chosen moment of time, as Borel had done.[4] In the Dällenbach lectures Einstein also deals with the probability *dW* in a second way, by taking an ensemble of systems, the conclusion being that 'time ensemble = system ensemble.'[5] Next he takes the combination of two rotations, noting that incommensurable circulation times give a dense motion for the combined system.

Part of the reason why Popper suggests an epistemic notion of probability in Einstein may be due to the latter's saying, in the letter quoted, that it is essential that one does not know the initial state of the rotating point precisely. It is easy to see the consequences here: If one knew the initial position approximately, but the circulation time exactly, one would be able to make predictions on the whereabouts of the point. But if the circulation time is also known only approximately, the initial indetermination will grow until nothing can be said except the probabilities of finding the point in different parts of the curve.[6] This brings us to the core of the argument for subjective probability: Two persons can know the *same* initial conditions with different accuracies. The motion is still the same, hence giving an initial probability distribution over the set of possible states seems like a purely subjective act. On the other hand, the modern theory of dynamical systems teaches us a fact taken for granted by Einstein and others, namely, that there exist cases of 'classical embeddings of probability and chance.' These are cases which from *any* macroscopic point of view appear as purely probabilistic processes, even if they admit of an underlying classical trajectory.

I shall next review the historical origins of the theories of objective probability as based on classical physics, and lastly discuss the most developed approach to it, the one worked out by Eberhard Hopf in the 1930s.

## 5.2 THE EARLY PHASE OF A THEORY OF OBJECTIVE PROBABILITY

In the late 1860s Boltzmann and Maxwell had certainly come to the conclusion that it is necessary to add some kind of probabilistic

[4] See above, p. 104. [5] His word for ensemble is Gesamtheit.

[6] It can be shown in the entropy theory of ergodic systems that Einstein's example has the following property. Repeated observation of the same accuracy gives more and more information on the initial state, which information in turn can be used for longer and longer predictions. In the limit observations do not tell us anything about the system we would not know already. One says the system has zero entropy. Such a dynamical notion of entropy was introduced by Kolmogorov in the late 1950s.

assumption to the mechanical basis in order to secure a foundation for statistical physics. Probability appears as a concept in the object theory of physics itself. In Boltzmann's great work of 1872, a new conception of physical state is reached: The distribution of molecular velocities gives a *complete description of physical state* (1872, p. 322). Maxwell, in his (1873), had introduced the idea of chance as the critical dependence of an event on the details in its conditions of occurrence. Such dependence is the case if the details are so fine as to lie beyond any possibility of control. Maxwell had criticized the causal law, 'from the same antecedents the same consequences follow,' as being empirically empty, and noticed the failure of the weaker form, 'from like antecedents like consequences follow,' if the physical process experiences a discontinuity in its evolution. Such points of discontinuity are the strongest kind of *instabilities*, and their recognition marked Maxwell's step toward indeterminism. These revolutionary thoughts were published in Campbell and Garnett's biography of Maxwell in 1882.[7]

The changing interpretation of the concept of probability was largely due to its role in statistical physics, as we have seen in Chapter 3. One of the first philosophers, if not the very first one, to introduce an idea of objective probability, inspired by the new science of statistical physics, was Johannes von Kries in his 1886 book *Die Principien der Wahrscheinlichkeits-Rechnung*. It has been considered the best philosophical book on probability of its time. One chapter is devoted to the old topic of games of chance, but from a new point of view. Von Kries argues that in a simplified version of roulette, red and black obey the same probabilities, by viewing the game as a mechanical system where given initial conditions lead through a mechanical motion into the final state. The simplification is that there is a slanting board on which a ball is set rolling, with the other end of the board divided into a great number of equally broad stripes of 'red' and 'black.' If any sufficiently regular probability distribution is assumed over the initial states, the total probability of red and black will be approximately equal. No details of how this should be brought about are worked out by von Kries. But his insight into the matter was correct, as is shown by his subsequent remarks. For example, looking backward at the initial conditions leading to the alternating red and black stripes, he notes that if the initial distribution has a suitable 'periodicity,' the total probabilities of red and black would differ (1886, p. 51). His mathematical condition for equidistribution of the results is that the initial distribution of probability should have nearly equal values for initial conditions

[7] For more details, see the discussion of Maxwell in Ch. 3.

leading to stripes of red and black located close to each other. In the work of von Kries, the idea of chance as instability of a mechanical system is also present. He requires that a small variation in the motion should suffice to produce a different end result (p. 58).

Henri Poincaré is generally recognized as the first one to apply probabilistic considerations to the theory of dynamical systems. This took place in his 1890 study of the three-body problem, in connection with the recurrence theorem.[8] Another application is found in his *Calcul des probabilités* (1896; 2nd ed., 1912), namely, what became to be called *the method of arbitrary functions.* As concerns the former application of probabilistic reasoning to the dynamical systems of celestial mechanics in Poincaré (1890), we have seen in Section 2.1(a) that a related application was made at the same time by Gyldén also. The method of arbitrary functions in turn is based on ideas very similar to those of von Kries, even though it is difficult to say whether his 1886 book was known to Poincaré. At the time it was not customary to make more than the most minimal references to the work of others. But Poincaré's treatment of the example of a rotating disc or wheel with alternating red and black arcs is very similar to von Kries'. To determine the probability of a fixed pointer showing at red or black after a rotation, Poincaré assumes an arbitrary distribution (fonction arbitraire) of the initial state, here the angular velocity, to be given. The motion itself is mechanical, so that the end position of the wheel is uniquely determined by the initial state. If it is assumed that the arbitrary function is continuous and has a bounded derivative, the final distribution of red and black is approximately independent of the initial one (1912, p. 148). The goodness of the approximation depends, for example, on how densely the disc is divided into red and black arcs.

In his book *Science et méthode* (1908), Poincaré departs from the idea that chance would only be an expression of ignorance.[9] Despite that 'we have become absolute determinists,' there is room for chance in nature (1912, p. 2). There is chance whenever a very slight change in initial conditions brings about a great change in the final state. An example is a cone standing on its sharp point (1912, p. 4). Another of the examples discussed by Poincaré concerns the minor planets which are evenly distributed along the ecliptic. He says that their motion replicates in the large the motion of his roulette example (p. 12). There is a certain difference, however, in that the motion of the minor planets

[8] See Sec. 3.1(b).

[9] These thoughts of Poincaré's are reprinted as an introduction to the second edition of the *Calcul des probabilités*, 1912.

is unique whereas the motion of the roulette ball is repeatable so that a distribution over initial conditions makes sense. But the difference is counteracted by the fact that there are a couple of thousand minor planets. Poincaré's application of the method of arbitrary functions shows that whatever the empirical distribution (in angular direction) was at some time in the past, it becomes uniform.

One of the first to use Poincaré's method was Emile Borel (1906a). He applied it to the kinetic theory of gases, showing that the Maxwellian distribution for molecules is reached also under small random disturbances. These latter are taken into account through the assumption of an unknown distribution over the initial conditions. Later Borel tried to dispense with the assumption of an ergodic path through the method of arbitrary functions.[10] Another early appreciation of the method of arbitrary functions was the doctoral thesis of the philosopher Hans Reichenbach in 1915. This work also contains a useful survey of all the philosophical literature on probability of the time.

The motion of the minor planets is an example of mechanical and, therefore, also supposedly deterministic behavior. The concept of probability should accordingly not play any other role there than the one relating to our ignorance of the true and deterministic course of events. But the uniform distribution of directions of some two thousand objects is an observed fact that should be somehow *explained*. A similar problem arises with games of chance of a mechanical type: how to explain the observed statistical laws? Poincaré knew three basic elements of such explanations: 1. Chance as unstable behavior; 2. the accumulation of very many small effects; 3. the coincidence at the same time and place of independent processes.

Ideas of the above kind were carried much further by Marian von Smoluchowski in his posthumous paper 'On the concept of chance and the origin of probabilistic laws in physics' of the year 1918.[11] First, chance is defined as instability, the typical element in many games of chance. Second, it is required that a physical and objective notion of probability be determined, not from our degree of ignorance concerning an event, but from the conditions that have an effect on its occurrence (1918, p. 254). Poincaré's method of arbitrary functions shows in what way physical chance can lead to a determinate probabilistic law (p. 256).

Von Smoluchowski defines chance as a *causal relation* of the following

[10] See Sec. 3.2(a).

[11] Due to wartime circumstances, *Die Naturwissenschaften* of 1918 had limited distribution. Von Smoluchowski's essay can now be found easily, thanks to the reprint in Schneider (1988).

kind: The effect $y$ is assumed to be a function $f$ of the 'variable cause' $x$, or $y=f(x)$, such that the effect $y$ depends on 'very small' variations in the cause $x$. Conditions imposed on the relation $f$ guarantee a unique probability distribution for $y$ as a consequence of instability. Von Smoluchowski lists as 'the essential features of a "regulated" chance' the following (p. 258):

1. Small variation in cause—great variation in effect.
2. An 'oscillative' causal relation: the same observable effect can be produced by different causes.
3. Approximately uniform distribution of chances within sufficiently small intervals of $x$.

After discussing the notions of chance and physical probability in mechanical games of chance, von Smoluchowski indicates briefly how these concepts relate to the systems of statistical physics.

Von Smoluchowski finds the explanation of chance and probability through the method of arbitrary functions unsatisfactory in the following respects (p. 259):

1. It is an assumption that the cause $x$ follows a probability law. The concept of probability is presupposed, and only the persistence of the probability law of the effect explained.
2. Properties of the probability law were assumed which were described as its 'regularity.'

Concerning the latter, it is an idealization that a continuous function describes the discrete individual events. But the empirical distribution of the effect is approximately the same for an arbitrary distribution over the cause $x$. Exceptional cases are possible, and von Smoluchowski illustrates them by the example of a rotating disc. If it is aimed at at random intervals, the points of hit are approximately uniform in direction. But if the intervals are commensurable to the time of rotation of the disc, there will be a concentration of points of hit in some places, while other places will remain free. Commensurability is a singular exception, for 'in set theory one proves, as is known, that there are popularly speaking infinitely many more irrational than integer numbers' (p. 260). It follows that 'times commensurable to the given time of rotation form only an infinitely small fraction.' Consequently, if the disc is aimed at at random intervals, 'it is infinitely improbable that one would meet ones that are commensurable to the given interval' (ibid.). A similar conclusion holds for the case of Lissajous figures.[12] In terms

[12] Discussed in Sec. 3.2(a).

of the causal relation between $x$ and $y$, it is required that the set of assignments of $x$ values leading to approximately the same assignment of $y$ values is 'immeasurably more numerous' than the set of assignments of $x$ values leading to markedly varying $y$ assignments (p. 261).

Von Smoluchowski's first objection to the explanation provided by the method of arbitrary functions concerned the origin or genesis of chance. How should one understand the origin of the random variation of the cause $x$? In experiments performed by man, chance is led back to psycho physiological reasons. Would probability find any application if man were excluded and the conditions of the physical process exactly determined, von Smoluchowski asks (p. 261). He then argues, against the usual assumption, that this would be the case: An assertion such as one claiming statistical equidistribution in time of directions of motion 'describes an objective fact independent of man' (p. 261). That fact is not changed even if we assumed the motion theoretically determined through an exact knowledge of the system under study. To illustrate more clearly 'the concept of objective probability completely independent of man', von Smoluchowski considers further radiation and molecular motions as sources of chance in nature. He construes a mechanical model which reproduces the statistical law of radioactive decay, thereby replacing chance in radioactive disintegration by what he calls 'molecular' chance. Though he in no way believes radium atoms to possess such a mechanical structure, he shows that a mechanical model displaying chance through instability is possible. 'It shows that the *apparent contradiction* [between chance and lawlike effects of causes] *does not exist* and that chance—in the sense of physics—can very well be brought by exactly defined lawlike causes' (p. 262). It remains an open question whether the 'physiological' and 'molecular' types of chance exhaust all possibilities, and whether the former is reducible to the latter. Von Smoluchowski summarizes his position on objective probability as follows (p. 262):

It seems to us *a result extremely important also for philosophers*—even if it can be shown only in the restricted field of mathematical physics—that *the concept of probability* in the usual meaning of *chance events following a frequency law, has a strictly objective meaning*; that one can make precise *the concept and genesis of chance,* also if one withholds to determinism, and that *the law of large numbers* is not here a mystical principle, neither a purely empirical statement, but really follows as a simple *mathematical consequence of the particular form representing the causal connection in this kind of cases.*

After von Smoluchowski's analysis of the philosophical foundation of objective probability, detailed studies of specific systems were conducted by Bohuslav Hostinsky. The first case he attacked was the famous needle problem of Buffon. A needle of length $2b$ is thrown in a grid of parallels having a distance of $2a$ from each other. What is the probability that the needle touches a parallel? Hostinsky (1920), criticizing the assumptions of the classical result $2b/\pi a$, is led to the same result by the method of arbitrary functions. Two further cases are treated in Hostinsky (1926). They are idealized descriptions of dice tossing where the symmetric probability distribution of the six sides is derived from a mechanical basis. The classical calculations assume equiprobability and just count the number of favorable cases, whereas in reality, says Hostinsky, there is a continuous number of different results. Every probability problem dealing with the motion of bodies can be treated as a problem of continuous probability, the probabilities over final states being determined from assumptions concerning the nature of the motion, and from a generally nonuniform initial distribution (1926, p. 113). The short monograph Hostinsky (1931) deals in three chapters with the method of arbitrary functions, with discrete Markov chains, and with Markov processes. He there adds a remark on the example of Buffon's needle, to the effect that the computation of the probability has to take into account the physical conditions of the experiment (1931, p. 8). Otherwise the probability is not uniquely numerically determined, as is indicated by some of the classic paradoxes of the calculus of probability. Different answers can be suggested because the trial or repetition is not determined well enough to yield unique probabilities.

In the 1930s the method of arbitrary functions, in combination with the recent advances of ergodic theory, served as a foundation of physical probability in the works of Eberhard Hopf and Alexander Khintchine. Hopf's work is the climax of this branch of probability theory, which later fell into oblivion.

## 5.3 THE THEORY OF HOPF

Equipped with the very recent measure theoretic probability and the probabilistic ergodic theorem, Hopf in his 1934 paper 'On causality, statistics and probability' asks for an explanation of the stable frequency phenomena found in nature. He wants to find out 'the true origin of the laws of probability' (1934, p. 51). He mentions two predecessors in such an attempt, namely, Poincaré and von Smoluchowski. The combination of measure theory and the probabilistic ergodic theorem leads to a systematic theory of abstract dynamical systems. In Kolmogorov's 1933 book presenting the measure theoretic approach, probabilistic independence was taken as the feature characteristic of probability as a branch of measure theory. There Kolmogorov mentions as one of the most important tasks of the philosophy of natural science, 'to make precise the conditions under which any given real phenomena can be held mutually independent (pp. 8–9). Obviously, Kolmogorov wanted to have a physical condition for the applicability of a law of large numbers. Hopf's aim is a broader one, an explanation of frequency phenomena as produced by 'strictly causal mechanisms.' The macroscopic observable events are represented as parts or subsets of a state space, say $H$ for the part of heads and $T$ for tails in coin tossing. The motion has to be unstable, and the measures of $H$ and $T$ are approximately equal. A continuous probability distribution over the initial conditions of a toss is assumed as the theoretical representation of the repetition of the experiment (1934, p. 52). The points leading to $H$ and $T$, respectively, should be 'practically uniformly distributed' in the state space $S$ for appreciable regions $A$ of S, for a frequency phenomenon to take place independently of the initial distribution.[13] Hopf wants to derive a 'physical law of large numbers,' founded on a physical notion of independence of events, for unstable causal mechanisms. Such mechanisms typically are dissipative, that is, the visible motion is turned into heat by friction and comes to a rest. Hopf in 1934 mainly studies conservative systems, for the reason that this makes the mathematical methods of ergodic theory applicable. He thinks though that the essential results for conservative systems will carry over to the dissipative ones (p. 64). In his later papers (1936, 1937b), dissipative mechanisms are studied.

Statistical regularity with respect to a conservative mechanism is defined in (1934) in the following manner. Let $A$ be a part of state space $S, m$ the Lebesgue measure over $S$, and $f$ a distribution over the states. Let $T(t, x)$ be the state after time $t$, starting from state $x$ at time 0. Since a conservative system preserves the measure $m$, one has for the measure of states in $A$ after time $t$, with $A_{-t} = T(-t, A)$,

$$\int_S I_A(T(t,x))f(x)\,dm = \int_S I_{A_{-t}}(x)f(x)\,dm \qquad (5.1)$$

If this number of states in $A$ after time $t$ tends to a limit $L(A)$ as $t \to \infty$ independently of the distribution $f$, the event $A$ is *statistically regular* (1934, p. 66). The independence over $f$ here expresses that the statistical law is reproduced independently of the way the experiment is performed,

[13] This latter remark is made in a short paper (1935, p. 6).

the way of choosing the initial conditions of an experiment being represented by the distribution $f$.

Hopf's explanation of true statistical independence is contained in his 'independence theorem' (1934, p. 76):

> If an event $A$ is statistically regular with respect to a conservative mechanism, and if the event $A'$ has the same property with respect to another such mechanism, the simultaneous event $A \times A'$ is always statistically regular with respect to the resulting product mechanism and its relative frequency equals $L(A \times A') = L(A)L(A')$.

Later in his paper Hopf tries to base his conclusions on ergodic theory, instead of assuming an arbitrary initial distribution. He also relates his approach to the theory of von Mises by giving a measure theoretic argument for the impossibility of a gambling system (pp. 97–98).

In Hopf (1936) the way statistical independence is derived from the physical description, is illustrated by an example of two roulettes. The physical independence of the operation of the two mechanisms is represented by the assumption that the probability distribution for the joint results is genuinely two-dimensional (p. 181). That same paper gives two general conditions for the equidistribution of final positions in dissipative motion, and contains the first treatments of physically unsymmetrical mechanisms. Both topics are presented by examples. In the first, there is a completely balanced roulette wheel. What is the probability that the wheel comes to a rest within given angular bounds $\alpha_1$ and $\alpha_2$? The equation of motion for a completely balanced wheel is, with the angular velocity $\omega = d\alpha/dt$ and a frictional momentum $r$,

$$\frac{d\omega}{dt} = -r(\omega) \tag{5.2}$$

that is, the rate of change of the angular velocity $\omega$ is the negative of the frictional momentum $r(\omega)$. The final position $\alpha$ is a function of the initial position for which one can set the value $\alpha = 0$, and of the initial angular velocity $\omega_0$. The solution of the equation of motion is

$$\alpha = \int_0^{\omega_0} \frac{\omega}{r(\omega)}\, d\omega = F(\omega_0). \tag{5.3}$$

The two conditions for the derivative of $F$ that guarantee an equidistribution are:

$$\lim_{\omega \to \infty} F'(\omega) = \infty \tag{5.4}$$

$$\lim_{\omega \to \infty} \frac{F'(\omega + s)}{F'(\omega)} = 1 \quad \text{uniformly for any finite } s. \tag{5.5}$$

The effect of (5.4) is that the end position depends sensitively on the initial angular velocity $\omega$. Both conditions are easily fulfilled if the friction is assumed constant over initial velocity. But a uniform distribution of end positions (between 0 and $2\pi$) also follows from other kinds of assumptions that satisfy the above two conditions. One is that the initial angular velocity is high, another that friction is low.

An example of an unsymmetrical mechanism is given by a vertically placed roulette wheel, with a constant friction term, but an eccentric placement of the wheel. The latter has the effect that the forces affecting the wheel depend on its position (1936, p. 187). The equation of motion is $d\omega/dt = -k(\alpha)$, where $k$ is a positive periodic function of position $\alpha$ whose density is proportional to $k(\alpha)$. The end position $\alpha$ is determined by the initial velocity $\omega_0$ through integrating the equation of motion, which gives

$$\frac{\omega_0^2}{2} = \int_0^\alpha k(\alpha)\, d\alpha. \tag{5.6}$$

A new angular variable $\beta$ is defined by setting $\beta = c\omega^2/2$, with $c$ given through

$$c = \frac{2\pi}{\displaystyle\int_0^{2\pi} k(\alpha)\, d\alpha}. \tag{5.7}$$

The angle $\beta$ is a function $F(\omega)$ of the initial angular speed fulfilling the two conditions (5.4) and (5.5) and is therefore uniformly distributed. The distribution for $\alpha$ can be determined from the relation $d\beta = ck(\alpha)\, d\alpha$. Hopf says modestly that 'this seems to us to be the first example that can really be discussed, in which probability cannot be guessed by symmetry or other plausibility considerations' (p. 188).

In 1936 Hopf had written: 'Processes of equidistribution and of mixing... are very numerous in nature, as already everyday experience teaches. It is to be hoped that the future will provide more cases to the few that can be discussed up until now, also from the field of biology' (1936, p. 195). The same year he had proved the ergodicity of what is known as geodetic motion on a surface of negative curvature. In 1937 in his book *Ergodentheorie* he saw the general importance of ergodic theory residing in its giving 'an understanding of the stable frequency phenomena in nature.' The solution of the problem of onset of turbulence on the basis of classical hydrodynamics, once it is achieved, will bring the importance of the classical approach 'in bright light,' he says (1937, p. iii). (No solution was forthcoming, though.) The theory of objective probability in classical systems was more or less forgotten until very recently. One factor certainly was the predominant role played by quantum mechanics, both in physics and in philosophical thought about probability. The foundational importance of the classical approach is now being revived in the theory of nonlinear dynamics.[14] A thread of continuation from the 1930s can be followed in some purely mathematical works: Kolmogorov's introduction of the notion of dynamical entropy in the 1950s and studies of billiard systems and others by his students in the 1960s. The other Moscow probabilist, Khintchine, advocated very clearly an approach to probability along the lines of this chapter, in his (1938) and (1954), but his views, too, remained without noticeable effect.

[14] See also the recent monograph Engel (1992) for a mathematical development of the topic.

# 6

# *Von Mises' frequentist probabilities*

## 6.1 MECHANICS, PROBABILITY, AND POSITIVISM

Richard von Mises was an applied mathematician. He first specialized in mechanics, hydrodynamics especially. By applied, he really meant it: A book of 1918, for example, dealt with the 'elements of technical hydromechanics.' Another related specialty was the theory of flight, much in vogue early on in the century.[1] His work on probability starts properly around 1918, and from the same time are his first writings on foundational problems in science: on foundations of probability in 1919, and on classical mechanics in 1920. Von Mises' philosophical book *Wahrscheinlichkeit, Statistik und Wahrheit* of 1928 was the third volume in the Vienna Circle series 'Schriften zur wissenschaftlichen Weltauffassung', edited by Philipp Frank and Moritz Schlick. The year 1931 marked the publication of von Mises' big book on probability theory, *Wahrscheinlichkeitsrechnung*, whose exact title adds, 'and its application in statistics and theoretical physics.' The posthumous *Mathematical Theory of Probability and Statistics* is based on lectures from the early 1950s.

Von Mises was a declared positivist, identifying himself with the philosophy of the Berlin group, the Vienna Circle, and the Unity of Science Movement. His *Kleines Lehrbuch des Positivismus* (1939) appeared in an English version in 1951 as *Positivism: A Study in Human Understanding*. It attempts to give a broad presentation of the logical empiricist world view, from foundations of knowledge and the sciences to morals and society. After having been professor of mathematics in Berlin until 1933, von Mises first went to Istanbul, and in 1939 to the United States, where he became professor of aerodynamics and applied mathematics at Harvard University. The other main representative of frequentist probability of the logical empiricist school, Hans Reichenbach, followed a similar path, although holding chairs in philosophy.

[1] Apparently von Mises was a remarkable specialist in the theory of air flow. His *Theory of Flight* is still in print.

The part of von Mises' work in probability that has always been associated with him, has been his foundational system. Right from the beginning, in his (1919b) he limits probability to apply to *infinite sequences* of 'labels.' These labels are points in a sample space, or 'Merkmalraum.' For each region $A$ of the sample space, the limit of relative frequency of labels in $A$ has to exist. Second, a requirement of *randomness* is postulated for the sequences, thereafter to be called *collectives.* Mainly von Mises' theory of random sequences has been remembered as something to be criticized: a crank semimathematical theory serving as a warning of the state of probability before the measure theoretic revolution. But it is striking that the greatest revolutionary, Kolmogorov himself, based the *application* of probability on von Mises' ideas. A similar role was granted to von Mises by Khintchine in the late 1930s. During almost sixty years of measure theoretic probability, Kolmogorov's explicit reliance on von Mises as far as the interpretation of probability is concerned in the *Grundbegriffe* of 1933, seems to have gone mainly unnoticed. Indeed, von Mises' approach is taken to be a relict of primitive times, to the extent that, as a recent article on Kolmogorov's early work put it, 'In view of Kolmogorov's high opinion of von Mises a few explanatory remarks are appropriate here.'[2] On the other hand, a revival of von Mises' theory was initiated by Kolmogorov himself.[3] That theory is a field of very active research today, with aims different from the 'probabilistic analysis' measure theoretic probability has served so well.[4]

The postulate of randomness in a collective was subject to great debates in the 1930s. Von Mises' idea was to require that the limiting frequency in the collective is preserved in appropriately chosen *subsequences* of the original one. But in a binary sequence, say, there would be an infinity of both 0's and 1's if the limiting frequency is strictly between 0 and 1. Therefore there would exist subsequences containing only 0's, or only 1's, and so on. It is astonishing how many mathematicians,

[2]This is J. L. Doob, who was among the very first to fully endorse the measure theoretic probability of Kolmogorov. The passage is from the Kolmogorov memorial issue of *The Annals of Probability* (vol. 17, No. 3, 1989). The words of explanation tell us of attempts at defining individual random sequences, the result of which, even if it were successful, 'would obviously be too awkward and too limited to be considered seriously as a useful basis for the extraordinary scope of modern probabilistic mathematical analysis.' Still, the theory of random sequences presents 'an appealing conceptual problem which Kolmogorov discussed on several occasions,' namely, that of formalizing the notion of a random sequence (p. 815).

[3]To be discussed in Sec. 7.5.

[4]See the recent review by Cover, Gacs, and Gray (1989) in the Kolmogorov issue mentioned in note 2.

logicians, and others gave this existence proof as a *serious argument* proving the contradictoriness of von Mises' notion of collective. Clearly the postulate of randomness is based on a notion of existence entirely different from the one absolutized by set theory. Von Mises gave several explanations of what he meant by the choice of subsequences. One intuitive idea is as follows: If in a gambling situation you had the ability to choose, *in advance* of each single trial to be considered for betting, a subsequence that produces a frequency different from the original one, you would be able to convert this ability into a *gambling system*. Several mathematicians devised alternative theories of random sequences in the 1930s. Abraham Wald provided a model for the von Mises theory by delimiting the choices of admissible subsequences to a denumerable set expressible in a logical system. But the theory was on the whole declining: The measure theoretic approach became the dominant one, as will be explained in Chapter 7. Alonzo Church's (1940) suggestion for admitting only recursive selections of subsequences in collectives was the last original contribution to the theory for the next twenty five years to come.

Von Mises' other work in probability has been overshadowed by the debates around his foundational system. But he was first and foremost an applied mathematician. Thus he invented the circular normal distribution, also called von Mises distribution, in 1918 in a study of the periodic system of the elements. The question is, whether results of experimental determinations of atomic weights can be taken as coming from integer-valued weights with normally distributed error terms. The observed values can be 'wrapped' on a circle of unit circumference through a Fourier transform (a technique we have met repeatedly in Chapter 2), with the consequence that only the deviations from an integer value are preserved. By a circular normal distribution one means that the deviations measured in arc lengths are normally distributed. (In case the radius is large in relation to the dispersion, the circular normal distribution approaches the ordinary Gaussian distribution.) The task in the specific problem von Mises studied was to determine the probability that the true value is an integer.

Von Mises' long survey (1919a) of the central results of probability theory was prompted by the 'lagging behind' of probability mathematics in respect of 'precision in formulation and proof that has become the obvious standard in other parts of analysis' (1919a, p. 1). What von Mises wants are rigorous conditions and derivations for the central limit theorem and analogous results. Polya's work on the central limit theorem (1920) seems to be mainly prompted by this work, and very soon there followed the now standard conditions for the central limit theorem of Lindeberg (1922). In von Mises (1921a) the Poisson distribution for events with a low probability is derived in a more general setting. The individual probabilities $p_i$ are allowed to vary, as long as their average remains the same.

At the time of the mathematical exercises described, von Mises was also involved in a far more radical project, namely, the creation of a statistical physics freed from mechanical assumptions. This line of research was relevant for probability mathematics, also, in its relation to the theory of Markov chains and stochastic processes more generally. From his empiricist and verificationist philosophy, von Mises had come to the view that classical mechanics cannot serve as a foundation for statistical physics. The assumption of a microscopic mechanical trajectory behind the observable becomes *idle*; the equations for a macroscopic number of particles are of no use in predicting any scientific fact, as he says. On the other hand, von Mises held the opinion that genuine probabilistic behavior, in an immediate sense, is not compatible with a mechanical description: One cannot derive probabilistic laws from mechanical premises. These thoughts led von Mises to formulate a program for a purely probabilistic statistical physics. As we have noted in Section 3.2(b), this was the earliest clearly stated program for creating an ergodic theory independent of its origins in classical statistical mechanics, as a branch of probability theory. The particular way von Mises tried to put up a system of statistical physics was based on the use of Markov chains. He assumed the observable macroscopic events form such a chain. As Khintchine (1933a) notes, this assumption unnecessarily excludes systems of statistical physics with long time correlations between states. On the other hand, the applicability of Markov chains to statistical mechanics was an important factor in the creation of a general theory of random processes in the late 1920s and early 1930s.

It is remarkable that von Mises' probabilism as such had nothing to do with the emerging quantum mechanics. (Indirectly, though, quantum phenomena played a role through the problem of specific heats.) As is shown by his early paper (1912), his predecessor rather was the *Kollektivmasslehre* of Theodor Fechner (1897).[5] The Gaussian normally distributed events form here an ideal type, as in the geodesic measurements of Gauss himself. The cases interesting the statistician often fail to be of this type, says von Mises, and therefore the objective of *Kollektivmasslehre* are results of repeated observation in general (1912, p. 10). The task is to

[5] Fechner's philosophical ideas about probability and indeterminism are studied in Heidelberger (1987).

characterize their distribution by parameters that have an intuitive meaning, such as mean value and dispersion. By 1919 von Mises was ready to introduce *randomness* as a basic concept in the theory of probability.

## 6.2 FOUNDATIONS OF PROBABILITY: THE THEORY OF RANDOM SEQUENCES

Von Mises' clarification of the analytical foundations of the central limit theorem and related results in (1919a) was followed by his foundational system (1919b). It announced a third paper on 'the principal problem groups of probability theory,' as a completion of the trilogy, but that never came out. Cournot, le Roux, and Bruns' 'Wahrscheinlichkeitsrechnung und Kollektivmasslehre' are mentioned as predecessors to this theory (p. 52, note 2). He begins by a criticism of the classical definition of probability as equipossibility that he sees as a *petitio principii*. And how should one incorporate *unfair* dice and the probabilities of vital statistics under this scheme, von Mises asks (p. 53). In these criticisms of classical probability he was well taken by many: Khintchine (1929), Kolmogorov, and even the subjectivist de Finetti (1936b) all mention them with approval. Von Mises sets as his task, 'through abstraction and idealization to represent the connections and dependencies of well determined, observable phenomena' (p. 53). Such a task is achieved through the axiomatic method. Von Mises prefers to proceed with a very informal axiomatization, calling his axioms alternatively conditions. Previous comparable attempts referred to are Bohlmann (1901), Broggi (1907), and Borel (1909a) . The philosopher von Kries (1886) is also mentioned, though von Mises says that what he himself presents has no immediate connection to philosophical discussions (p. 55). In passing, he wants to bring forth 'the remarkable *physical* explanation of the appearance of probabilistic phenomena as recently given by M. v. Smoluchowski.'[6]

A *collective* is an infinite sequence of elements $e = (e_1, e_2, e_3, \ldots)$ such that to each element a *label* ('Merkmal') is ordered. A label is a point in a $k$-dimensional real space (the 'Merkmalraum'). There must be at least two labels, to both of which an infinity of elements is ordered. The word 'collective' conveys the idea of putting together elements into an entity ('Sammelgegenstand') and stems from Fechner's 'Kollektivmasslehre.' The first axiom or condition for collectives is (p. 5):

[6]That is of course von Smoluchowski (1918), discussed in Sec. 5.2.

Condition I: Existence of limits. Let $A$ be an arbitrary set of points of the label space that can be given (= *angebbar*). Then the limit of relative frequency of $A$ in the sequence of elements e exists: $\lim N_A/N = W_A$.

If the second condition, to be stated shortly, is fulfilled, $W_A$ is defined to be the probability of $A$ in the collective. Notice that probability is defined relative to *one single* collective. It follows that probabilities are numbers between 0 and 1, and that they are finitely additive. The probability values $W_A$ for all the point sets $A$ of the label space form the *distribution* of a collective. Note also that distribution is a set function defined over subsets of the label space.[7] It is a more general concept than the probability distributions encountered in classical probability. These can be recovered from the special case that $A$ has only one point.

Next we come to the second condition that postulates *randomness* (p. 5):

Condition II. Irregularity of the ordering. Let $A$ and $B$ be disjoint sets, with the nonzero probabilities $W_A$ and $W_B$ in collective $K$. Drop all the elements of $K$ whose labels do not belong to $A$ or $B$. A subsequence of this sequence is formed without making use of differences in labels. The limits $W'_A$ and $W'_B$ of the subsequence exist, and $W'_A/W'_B = W_A/W_B$.

Particularly, if $B$ is the complement of $A$, $W_A = W'_A$. As simple examples of methods of choosing subsequences, von Mises gives arithmetical laws, and the ordering of labels in another collective (p. 57). That is to say, neither *lawlike* nor *random* choices can produce a different probability of a label. The simple objection that there is a subsequence with only zeros is refuted, since its rule of choice uses the labels. That objection would not suffice for showing that von Mises' axiom system specifies an empty set of collectives. Various intuitive renderings of axiom II are given, for example, one stating that place selections not using label differences are 'admissible'; they do not change probabilities. The axiom states 'the impossibility of a gambling system' (p. 58).

In choosing a subsequence, it is permitted to make selection depend on the previous labels, or results of previous trials in more common terminology. However, condition II excludes the possibility of obtaining statistically differing results. In the simple case the labels would be 0 and 1. Taking the subsequence of those elements following a 0, and another following 1, must by condition II lead to the same limiting distribution. Therefore von Mises' randomness postulate serves the purpose of the

[7] Von Mises was not very specific at this point. He assumed the probability function would be defined for all subsets. Very soon, he was notified of the need for restrictions by Hausdorff. See his Selected Papers, vol. 2, p. 71, note (c).

assumption of probabilistic independence in the usual formulation of probability theory. This is a feature von Mises does not explicitly recognize. He thinks *chance* amounts to the fulfilment of his condition II. But obviously there is chance operating also in cases with dependence; only, there is perhaps *less* chance. In von Mises, probabilistically dependent events are in fact not excluded, but appear in a certain way in collectives derived from given ones.

There are four basic ways of deriving new collectives from one or several assumed to be given. The task of probability theory is to show how to combine these basic operations on collectives, and how to calculate the distributions of the new collectives from those of the given ones. But it would be erroneous really to ask for a collective to be given, in the sense that there would be a complete specification of the ordering of labels for each index $i$ of the elements $e_i$ (p. 58). This remark clearly indicates that collectives in von Mises are not constructive objects. His proposition 5 (p. 59) says: 'A collective is completely determined by its distribution; it is impossible to give the complete ordering one by one.' In fact, he soon goes on to add that one cannot construct collectives analytically. 'One has to remain satisfied with an abstract *logical existence* which only contains that one can operate without contradiction with the concepts defined' (p. 60). These remarks on collectives are rather similar in character to those made by Borel on 'sequences due to hazard' (see Section 2.2(c)). There, too, the only way to study the sequences was by probabilistic means.

A careful reading of von Mises reveals the following fact. He invariably talks about probability in *a*, that is, in *one* collective. This is so natural that it is not even mentioned, though we may find it strange. Here there is no plurality of possible worlds, no choice of path in the branching of the present into all the possible future courses of events as dictated by chance. More technically, the probability of an event is not the measure of (the cylinder set of) all the worlds which share the event of interest, relative to all possible worlds. Instead chance somehow resides in the only and actual world where every event is singular, and where the unending succession of events is unique and unrepeatable. Events have properties; these are designated by their *labels* which are the bearers of the probabilities. This is a minimalist ontology of an empiricist.[8]

Von Mises has been criticized often for his limiting frequency interpretation of probability, for the limits never are there really. This

[8] Echoes of similar thinking can be found in Gardner's (1987) 'nonprobabilistic' spectral analysis; cf. his introduction especially.

criticism is somewhat superficial as such, as von Mises explicitly took the limit as an idealization and mathematical abstraction (1919b, p. 60). He admits readily that in the world of experience, only finite sequences are observed. He says that to the two properties of a collective, there correspond definite *observable phenomena.* Say, in throwing dice, experience teaches us that the relative frequencies are approximately constant, so that 'we recognize a behavior which finds its appropriate expression in the assumption of *existence of limit* for a sequence of tosses extended into infinity' (p. 66). Experience also teaches that choices of subsequences do not alter the frequencies, this corresponding to condition II in the empirical world.

Von Mises compares the task of probability to that of mechanics: The given probabilities correspond to initial conditions, and the task is to determine other probabilities from the given ones. Just as it is not the task of theoretical mechanics to determine any initial conditions, it is not the task of probability theory to determine any probability distributions as such (p. 67). According to von Mises, thinking that probability theory could accomplish such a task, has been a major source of confusion. For example, he contends that Bertrand's paradox vanishes once one recognizes that the sought for probability depends on the probabilities assumed. A similar fallacy is to take an assignment of equal probabilities in games of chance as the only and a priori correct one, when in fact it is an assumption defining fairness of the gambling device (p. 67).

The application of probability is in von Mises analogous to the application of other physical theories such as mechanics. It is a 'two-way bridging' of theory and reality: First initial conditions are determined, here the distributions of the given collectives. Second, conclusions are made from computed results to the course of external phenomena (p. 68). The three main fields of application are: 1. games of chance, 2. statistics, and 3. physics. In the last case, which von Mises says is the easiest one, a hypothetically taken distribution is introduced. The justification of the hypothesis is tested through the results of the ensuing theory. An example would be the hypothesis of uniform distribution of velocities of the molecules of a gas. The thermodynamical consequences of the probabilistic hypothesis would be tested against observations.

In the second field of applications, namely, statistics, initial values for probabilistic computations come from counting. It is useful to broaden the perspective a bit here, in order to bring von Mises' ideas closer to other frequentist thinkers such as R. A. Fisher. In von Mises, *sequences* of elements are considered, but we can consider more generally a class of elements with or without any intrinsic ordering. Sometimes there is the natural time ordering of consecutive trials, as in von Mises. At other times we can think of a random drawing of members from a population. It suffices to concentrate only on the distribution given by the random sampling, just as it suffices to consider only the distribution in a collective. Probability obviously is determined as a fraction in a finite class in a statistical application. Since probabilities usually are assessed through the examination of a part of the class only, condition II is taken to be valid. It expresses the condition of representative, random sampling.

Von Mises mentions insurance and vital statistics as examples of the tacit acceptance of conditions of randomness. In the more general terms introduced here, this amounts to saying that whenever an inference from a sample to a population is accepted, an equivalent to his condition II is also required. There is no mention of the tremendous problem of finding 'the right reference class' in von Mises in this connection. For example, an insurance company might want to classify the persons insured into classes with different risks. Doing so, it could not treat the class of all clients as a collective insensitive to selection of subclasses. But von Mises only admits that the finiteness of the situation leads to a certain inexactness (p. 69).

Games of chance, the first field of application, are according to von Mises the most difficult cases. The classical requirement of equiprobability only defines fairness of the gambling device. To check this in practice, one has to perform a great number of trials with it. But 'it remains always uncertain what an effect the smallest deviations from symmetry have, deviations so small as to escape immediate observation' (p. 70). Earlier, von Mises had referred to what is known as Wolf's dice data. The difficulty is that it may be next to impossible to produce fair dice, and at least equally difficult to detect one from dice data.

A fourth field of application of probability is pure mathematics. Von Mises says that in cases such as the decimal expansions studied by Borel (1909a), the randomness condition II is not fulfilled. If one wanted to incorporate these cases into probability theory, the multiplication theorem (that is, the rule of multiplication for independent events) would have to be dropped (p. 71). Von Mises' thought seems to be that the decimals are given by a law and therefore they are not independent. He adds the remark that Hausdorff (1914) increases the load the notion of probability has to carry, through the use of the word probability for the relative measure of a set. Notice though that in von Mises, too, probability is a function defined over certain subsets of a space. He sketches in fact an integration theory for his notion of probability in

§3 of his (1919a) paper. It has been identified as coinciding with the Peano–Jordan concept of measure.

The form the calculus of probability takes in a von Misesian clothing is somewhat complicated. As Kolmogorov (1933, p. 2) indicates, the objective of having a theory that is as close as possible to the applications of probability in statistics, would not permit the simplest axiomatization. Once the concepts are explained in a terminology familiar today, it does not look that strange, though. But with a different set of basic concepts and principles, different things can be (or have to be) proved, whereas others, like the law of large numbers, are incorporated in the definitions, a surprising and educational experience. The basic operations on collectives are as follows:

1. *Subsequence selection.*
2. *Mixing.* A collective, that is, the ordering of labels to a sequence of experiments, is produced by a noninvertible function of the label space. A simple example is offered by the sum of the faces from a toss of a pair of dice.
3. *Division (Teilung or Aussonderung).* A collective $K'$ is formed from a given collective $K$ by taking only those elements $e_i \in K$ whose labels belong to a subset $M'$ of the original label space $M$. $M'$, like any proper label space, has to have at least two members, both with an infinity of labels ordered to the elements. Division does the job of conditional probability in von Mises' theory.
4. *Combination (Verbindung) of two collectives.* The elements of the new collective are pairs of elements of the given collectives. The distribution of the new collective is a product formed by the distributions of the given collectives. This multiplication rule of probabilities is proved from the requirement that combination produces a collective, in particular, fulfils condition II (1919b, p. 85). Combination generalizes inductively to any finite number of given collectives.

After the four basic operations on collectives have been introduced, probability theory can be defined as the study of how these operations can be combined, in various interesting or practically relevant ways, to produce new collectives (p. 88). Anyone who wants to see how it goes can turn to the nearly 600-page *Wahrscheinlichkeitsrechnung* of 1931.

A von Misesian probability theory applies exclusively to events that can be thought repeated a great number of times. Problems are dismissed as not well defined if they do not fulfil this condition. As an example, the St. Petersburg paradox is mentioned. In it, a bet of the following kind is proposed: A sequence of coin tosses is performed until heads appears for the first time. If this takes place at the $n$th toss, the gambler wins $2^n$ rubles. What is the fair amount to pay for playing one round of such a gamble? The expectation of gain is $\sum(\frac{1}{2})^i 2^i = \infty$. But no one is willing to pay just any amount for the right to play this game only once. Von Mises says no collective is defined here, since no indefinitely proceeding sequence of gambles is proposed (p. 91).[9] The idea of subjective probability is dismissed with, as one not amenable of a quantitative treatment. In questions of statistical inference, emphasis on quantitative treatment requires large samples. This is in complete accordance with von Mises' identification of probability as a frequency in an unlimited sample.

## 6.3 A PURELY PROBABILISTIC PHYSICS

We see that in von Mises an element of randomness is postulated into the theory. The question naturally arises, whether that randomness is compatible with the supposedly deterministic classical dynamics that still was the accepted basis of physics. Von Mises gives an answer in his paper 'On the present crisis of mechanics' (1921b). The basic question is, can Newtonian mechanics give an explanation of the motions of all bodies? He addresses relativity, hydrodynamics, and statistical mechanics (pp. 429–31). In his *Wahrscheinlichkeitsrechnung*, of the year 1931, he says that in this paper, the statement is made that it is not possible to have a satisfactory theory of physical phenomena without the use of probabilistic basic concepts (p. 561, note 56). His reason is the following: Undeniably there exist genuinely statistical sequences in nature, and these are such as cannot have any 'algorithm.' What he exactly meant by the inexistence of an algorithm is not so clear, but at least it should contain that there is no method of computing, even in principle, the sequence. As a last step of his argument, von Mises states without any question that classical mechanics, on the contrary, does have an algorithm. It follows that probabilistic and mechanical behavior are incompatible.[10] When one reads the 1921 paper, one has nevertheless some difficulty in locating an exact statement to that effect. A much later paper, a talk on causal and statistical laws in physics, von Mises (1930) does not make the strong claim, but gives instead a more sophisticated opinion. The basic question is again, can Newtonian mechanics explain the motions of all bodies? Von Mises concentrates on

[9] Much more interesting solutions have been proposed, all surveyed in Jorland (1987).

[10] Such a conviction was one of the principal starting points of the work of N. S. Krylov from the late thirties, now available in English as *Works on the Foundations of Statistical Physics*, Krylov (1979).

explanation. To explain, for him, is to reduce the more complicated to the more simple (p. 146). For example, one might reduce complicated planetary motions into a simple gravitational law of force. But if we consider the motion of matter on the atomic scale, the laws of force operative are inaccessibly complicated, so that classical mechanics becomes idle ('leerlaufend'), even if it is not contradicted (p. 147). The hypothesis that trajectories follow classical orbits has to be discarded, for von Mises concludes that 'an assumption from which we cannot determine whether it is right or not is not scientific' (p. 148). The methods of classical mechanics leave the problem of motion on an atomic scale completely untouched, whereas those of probability theory give a result that is in accordance with experience.

Next we come to von Mises' treatment of the ergodic problem. He was naturally aware of the earlier attempts of Einstein and others at founding statistical physics on classical dynamics. Contrary to these attempts, he had concluded that classical mechanics is not a relevant framework. In 1920 he published his 'Exclusion of the ergodic hypothesis from physical statistics,' in which he replaced the ergodic hypothesis of classical dynamics with a probabilistic formulation. The traditional dynamical version of the ergodic (or quasi-ergodic) hypothesis is the assumption that a mechanical system goes through all states (comes arbitrarily close to all states) in its evolution. As a consequence of Liouville's theorem, this leads to giving the same average kinetic energy to all degrees of freedom of the system (equipartition of energy). Von Mises emphasizes that in the case of specific heats, such equipartition is contradicted, so that the ergodic hypothesis had to be given up in many cases.[11] He formulates the task of the ergodic hypothesis differently, in a general way that avoids the difficulty: Given a microcanonical ensemble, one ought to be able to conclude that the microcanonical probabilities give the relative times that each of the systems of the ensemble spends in different regions of the phase space. Von Mises admits that the ergodic hypothesis makes this inference possible, but he contends at the same time that it is not at all necessary for the inference. He calls the classical ergodic (or quasi-ergodic) hypothesis an 'ergodic hypothesis in the strict sense' and offers anything that does the following task as an 'ergodic hypothesis in the wide sense': namely, the task of allowing an interpretation of a combinatorially determined probability as the relative time in the actual motion of a specific mechanical system (1920, p. 226). However, even this formulation does not satisfy him. It

[11] The matter is explained by quantum theory, but von Mises does not mention quantum effects in this context.

dispenses with the original mechanical ergodic hypothesis, but the conclusion is still unacceptable. He makes the point that a theory can lead to definite statements about the actual course of single systems only through mechanical equations, but never from probabilistic considerations.

Von Mises gives Brownian motion as an example of an unstated use of the ergodic hypothesis. The probabilistic calculations of the theory give a virtual ensemble (uninterpreted probability density); however, one tests the theory without hesitation by time averages. Therefore this approach is based on the ergodic hypothesis in the wider sense. But von Mises is not a supporter of the method of ensembles, for he says that with virtual ensembles, 'there would not be any physical content in our considerations, for physics is concerned with the prediction of phenomena occurring in time' (1920, p. 227).

Von Mises comments on the Boltzmannian approach as represented by Einstein as follows. Here one starts with a time ensemble, defined through the relation between entropy and probability (entropy proportional to logarithm of probability). 'But a deep contradiction remains in physical statistics, one that has not been conquered yet, namely, that one takes from a certain point of view the course of events as completely determined through the physical equations, the differential equations of motion of the system; yet one thinks one can make definite statements about this course of events, from a completely different point of view' (1920, p. 227). This, apparently, is the statement that classical dynamics and genuinely probabilistic phenomena are in contradiction with each other.

Von Mises' own purely probabilistic approach to statistical mechanics started from the requirement that measurements always have to be of finite accuracy only. For von Mises, this is a consequence of the atomic structure of matter. The assumption of exact microstates behind observable macrostates or coarse-grained states is idle, 'leerlaufend.' By accepting only macrostates as empirically meaningful, one sees that the determination of the evolution of states in time of classical dynamics becomes empirically meaningless. Probability is introduced as a hypothetical element. It should come in a form which leads to probabilistic laws about the time development of the system. This makes it possible to do the job of the ergodic hypothesis in its classical formulation: the replacement of virtual ensembles (uninterpreted probabilistic assumptions) with predictions of behavior in time. The specific structure that von Mises gave his theory can be condensed into the following (1931, §16). Only a finite number of macrostates are physically observable states. It is assumed that the course of events forms a Markov chain: Given the present macrostate of the system, the transition probability to the next macrostate depends only on what the present state is, but not on the previous 'history' of the system. If the transition probabilities remain the same in time and if they fulfil certain additional conditions, there will be a unique limiting distribution for the macrostates of the system.

Von Mises' work in statistical physics has mainly contributed to the mathematical theory of Markov chains. His probabilistic ideas influenced the transition from ergodicity of dynamical motions to ergodicity as a purely probabilistic and very general concept, as has been explained in Section 3.2(b).

## 6.4 THE FATE OF COLLECTIVES

The 1920s were a brooding time for foundations of probability. Interest in mathematical probability was accumulating, with developments in the limit theorems and random processes being the foremost fields of research. Then, from around 1930 on, different approaches to foundations of probability started appearing.[12] Von Mises himself had remained silent on foundational questions from 1920 to the appearance of his 1928 philosophical book. The foundational debate saw a refinement both in the measure theoretic approach to probability and in the approach through random sequences. The fate of collectives was to be the loser in this debate. We have discussed early measure theoretic probability in Section 2.1(a), and shall resume that line of development in connection with the work of Kolmogorov in Chapter 7. But the defeat of collectives was not final, as we shall see in our closing up on Kolmogorov. The von Mises theory was given a new start in the 1960s, by no one else than Kolmogorov himself.

One of the first criticisms of von Mises' theory was made by Khintchine in 1929, in his 'Von Mises' theory of probability and the principles of physical statistics.' He says that the constructive part of the theory, with its four basic operations, does not raise suspicion. It is the foundations that are problematic. The idealization of limit of relative frequency is not fruitful, for it does not fulfil the methodological principle that the properties of an idealized object should be approximately realized in a real object. Khintchine dismisses with the notion of collective: He says that contemporary mathematics knows only two kinds of binary sequences, those following a law, and the choice sequences of Brouwer. The latter ones cannot be thought of as ready-made, and they only have properties which are based on finiteness.

[12] See, for example, the bibliography of Kolmogorov 1933.

The existence of a limit is not a property of a choice sequence; one cannot apply the disjunction, either there is a limit or there is not a limit to it.[13]

From the late 1920s on several alternatives to von Mises' notion of collective were proposed.[14] The first one was the theory of *admissible numbers* of Arthur Copeland. It generalizes the notion of normal numbers proposed by Borel.[15] In 1932 Reichenbach suggested another definition of what he called normal sequences. Finally Popper, in his famous book on the logic of scientific discovery (1935), suggested a notion of sequences without aftereffect. The three definitions can be shown equivalent, as in Jean Ville's study of the notion of collective (1939, Ch. 3). Admissible numbers are obtained from the definition of a collective by requiring the selection of subsequences to be of the following kind: Let $a = (a_1, \ldots, a_m)$ be an arbitrary sequence of labels. Given an infinite sequence, choose only those members which immediately follow the sequence $a$, and require this subsequence selection rule to preserve the limits of relative frequencies. A defect of the definition is that there are admissible numbers that can be constructed, like there are absolutely normal numbers that can be constructed. The latter was shown by Sierpinski (1917) and Lebesgue (1917), the former follows from Copeland (1928).[16] A simple example of an absolutely normal number, sometimes called Champernowne's number, was discovered by Champernowne (1933). In the scale of ten, it is obtained by writing the decimal 0.123456789101112..., that is, by writing all the natural numbers in succession as the decimal expansion. Von Mises did not accept admissible numbers as a basis for collectives, because they may be obtainable through a law. This, in turn, would lead to a possibility of selecting a subsequence of an admissible number where the limit frequency is different. Another frequentist theory of probability was formulated by Tornier (1929). It contains a denumerable repetition of denumerable sequences where von Mises operates with just one collective.[17]

The simplest objection to the theory of collectives was that each proper (binary) collective has an infinity of 0's and 1's, and therefore admits of a subsequence with only 0's, and another with only 1's. For von Mises this argument only proves that collectives cannot be the kind of explicitly given mathematical objects that would permit determining

[13] By the late 1930s, Khintchine's tone became much more critical, as is shown by the posthumously published paper (1961).

[14] These theories have been reviewed in Martin-Löf (1969) and van Lambalgen (1987).

[15] See Sec. 2.2 (c) for normal numbers.

[16] See Martin-Löf (1969, p. 23).

[17] See Martin-Löf (1969, pp. 25–27).

their members one by one. A second line of objections to von Mises' theory was that it excludes sequences that are logically possible. The objections called for a refinement of the basic concepts of the theory. Von Mises himself did not work much on such improvements, even if he repeatedly defended his theory on intuitive grounds. The main contributor to his theory was Abraham Wald, who made the decisive steps toward a logically consistent formulation. Claims to the emptiness of the notion of collective were finally removed by his (1937) proof that for any denumerable set of functions performing the subsequence selection, there exist collectives.

Wald (1937) removes the separation between elements and the ordering of labels to these in von Mises, dealing directly with a sample space $M$ where the random elements take their values. Selection functions $f_n$ are 0–1-valued functions operating on sequences $(m_1, \ldots, m_n)$ of the product set $M^n$. For a denumerable sequence $m$, $m_i$ is selected to the new subsequence if $f_{i-1}(m_1, \ldots, m_{i-1}) = 1$. The probability of a set $L \subset M$ is defined to be the limit of relative frequency, provided it exists. The principal result of Wald is the following: Let $M$ be any set and let $J$ be the set of Jordan measurable subsets of $M$, with $\mu$ the normalized Jordan measure. Then if $S$ is a denumerable system of selection functions, there exists an infinity of collectives for which $L \in J$ and the probability is $\mu(L)$. Wald defends his assumptions as follows: First, in an application one would never encounter more than a denumerable infinity of selection functions. He in fact suggested that one could take as $S$ those functions that are expressible in a formal system for mathematics, like the one of Russell's Principia. The second restrictive assumption, to Peano–Jordan measurable sets, stems from the following result. There exists no collective that would give as limits of relative frequencies the measures $\mu(L)$ for a larger collection of sets than the Peano–Jordan measurable ones (Wald 1937, p. 46). The same restriction had been arrived at by Copeland (1937, p. 339; also 1931) and by Tornier (1929). The restriction is naturally of interest only if $M$ is continuous. The defense of the restriction is based on the interpretation of probability as a limiting frequency in a collective. If a Lebesgue measurable subset of $M$ is not Peano–Jordan measurable, its measure cannot be approximated by a finitely additive measure. The event is essentially infinitistic. An example of such an event is provided by the strong law of large numbers and the law of iterated logarithm. The heart of the problem is that the limit of relative frequency, while fulfilling all the other properties of probability, fails to be denumerably additive. Limit of relative frequency can only be used for interpreting measures of events which can be arbitrarily well approximated by a finitely additive measure.

The theory of collectives was subjected to a penetrating critique by Jean Ville in 1939. The introduction of his small book, *Etude critique de la notion de collectif*, contains discussions of foundations of probability from a time when the measure theoretic approach had not yet quite reached its position as a standard. The latter gives an axiomatic (implicit) definition of probability. In the frequentist theory, on the other hand, one aims at giving an *explicit* definition of probability on the basis of other basic concepts (Ville 1939, p. 5). The frequentist or statistical approach is also the empirical one which takes probabilities simply as an expression (constatation) of stable frequencies, whereas the axiomatic approach tries to *explain* the latter (p. 6). Ville open-mindedly discusses subjective probability as represented by de Finetti. A subjective probability is a prevision (expectation) of frequency (pp. 11–13). He further addresses the problem of probabilities of singular events, an inexistent concept according to the standards of the frequentist philosophy of probability. He says that in most cases, a subjective probability estimate would be judged as follows. 'If you make notice of all the events to which I have given the probability 3/4, and if you observe for a large number of cases the relative frequency of times the event has in fact been produced, I predict that this frequency could be close to 3/4' (p. 17). Judging the quality of an assessment of probability in this way comes very close to the idea of calibration, or measurement of deviation between probability assessment and observed frequency.

Turning now to the frequentist theory, we noted that a frequentist interpretation in the style of von Mises has to abstain from accepting events relying on denumerable additivity. The strong limiting theorems are good examples. More generally, events of the type, '$A$ occurs infinitely often,' are of this kind. Specifically, the behavior of limiting frequency in the law of iterated logarithm is an 'i.o.' result. Accordingly, Ville was able to prove his main objection in the form of a theorem. Consider a binary alternative with an arbitrary probability $p$, and any denumerable class of selection rules $S$. There is a collective such that the frequency of 1's in the sequence is always greater than or equal to $p$ (1939, p. 63). The law of iterated logarithm gives exact bounds for the oscillation of finite relative frequency, but the definition of random sequences as collectives is not even able to secure that the frequency ever has a value below the limit. Ville does not explicitly refer to the law of iterated logarithm here. But he observes (p. 67) that in the 'modernized classical theory' (as he calls the Kolmogorovian or similar measure theoretic approach) one proves the following: If we have simple events with independence and success probability $p$, the relative frequency $k/n$ fulfils with probability 1 infinitely often the inequality $|k/n - p| > 1/\sqrt{n}$.

This latter value of course was the earliest lower bound for the oscillation of frequency obtained for strong laws of large numbers, to be perfected into the exact value given by Khintchine's law of iterated logarithm.[18] Fulfilment of the inequality infinitely often also gives an example of an event whose occurrence is only decided after the production of the whole denumerable sequence is in some sense completed. Therefore, to prove a similar result for collectives, one would have to imagine an infinite sequence of collectives and show that they all have the probability 1 property of the measure theoretic approach (p. 68). But instead, Ville proves that for any $p$ and any denumerable set of selection functions, with $P$ denoting the property that the above inequality holds only for a finite number of times (p. 69):

α. There is a sequence of collectives each with limiting frequency $p$ such that the limit of relative frequency of collectives for which $P$ holds has value 1.
β. Same as in α, except that the limit in the sequence of collectives is 0.

The conclusion is that results making essential use of denumerable additivity are outside the reach of the frequentist theory.

Ville's own approach to von Mises' idea of random sequences builds on the intuition of the impossibility of a gambling system. In the simplest case, there is a symmetric coin-tossing process used for betting, and the gambler can only choose the sum that is to be doubled or lost at each repetition. His sum $s$ gained after $n$ gambles, starting with a unit sum, is a function of the previous outcomes $x_1,\ldots,x_n$. Given that the gambler has the sum $s(x_1,\ldots,x_n)$ at stage $n$, the fairness of the game implies that his expectation of future gain is this same sum. Such a sequence of gains $s(x_0)=1$, $s(x_1)$, $s(x_1,x_2),\ldots$ is called a *martingale*. A general result says that the supremum of $s(x_1,\ldots,x_n)$ is with probability 1 finite. This says in other words that the probability of making an indefinitely large gain is 0, so the gain becomes infinite only in a set of measure 0. The definition of irregularity through martingales says collectives do not belong to such null sets. Equivalently, collectives belong to the complements of these measure 0 sets. But one cannot require them to belong to *all* such sets (Ville 1939, p. 93). Excluding all probability 0 properties from collectives would in the end exclude all logically possible sequences (p. 136). Ville's overall conclusion was that although the definitions of random sequences by Wald and by the martingale approach are intuitively satisfactory, one cannot use the definition of probability as an ordinary limit in the mathematical sense. To avoid contradictions,

[18] See Sec. 2.4.

restrictions are imposed on the selections and on the events for which probability is defined. They are not imposed by the nature of the problem however, in Ville's opinion (p. 140).

In his conclusions Ville again returns to the problem of singular events. The probabilities of isolated (singular) events are incorporated by the introduction of 'nonhomogeneous' collectives, namely, those consisting of the repeated judgments of in what category the event is to be placed. Probability is the frequency within this category. Ville says the difference of these sequences of judgments and von Mises' collectives is a linguistic one (p. 139). His final verdict on collectives is as follows: An expression such as 'the frequencies have a tendency to group around a certain value' seems impossible to avoid, even if it is not mathematically precise. Such expressions are, however, sufficiently precise for a justification of the axioms of probability by 'a certain kind of empirical passage to the limit' (p. 140). Obviously, a great number here is 'empirically' the same as a limit. But Ville believes to have shown that 'a passage to the limit in the sense of analysis cannot be used for the definition of probability' (ibid.).

We shall resume some of the discussions and debates on the nature of collectives and the explicit definition of probability with their help in Section 7.5 on finite random sequences. Ville's book practically was the last contribution to the von Mises theory in a long time to come, with the war setting in and all. As mentioned, one small piece was added by the logician Alonzo Church in 1940. He suggested that the idea of effectively choosing a subsequence be made precise with the newly formulated concept of recursiveness. That concept, according to the well-known Church thesis, was meant as a mathematically precise formulation of 'mechanical,' algorithmic computability.[19] The class of recursive selections is denumerable so that the conditions of Wald's result are fulfilled. But the objection of Ville, that the finite relative frequency could still fail to oscillate on both sides of the limiting value, also held for the Church suggestion. That problem was only resolved in the mid-1960s, as we shall see in Section 7.5.

[19] Some of the original papers by Gödel, Kleene, Church, Turing, and Post are reprinted in Davis (1965).

# 7

# *Kolmogorov's measure theoretic probabilities*

## 7.1 FOUNDATIONS AND PHILOSOPHY OF MATHEMATICS

*Grundbegriffe der Wahrscheinlichkeitsrechnung* by Andrei Kolmogorov is the book which has become the symbol of modern probability theory, its year of appearance 1933 being seen as a turning point that made earlier studies redundant. In mathematics, it is fairly common to take a field of study as given, as being defined by a set of commonly accepted postulates. Kolmogorov's presentation of probability in terms of measure theory serves well to illustrate this supposedly ahistorical character of mathematical research: With some knowledge of set theory, one can take the book, and learn and start doing probability theory. In such an approach, the concepts and the structure of probability theory appear fixed, whereas the experience of those who built up modern probability must have been very different. There were many kinds of approaches to the foundations of the subject. The idea of a measure theoretic foundation was almost as old as measure theory itself, and it had been repeatedly presented and used in the literature. Therefore the mere idea was not the reason for the acceptance of Kolmogorov's measure theoretic approach, but rather what he achieved by the use of measure theoretic probabilities. The change brought about by Kolmogorov was a big step, but not the kind of dramatic revelation some later comments suggest. The two essential mathematical novelties of *Grundbegriffe* were the theory of conditional probabilities when the condition has probability 0, and the general theory of random or stochastic processes. The latter became the most essential object of study in modern probability.

The mathematical study of stochastic processes with continuous time was begun in the late 1920s. It proved enormously important for the development of probability, and such processes have stood at the center of research ever since. The first systematic, extensive development is given in the long paper by Kolmogorov (1931), 'Über die analytischen

Methoden in der Wahrscheinlichkeitsrechnung.' It was motivated from problems in statistical physics, and used the measure theoretic formulation of probability. Another paper (1929b), also explains the measure theoretic approach to probability prior to the book of 1933. In view of the unique role this book has had in the development of modern probability, one is surprised by the relative lack of detailed commentaries. For example, a logically trained eye easily finds some of the principal questions on foundations of mathematics behind the presentation. In this respect it is very much a product of its times. Kolmogorov was in fact deeply interested in such foundational problems. Some of his very first works were concerned with the rules of correct reasoning with infinity. He put up an intuitionistic or constructive theory that would explain the place of infinity as an ideal notion in mathematics.

At present, after more than fifty years of exposure to measure theoretic probability, it may be difficult to conceive of any alternatives. Such was easier coming to some of those who created the theory. In what may seem like a spectacular turn of minds, Kolmogorov began developing a new approach to probability in the 1960s. This was a turn back to a theory of the type created by Richard von Mises in the 1920s. Without historical understanding of the problem situation in the early 1930s, it may seem like a step in the void.

The theory of probability forms only a small part of the published work of Kolmogorov which runs to hundreds of items.[1] His other mathematical works belong to the fields of logic, real and functional analysis, the theory of turbulent flow, classical dynamics, and so on. Kolmogorov's first work on probability theory was a joint paper with Khintchine in 1925. From then on he appeared as the ingenious proof-maker, many times strengthening the limit theorems obtained by Khintchine. The latter had started the study of probability theory in Moscow, with his result known as the law of iterated logarithm, in 1923.[2] Probability theory became a field of very active research. The publications of Khintchine and Kolmogorov show that a close collaboration often stood behind the new developments. They also shared interests and viewpoints in the foundations of mathematics. Research on probability theory was conducted contemporaneously

[1] A bibliography can be found in the special issue on Kolmogorov of the *The Annals of Probability*, vol. 17, No. 3 (1989). It also contains an account of his life and works by Shiryaev.

[2] See Khintchine (1923, 1924) and Kolmogorov (1929a).

by S. N. Bernstein, E. E. Slutsky, and V. Glivenko among others.[3] Earlier tradition included the works of Chebyshev and Markov.

Kolmogorov's mathematical education was in the spirit of set theory and modern real analysis of the school of Lusin.[4] His first paper, written in 1922 but published as Kolmogorov (1928b), deals with what is today called descriptive set theory. In his first publications in 1923 and 1924 he studied the properties of Fourier series, giving in (1923) an example of a series that diverges except for a set of measure 0. In (1926) he gave an everywhere divergent series.[5] In 1925 he published the first of his early papers on logic and foundations of mathematics; a belated English translation appeared only in 1967. Reading it, one finds out how unfortunate it was not to give it out in a more accessible language and place. It is the first publication ever on a constructive system of logic, Kolmogorov's main aim being to 'save' classical mathematics by showing that it has *formal validity*. He gives a translation of classical formulas to constructive ones, an invention usually attributed to Gödel (1933) and Gentzen (1933), but besides the technical results it also contains a general approach to the philosophy of mathematics.

The 1920s saw the great debates between Hilbert the formalist and Brouwer the intuitionist. Hilbert's programe, deriving from the model of *Grundlagen der Geometrie*, aimed at establishing the consistency of classical mathematics. Only the finite has immediate meaning. Mathematics is to be formalized into a system of finite strings of symbols manipulated according to mechanical rules. Infinitistic notions and principles are *ideal*: One should show that they can be safely used for deriving results concerning the *real* finitistic part. For each such derivation, there should be a (usually more complicated) finitistic derivation. Mathematical existence is inferred from consistency. Specifically, if an assumption of nonexistence of an object leads to a contradiction, one is allowed to infer the existence of the object. Properties of formal systems are studied in *metamathematics* which is nonformalized and uses intuitive ('inhaltlich') finitary reasoning.

The starting point of Brouwer's criticism of classical reasoning is the indirect derivation of an existential statement, which in no way recognizes the object claimed to exist. The weak point which has no convincing

[3] See Alexandrov et al. (1969) for Bernstein. Kolmogorov has written obituary notices of Slutsky and Glivenko (1948, 1941). A discussion of the development of Russian probability in this century is contained in Maistrov (1974), but it is unfortunately brief and superficial.

[4] See Phillips (1978) for the school of Lusin.

[5] An explanation of this remarkable result can be found in Zygmund's book (1935, p. 175).

justification is captured in *the principle of excluded middle.* It asserts the truth of $A \vee \sim A$ for any $A$. An equivalent formulation is the law of double negation: $\sim\sim A \rightarrow A$. In (1924a) Brouwer gave the constructive notion of negation: If $\perp$ is an impossible proposition (an *absurdity*), $\sim A$ is defined as $A \rightarrow \perp$. The law of double negation says that from the impossibility of the impossibility of $A$ follows the truth of $A$. If $A$ is an existential proposition, the law of double negation would permit precisely the kind of inference Brouwer set out to criticize: from the impossibility of nonexistence, to infer existence. One way of seeing the implausibility of the universal validity of the principle of excluded third is offered by Kolmogorov (1932c). He interprets the propositions $A, B, C, \ldots$ as expressing problems. Thus, $A \& B$ is a problem that is solved by solving both $A$ and $B$, $A \vee B$ is solved by solving at least one of $A$ or $B$, and $A \rightarrow B$ is solved by reducing the solution of $B$ to that of $A$. This gives also negation: $A$ is impossible if it leads to the solution of the absurdity $\perp$. The problem $A \vee \sim A$ would be solved if we had a way, applicable to just any problem $A$, of solving $A$ or showing $A$ impossible, which clearly is not the case. The usual propositional logic comes out by saying the problem is to establish the truth of a proposition $A$.

Kolmogorov's great objective in the 1925 paper 'On the principle of excluded middle' was to show that 'all the finitary conclusions obtained by means of a transfinite use of the principle of excluded third are correct and can be proved even without its help.' To show this, he notes that $\sim\sim\sim A \rightarrow \sim A$ holds constructively. Then he replaces judgments of classical logic with their *double negation translations*: All its formulas and subformulas can in fact be so transformed that the principle of double negation elimination only finds application to negative formulas. The upshot is that for any classically provable formula, there is a constructively provable formula *classically* equivalent to it. After this result, only prejudice can permit claiming that constructive reasoning is weaker than classical. Kolmogorov's paper, though, was not noticed, but it seems to me that it did not go completely lost: For its basic idea, namely, that a double-negated form of a theorem of classical logic holds constructively, was used by his Moscow fellow probabilist Glivenko in (1929). Heyting's well-known formalization of intuitionistic logic, the basis of all subsequent work on the topic, refers the idea to this paper (cf. Heyting 1930, p. 43). Later, in Gödel (1933) and Gentzen (1933), similar double negation transformations were applied to classical first-order arithmetic.[6] These results show that the consistency of the

[6]Gentzen withdrew his paper at proof stage upon hearing of Gödel's publication. The text first appeared, in English translation, in his collected papers in 1969.

classical theory reduces to that of the constructive one. Gödel's second incompleteness theorem of 1931 stated the unprovability of consistency of arithmetic. The importance of the relative consistency for Gödel was that constructive arithmetic was founded on more intuitively reliable principles than the classical one.

Both Hilbert and Kolmogorov maintain that only the finitary part of mathematics has meaning. Hilbert locates the meaningful part in the finitist metamathematics, in which one speaks of the properties of formal systems, such as consistency or completeness. But Kolmogorov is an intuitionist and wants to put up directly a system of constructively valid principles. He finds that a formalist is driven into arbitrariness as his only criterion is consistency. If there is an unprovable formula one can add it or its negation to the system. The choice is purely conventional (1925, p. 417).

Kolmogorov attempts to legitimate classical mathematics, 'alongside the development of mathematics without the principle of excluded middle.' The double-negated results are 'pseudotruths' which coincide with ordinary (constructive) truth in a finitary case; in that case one can use classical logic. (I shall return to this in a while.) A specific consequence of the double negation translation is that the consistency of a classical result reduces to the consistency of the corresponding translated one (1925, p. 431).

More light on the circumstances of Kolmogorov's early paper on constructive logic is thrown by Khintchine's 1926 'Ideas of intuitionism and the struggle for content in contemporary mathematics.' There we find that a seminar on intuitionism was being held in Moscow at the time. Khintchine's paper criticizes strongly Hilbert's formalism, and defends Brouwer's intuitionism as a 'mathematics with content.' It is interesting to notice that the criticism of Hilbert's philosophy of mathematics relates to more general tendencies against 'empty formalism' in Soviet cultural life of the 1920s.

As mentioned, Kolmogorov's second early paper on constructive logic (1932c) gives the problem-interpretation of the newly formalized system of intuitionistic logic of Heyting (1930). The interpretation makes constructive logic understandable for those who 'do not share the intuitionistic epistemological preconditions' (1932c, p. 58). Indeed, constructive logic gives a *different meaning* to the logical operations, and once these are explained, it is easy to see what is meant by the 'failure' of old-established logical principles such as the law of excluded middle. The second part of Kolmogorov's paper goes beyond the level of such logical explanations, into the heart of the disagreement between intuitionism and the classical views. Kolmogorov states the basic principle of the intuitionist criticism of logical and mathematical theories as follows: 'Every assertion not without content must refer to one or several fully determined matters of fact [Sachverhalte] accessible to our experience' (1932c, p. 64). If a set is infinite and if $A$ says that every element of the set has a certain property, the claim of $A$'s falsity does not fulfil this principle. That is to say, we may be in a situation of checking indefinitely one after the other the individual cases, there being no guarantee that a falsifying instance will be found in the future.

Kolmogorov is one of those thinkers who, once having taken a stand, would not easily change their minds. Thus a passage from the *Grundbegriffe* of 1933 (G.W. for short) reminds us that Kolmogorov maintained his intuitionist philosophy of 1925: Referring to sets (events in probabilistic terminology) that are infinite unions or disjunctions of other sets, he writes (G.W., p. 16) that 'we consider these sets in general only as "ideal events" to which nothing corresponds in the world of experience. But if a deduction uses the probabilities of such events, and if it leads to the determination of the probability of a real event, this determination will obviously be unobjectionable also from an empirical point of view.' Like another of the relatively few followers of Brouwer's ideas, Hermann Weyl, Kolmogorov also went on with studies of classical mathematics as his main activity. But for him there was a special reason for this attitude, as we have seen.

In the 1950s Kolmogorov began studying the theory of algorithms. Church's (1940) definition of random sequences gave a connection between this field and probability theory. It was used by Kolmogorov in the 1960s, when he formulated anew some of the ideas of von Mises. In the address at the international congress of mathematicians (ICM) in Nice in 1970, published as Kolmogorov (1983), he gave some of the philosophical reasons: Hilbert's formalization of mathematics does not address 'the real, intuitively accepted content of mathematics.' Mathematics is a 'science of the infinite' (1983, p. 30). Further, 'this state of affairs is deeply grounded in the structure of our consciousness, which operates with great facility with intuitive ideas of unbounded sequences, limit passages, continuous and even "smooth" manifolds, and so on' (ibid.). Kolmogorov thinks the way the brain processes information is similar to the computer: 'It is a paradox requiring an explanation that while the human brain of a mathematician works essentially according to discrete principles, nevertheless to a mathematician the intuitive grasp, say, of properties of geodesics on smooth surfaces is much more accessible than that of properties of combinatorial schemes capable of approximating them' (pp. 30–31).

In the philosophy of probability proper, Kolmogorov represented a finite relative frequency interpretation. The problem of interpretation concerns the way the mathematics of probability is combined with the empirical world. His changing ideas on how probability theory is applied meant changes in the interpretation. The first indications in the direction of interpretation were in the *Grundbegriffe*, and I shall therefore address them later, in Section 7.4.

In the mid-1950s Kolmogorov published his most extensive clarification of the methodological role of probability, with English translation in Kolmogorov (1963a). A probability is a number applying to repetitive events. It is 'a constant $p = P(A/S)$ (objectively determined by the connection between the complex of conditions $S$ and the event $A$) such that the frequencies $v$ get closer "generally speaking" to $p$ as the number of tests increases' (1963a, p. 231). The qualification 'generally speaking' leaves room for imprecision which is unavoidable according to Kolmogorov. Laws of large numbers apply in the following way. Let $P$ be the probability that a relative frequency deviates from the theoretical probability $p$ of the event under interest by more than $\varepsilon$. A law of large numbers allows one to determine how long a series of experiments has to be so as to have the probability $P$ below a preassigned level, say 0.01. Let this number be $n$. The theoretical probability, typically with the assumption of independent repetitions, forms a probabilistic model. $P$ is also interpreted frequentistically: In a long series of $n$ trials each, the frequency is within $\varepsilon$ of $p$ in 99 percent of cases. But here one has to stop and remain content with the imprecision of the notion of probability (1963a, p. 239).

Kolmogorov (1963a) repeatedly comes to the question, 'does the randomness of the event $A$ demonstrate the absence of any law connecting the complex of conditions $S$ and the event $A$?' (p. 229). He thinks that statistical laws can have objective validity independently of the question of ultimate randomness. Such statistical or probabilistic laws are given relative to some specific conditions $S$. These conditions might be such as to make deterministic prediction of individual results possible. The statistical distribution, on the other hand, is unaffected by that possibility: It is brought about by a certain process independently of our knowledge (p. 249). One can see that Kolmogorov is at pains at allowing determinism to be compatible with an objective notion of statistical probability. He devotes a whole section (§5) to questions of determinism and randomness; still one cannot say on its basis that he would have denied the theoretical possibility of determinism in the style of Laplace. A related question concerns the notion of probabilistic independence. As we shall see later, Kolmogorov found probabilistic independence to be the notion that distinguishes probability theory from measure theory in general. It is interesting to note how he links independence directly to causality (1963a, p. 251): The practical meaning of independence is 'absence of causal connection.' The latter would also usually be the means for inferring independence. But the mathematical definition of probabilistic independence is formal and broader than the practical criterion, so that there would be cases where independence is simply proved by computation of the relevant probabilities which are seen to multiply in the right way. For example, the digits of a decimal expansion may be independent in this sense, even if they 'are interconnected by the way they are produced' (p. 252). Probability seems at least partially to overlap with some sort of determinism.

In methodological questions, or the philosophy of science, Kolmogorov appears as a tough inductivist. He suggests that statistical laws are discovered by an empirical study of frequencies. That study is followed by a computation of other probabilities, from the empirically determined ones using the rules of the theory (1963a, p. 231). This simplistic description is qualified in the sequel: 'Only in the first stage of the penetration of probabilistic methods into a new domain of science has the work consisted of purely empirical observation of the constancy of frequencies' (p. 254).

The 1960s brought new ideas. These were still changes within the general frequentist conception, as we shall see. It seems that Kolmogorov never expressed his opinion on the subjectivist or Bayesian view of probability and statistics.

In concluding this review of philosophical thought let us be reminded of the accepted philosophy of the 1920s and 1930s in the Soviet Union, dialectical materialism. It is a philosophy formulated in the nineteenth century, its materialism stemming from the belief that the world process in the large and small is the same as the material motions within it. It was tied to a Newtonian absolute and infinite space and time. Matter was the indestructible basis to which all existence is to be reduced. Such a philosophy could not be reconciled easily with relativity theory. It was in fact opposed as late as the 1950s. Materialism also professed a doctrine of determinism, of the 'necessity' of motion and of change more generally.[7] Acceptance of the indeterminism of modern physics came with a lag almost as great as with relativity theory. Kolmogorov

[7] A late reflection of Aristotelianism: Local motion is a special case of change, both abide with necessary laws.

in his active years never related his ideas on probability to the philosophy of materialism.[8] Probabilistic schemata were to be taken as mere models, he carefully suggested. In this way he remained noncommittal with respect to the question of chance in nature.[9] Another ideological problem that probability theorists encountered was Lysenkoism, especially after its acceptance as the official doctrine in 1948. For example, Serge Bernstein's book on probability theory was not reprinted after his refusal to withdraw the parts on Mendel's laws. Kolmogorov also had worked with biological applications and had published in 1940 a paper 'On a new confirmation of Mendel's laws.' In 1953 the paper was censored from his list of published works.[10] The 'new confirmation' were data – obtained and misinterpreted by a student of Lysenko – that proved to be a new brilliant confirmation of Mendel's laws,' as he put it (1940, p. 37).

Let us now turn to a brief description of some of Kolmogorov's earliest mathematical results on probability. The very first one is in the joint paper by Khintchine and Kolmogorov (1925) on 'the convergence of series whose members are determined by chance.' It is obvious from a comparison of the works of Khintchine and Kolmogorov of the 1920s that very often they shared research interests and results. Kolmogorov, by close to ten years the younger, is the one who invents short and powerful proofs, and who generalizes the results. This is already clear in the first and only joint paper. Khintchine proves in §1 in over four pages the following: If the sums of expectations and of variances of a sequence of random quantities $y_i$ both converge, the sum of the $y_i$ itself converges with probability 1. Kolmogorov gives the same result in less than a page. Then he finds a special case where the conditions are also necessary, to end with a general necessary and sufficient condition. The sense of probability 1 in the result is that the set of exceptions forms a set of Lebesgue measure 0.

From the beginning, the two Moscow probabilists were mainly concerned with the limit theorems of probability, the laws of large numbers, the law of iterated logarithm, and the central limit theorem. They aimed at strong versions of such limit results, where the very expression 'strong' indicates a formulation in which the exceptions

[8] Some much later items, from the 1970s [312] and [317] in the list of Kolmogorov's publications (note 1 above), deal with dialectical materialism and its relation to mathematics teaching.

[9] Quantum mechanical indeterminism was an especially touchy subject until the 1950s.

[10] The list was published in *Uspekhi Matematicheski Nauk*, **7** (1953), pp. 194–200, with continuation in *Russian Mathematical Surveys* in 1963, 1973 and 1983. For Bernstein, see Aleksandrov et al. (1969, p. 172).

to the limiting behavior predicted have measure 0. After Lindeberg (1922), conditions for the central limit theorem were actively studied. Kolmogorov's and Khintchine's occupation with this problem proved rather essential in the creation of the theory of stochastic processes with continuous time. In 1928 Kolmogorov gave a proof of an inequality now bearing his name, generalizing the one of Chebyshev.[11] In 1929 he strengthened the law of iterated logarithm and formulated what are known as Kolmogorov's zero–one laws. They are included as an appendix in the *Grundbegriffe*. The last time that Kolmogorov showed his ingenuity at improving upon Khintchine's proofs was in the late 1930s in connection with the ergodic theorem. As we saw in Section 3.2(b), Khintchine had made the ergodic theorem of Birkhoff into a purely probabilistic statement.

## 7.3 RANDOM PROCESSES AND STATISTICAL PHYSICS

Until about 1930 the random processes studied in probability theory had been almost exclusively discrete in time. The few exceptions were Bachelier from 1900 on, works on Brownian motion by Einstein and others in and after 1905, and work by Wiener on the same topic starting in 1921.[12] Then, in 1929, de Finetti published the first systematic papers on stochastic processes with continuous time.[13] At the same time Kolmogorov was creating a similar theory for what are today called Markov processes.[14] The only precursor he mentions is Bachelier, whom he credits with having been the first to systematically study probabilistic schemes with continuous time. But Bachelier's work is 'entirely devoid of any mathematical rigor.'[15] Other simultaneous studies of continuous time random processes were those of Hadamard and Hostinsky. On the physical side, it is worth mentioning Sidney Chapman's (1928) generalization of the continuous time distribution law of Brownian particles to the case of a nonuniform fluid.

Kolmogorov's first and very long paper of 1931 on continuous time random processes, 'On the analytical methods of probability theory' (A.M. for short), establishes a special measure theoretic framework for

[11] Cf. Feller (1968, p. 234) for the result.

[12] See Sec. 3.4 for the physical tradition in random processes, and for the beginnings of the mathematical theory of continuous time random processes.

[13] The first is de Finetti (1929a). Details are presented in Sec. 8.3.

[14] It is mentioned in Shiryaev (1989) that Kolmogorov had been working on the theory of continuous time processes since the summer of 1929.

[15] Cf. Kolmogorov (1931, p. 417, note 2).

studying continuous time stochastic processes. Only in the *Grundbegriffe* do we meet the now familiar construction of an infinite product space. In (A.M.) the main result is, in modern terms, the Chapman–Kolmogorov differential equations for a stationary Markov process. Two years later, in Kolmogorov (1933), these are derived for the general $n$-dimensional case. In (1937b) he treated what he called 'the inversion of the statistical laws of nature,' a problem in the theory of Markov processes stemming from statistical physics. Several other papers related these mathematical studies to physical problems.

Kolmogorov's first paper on Markov processes already studied the special case of discrete time Markov chains. It was in itself a field of active interest, and Kolmogorov devoted separate papers (1936a,b) to the creation of this attractively simple theory. Markov chains had been studied previously by Hadamard and Hostinsky among others, and they were also the central mathematical tool in von Mises' approach to statistical physics with which Kolmogorov was thoroughly acquainted.[16] His first publication on continuous stochastic processes, (A.M.) is interesting also more generally. It is based, as mentioned, on measure theoretic probability, and its introductory section contains some of Kolmogorov's rarely expressed opinions of foundational interest:

The mathematical treatment of natural and social events in based on *schematization.* A complete description of the state of a system, using a definite number of parameters, is required.[17] Such a description gives only a model, it is not 'reality itself' (A.M., pp. 415–16). Through a recourse to 'models,' Kolmogorov possibly wanted to clear space for an instrumentalist reading of his theory, as this would be a way of avoiding confrontation with the materialist ideology tied to the physics of a previous century. Within classical mechanics the schemes used are of the following character: The state $y$ of a system at time $t$ is uniquely defined by its state $x$ at an arbitrary previous moment $t_0$ through a unique function $f$ such that $y=f(x,t_0,t)$. Kolmogorov calls situations of this general type *schemes of a well-determined process.* It would be possible to consider systems where the state depends also on states previous to $x$, but such situations can in fact be avoided by adding parameters to the description of the present state. A familiar example

[16]The early history of Markov chains and processes is discussed in detail in Antretter (1989). A succinct exposition of the theory of Markov chains in given in Feller (1968).

[17]In Kolmogorov (1933a, p. 149) this is very explicitly formulated as the requirement of having a physical system with a finite degree of freedom n. The state space is a subset of $R^n$.

is to take instantaneous velocity as part of the description of state of a moving body (A.M., p. 416, note 1 especially). Next Kolmogorov says that outside classical mechanics, one often considers schemes where the state $x$ at time $t_0$ determines only a probability distribution for the possible future states $y$. These are called *schemes of a stochastically definite process.* Again, instead of considering also cases where the probabilities depend not only on $x$, one can avoid the influence of previous history by adding parameters (A.M., pp. 416–17).[18]

The preceding shows first of all that the idea of continuous time stochastic processes in Kolmogorov is directly based on the model of classical physics. Second, the notion of a Markov process stems directly from how the evolution of state proceeds in classical mechanics. The specific form in which determinism appears there is shifted to a probabilistic level of determination of probability law for the future on the basis of the present state. Kolmogorov's thinking in terms of models is clear when he says that 'the possibility of using, in the treatment of a real process, schemes of well-determined or of only stochastically definite processes stands in no relation to the question whether the real process itself is determined or random' (A.M., p. 417).

In §1, instead of a 'complete axiomatic system of probability theory,' Kolmogorov gives a 'general scheme' for a stochastically definite process: There is a collection of subsets of the state space of the system. It is a $\sigma$-algebra; in addition, it is supposed to contain every one-element set of the state space. The notation $P(t_1, x, t_2, A)$ denotes the probability of finding the system in the set of states $A$ at time $t_2$ under the hypothesis that it was in state $x$ at time $t_1$. The probability measure $P$ is $\sigma$-additive, and $P$ is further supposed to be measurable as a function of $x$. Random variables in general are introduced as measurable functions of the state space. Their expected values are Stieltjes integrals.

Kolmogorov assumes the transition probability function $P$ to be continuous in $t_1$ and $t_2$. This assumption is essential for the central results of the paper, the differential equations for stationary Markov processes, or the Chapman–Kolmogorov equations. The fundamental equation of Markov processes says, for a discrete set of states $\{x_1, x_2, \ldots\}$ and for $t_1 < t_2 < t_3$,

$$P(t_1, x, t_3, A) = \sum_i P(t_2, x_i, t_3, A) P(t_1, x, t_2, x_i). \tag{7.1}$$

[18]This method or trick suggested by Kolmogorov would not cover all cases. It would not always be possible to describe the effect of past history by $n$ real parameters. Khintchine's (1934) general stationary process, for example, cannot be covered in this way.

This comes from total probability: The probability of going from $x$ at $t_1$ to $A$ at $t_3$ is for any given $t_2$ obtained as a sum of the probabilities of all the possible paths. The Markov property makes the probability of passing from $x_i$ at $t_2$ to $A$ at $t_3$ independent of the probability of passing from $x_1$ at $t_1$ to $x_i$ at $t_2$, which proves the formula. In the case of a continuous state space the analogous integral expression appears as a postulate (A.M., p. 420). We shall give it in the form it appears in Kolmogorov (1933a). In this paper, a sequel to (A.M.), he explicitly assumes a physical system with $n$ degrees of freedom, that is, one with a state space $X \subset R^n$. With $V_y$ the volume measure in state space, he assumes $P$ has a density $f$ so that

$$P(t_1, x, t_2, A) = \int_A f(t_1, x, t_2, y)\, dV_y. \qquad (7.2)$$

The fundamental equation reads (1933a, p. 149)

$$f(t_1, x, t_3, y) = \int_X f(t_1, x, t_2, z) f(t_2, z, t_3, y)\, dV_z. \qquad (7.3)$$

In statistical physics this is called von Smoluchowski's equation.[19]

Next Kolmogorov goes on to discuss special cases, first the discrete time process (A.M., §3) obtained by relaxing the continuity assumption of $P$. It is supposed to change only at discrete times $t_0, t_1, \ldots$. Let $P(t_m, x, t_n, A) = P_{mn}(x, A)$ and $P_{n-1,n}(x, A) = P_n(x, A)$. If the distributions $P_n(x, A)$ are the same for all $n$, we have what Kolmogorov calls a *homogeneous scheme*. In the case of continuous time, the analogous condition is that $P$ can be expressed as a function of the time difference of $t_1$ and $t_2$:

$$P(t_1, x, t_2, A) \equiv P(t_2 - t_1, x, A). \qquad (7.4)$$

Kolmogorov says these schemes, the *time homogeneous schemes*, are the most important ones among continuous time schemes (A.M., p. 423). The general idea of stochastic processes with a time-invariant probability law was introduced by Khintchine in 1932. His idea of a *stationary process* was suggested directly from physical considerations. In both cases, Kolmogorov (1931) and Khintchine (1932a), the (unstated) physical analogue goes as follows. In general, the law of time development of a mechanical system is of the form mentioned by Kolmogorov: a unique function $f$ such that $y = f(t_1, t_2, x)$, where the development of the state $x$ depends on the initial time $t_1$. In case a

[19] Von Smoluchowski presented his equation in his (1915).

system is energetically isolated, that is, a conservative system, its law of motion is the same for different times. Such systems were called stationary a long time ago.[20] In these cases, the law $f$ acts in the same way for different times, and the time evolution depends only on the length of the time interval $t_2 - t_1$, exactly as Kolmogorov puts it for time homogeneous stochastic schemes.

Before Kolmogorov goes into the heart of the matter, the differential equations for continuous time Markov processes, he makes one more general observation. He says (A.M. §4) that without further specifying the structure of the set of states, one can only prove results concerning the 'ergodic principle,' a principle suggested by Hadamard's note (1928). A stochastically definite process obeys the *ergodic principle* if for any $t_0, x, y$, and $A$,

$$\lim (P(t_0, x, t, A) - P(t_0, y, t, A)) = 0 \quad \text{for} \quad t \to \infty. \tag{7.5}$$

The condition for discrete time reads

$$\lim((P_{mn}(x, A) - P_{mn}(y, A)) = 0 \quad \text{as} \quad n \to \infty. \tag{7.6}$$

The probability of entering the set of states $A$ becomes asymptotically independent of the present state. Why such a condition is called an ergodic principle can be seen from the following. Ergodicity appeared as a condition for statistical mechanical systems, its task being to guarantee that all the trajectories of a system behave in the same way, seen macroscopically. If that was so, that behavior could be determined from the physical description of the system, so that its time behavior became macroscopically predictable. Conditions (7.5) and (7.6) say that the probability law is asymptotically the same for the trajectories, or sample paths in statistical terminology, of a stochastically definite process obeying the ergodic principle. Allowing a frequentist meaning for probability, the sample sequences exhibit the same statistical behavior asymptotically. Thus the ergodic condition leads to a law of large numbers. In the case of Markov processes, instead of a physical equation of motion there appears a system of transition probabilities, from which the asymptotic probability law can be determined.[21]

Next I shall consider Kolmogorov's construction of a continuous stochastic process. In the *Grundbegriffe* a very general way of defining such processes is given, whereas in (A.M., §12) he follows a more particular path. Instead of Kolmogorov's rather involved discussion, I

[20] The term appears at least in Boltzmann, as well as in Zermelo (1900) and in Einstein's papers on statistical physics (Sec. 3.3).

[21] See Feller (1968, p. 393).

shall try to make the idea clear by the simplest possible case, as in Khintchine (1933, p. 2ff.).

Results stating that a limiting distribution follows the Gaussian normal probability law are called central limit theorems after Polya (1920). The classical Laplacian case holds for simple events $x_i = 0, 1$. Let $x_i = 1$ with probability $p$ and independent repetitions. The expected value of $x_1 + \cdots + x_n$ is $np$, and dispersion $\sigma = np(1-p)$. The central limit theorem says the limiting distribution of

$$\frac{\sum_i x_i - np}{\sigma} \tag{7.7}$$

is normal. The general heuristic behind central limit theorems is that the individual contributions $x_i$ to the sum (7.7) become smaller and smaller. Generalizations of the theorem apply to continuous random variables and do not require a common distribution for the summands. The following version is given in Khintchine (1933, p. 3): Let $x_1, \ldots, x_n$ be continuous random variables with zero expectation, variances $\sigma_i^2$, and distributions $F_1, \ldots, F_n$. Let further $F$ be the distribution of the sum $x = \sum_i^n x_i$, and $\Phi$ the normal distribution. It follows that for any $\varepsilon > 0$, there are $\tau, \lambda > 0$ such that

$$\frac{1}{\sigma_i^2} \int_{|x| > \tau} x^2 \, dF_i(x_i) < \lambda \quad \text{implies} \quad |F(x) - \Phi(x)| < \varepsilon. \tag{7.8}$$

The condition of the result expresses the idea of small individual contributions which together have a marked effect. Khintchine (1933, p. 7) takes this as the simplest case of the diffusion of a particle in one dimension. It starts from the origin, and each step adds a positive or negative term $x_i$ to the position $\sum_i^n x_i$ appearing as the sum of previous moves. First time is discrete, with one step taken at each unit of time. Next one lets the units of time become smaller and smaller.[22] Thereby one is led to the idea of considering, instead of the limit with an infinity of steps infinitely close each other, directly a random variable experiencing changes continuously in time. What one obtains as a limit becomes the 'exact probability law' holding at a given time $t$, as Khintchine says. If one sends an ensemble of particles, each moving according to this law, the analogy to diffusion becomes clear.

The foregoing is, in simple terms, the content of Kolmogorov's 'transposition theorem' (Übertragungssatz; A.M., p. 442) for moving from discrete to continuous stochastic processes. The laws of the

[22] This kind of procedure is suggested in de Finetti (1929a, p. 167).

processes studied in (A.M.) are those which fulfil the fundamental equation (7.3).

Kolmogorov's next work (1932a) on continuous time stochastic processes and its continuation (1932b) also proceed from this equation. This two-part note, building on the results of de Finetti from 1929, was presented to the Accademia dei Lincei by Castelnuovo. De Finetti had, apparently independently, started the study of continuous time stochastic processes in his (1929a,b). His mathematical methods were mainly based on the use of characteristic functions, as presented in Levy (1925) and Castelnuovo (1925–28). He defined in (1929a) what he called *processes with independent increments*, assuming straightaway that $X$ is a random variable for continuous time and that its change from time $t_0$ to $t$ is independent of its change on preceding intervals. He wanted to study the case where the probability law for $X$ is given via a differential condition for its change in time. To give the condition, he invented the concept of *divisibility* of a probability law. Divisibility requires that for any $n$, the probability law can be given as the sum of $n$ equally distributed independent increments. With this concept the time derivative can be defined by a limiting process, where $t = 1/n$ goes to 0 with $n \rightarrow \infty$. We see a certain similarity with the argument leading to the diffusion process above. De Finetti's first paper (1929a) treats continuous changes of $X$. In (1929b) discontinuities are allowed, an example mentioned being the instantaneous changes of velocity of a molecule of gas. De Finetti's discussion, as Kolmogorov's, is motivated by the example of physical systems.[23]

In Kolmogorov's two-part note (1932a, b) on processes with independent increments, the independence condition is written as follows[24]: We have a real random variable $X(t)$ with the probability law

$$F_t(x) = P(X(t_2) - X(t_1) < x) \quad \text{with } t = t_2 - t_1 \tag{7.9}$$

depending only on $t$, but not on $t_1$ or $X(t_1)$ or previous values of $X$. Following (A.M., eq. 21, p. 424, and eq. 8, p. 421) one obtains the functional equation

$$F_{t+t'}(x) = \int F_t(x-y)\, dF_{t'}(y). \tag{7.10}$$

Kolmogorov (1932a) gives a general solution to this equation under the assumptions that the distribution $F_t(x)$ is continuous and has finite

[23] See Sec. 8.3 for this aspect of the work of de Finetti.

[24] With slight changes in typography, as already indicated as general policy in these matters.

first and second moments. The second part (1932b, p. 868) of Kolmogorov's note contains a result belonging to the natural philosophy of chance, so to speak: If a process $X(t)$ with independent increments is continuous in time, the solution of equation (7.10) follows a normal distribution, and if $X$ changes discontinuously, it follows a Poisson law. The general case is a combination of the two.[25]

As we have seen, Kolmogorov's 1931 theory of stochastic processes with continuous time was built up by giving a probability law for the time development of a physical system. Two years later there followed a paper on continuous time processes by Leontovich and a joint paper by Kolmogorov and Leontovich (1933). These two papers are mentioned as the physical applications of the general theory of random processes in the preface of the *Grundbegriffe*. Two further works treating problems from physics are Kolmogorov's note (1934) and the paper (1937b). The latter is concerned with the following situation: Let $f(t, x, y)$ be a stationary probability density for obtaining state $y$ at time $t$ starting from state $x$ at time 0. Under certain conditions, the probability $h(t, x, y)$ of state $x$ at time 0, conditional on state $y$ at time $t$, is well defined. The 'question of the inversion of statistical laws of nature' is to find conditions under which $h(t, x, y) = f(t, x, y)$. One physical background for this question obviously is the invertibility of the laws of motion of isolated systems of classical mechanics. We have a time-invariant law of motion $T_t(x) = y$ which gives the state $y$ at time $t$ starting from $x$ at time 0. The time evolution is unique also toward the past: There is a function $T_t$ such that $T_{-t}(y) = x$. If a system is dissipative, its law of motion comes from a partial differential equation and is not in general invertible. In Schrödinger (1931), to which Kolmogorov refers in his (1933a, p. 155), the inversion problem is given together with an example of a derivation of an inverse probabilistic law.

## 7.4 THE GRUNDBEGRIFFE

### *7.4(a) The background*

David Hilbert presented his list of mathematical problems at the international congress of mathematicians in Paris in 1900. Hilbert's sixth problem is, following the example of the *Grundlagen der Geometrie*, to treat axiomatically those physical disciplines in which mathematics plays a predominant role. These are in the first place the calculus of

[25] The general case was later studied by Lévy (1934) in a paper that has become classic.

probability and mechanics. Hilbert adds that it would be desirable to have, together with the logical investigation of the axioms of probability theory, a rigorous and satisfying development of the methods of determining averages in physics. This goes specifically for the kinetic theory of gases. We have reviewed the problem and the early attempts at axiomatization by Laemmel, Broggi, and others (Section 2.1(b)). Kolmogorov's suggested solution to Hilbert's sixth problem had also later predecessors. He himself mentions von Mises and Bernstein as exponents of axiom systems with interests different from his.[26] In these, the concept of probability is a defined notion, and the attempt is 'to establish a connection as close as possible to the empirical origin of the concept of probability' (G.W., p. 2). But Kolmogorov concludes that for the sake of simplicity of the theory, 'it seems most appropriate to axiomatize the concepts of a chance event and its probability' (G.W., p. 2).

The idea of basing probability theory on measure theory is by no means original in Kolmogorov, as we have amply seen (Sections 2.1(b) and 2.3). Kolmogorov says himself that after Lebesgue, 'the analogy between the measure of a set and the probability of an event, as well as the integral of a function and the mathematical expectation of a random quantity, lay at hand' (G.W., p. iii). Subsequently, Fréchet formulated measure theory in an abstract way, so as to make it independent of its origins as a generalization of geometric measure in such spaces.[27] Kolmogorov says that this abstraction made it possible to found probability on measure theory, that 'the construction of probability theory according to these points of view has been current in the appropriate mathematical circles' (ibid.).

We have seen that Kolmogorov was an intuitionist. For him, and here he shared Hilbert's position, infinity is an ideal notion, whose use is only warranted by the requirement of consistency. Like Hilbert, but maybe with more reason after the double negation translation, Kolmogorov believed that the infinite is a conservative extension over the finite; that no finitary statements can be proved with infinitary notions that could not be proved with finitary means. This idea is now rather generally thought to have been refuted by Gödel's (1931) discovery of the existence of finitary arithmetic statements not having formal proofs within arithmetic. However, this state of matters does

[26] Bernstein (1917) contains an axiomatization of qualitative probability for which see below, p. 275.

[27] This is Fréchet (1915). A history of these developments can be found in Hawkins (1970).

not find expression in the *Grundbegriffe*. The advantage of the infinite is, beyond the extra proof theoretical strength shown by Gödel, that it is often more intuitive to think in terms of the infinite: for example, in terms of a unique limit instead of the finite approximations given by a converging sequence. Kolmogorov also wanted to follow the example of Hilbert's *Grundlagen der Geometrie* in the questions of formalization. Probability theory is to be formalized in exactly the same abstract way as geometry or algebra. As a consequence, the formalism has several other interpretations in addition to the one from which it grew. Thus probability theory can be applied to cases which 'do not have anything to do with the concrete sense of the notions of chance and probability' (G.W., p. 1). Behind this statement there is an application of probability to a purely infinitistic situation.

Two works precede the measure theoretic axiomatization of the *Grundbegriffe*, Kolmogorov (1929b) and (1931). In the latter, as we have seen in the previous section, there was a physical motivation for building a theory of probability. The need was to handle schemes of statistical physics where time and state space are continuous. Probability was introduced as a $\sigma$-additive measure over the state space, as the transition probability $P(t_1, x, t_2, A)$ for going from state $x$ at time $t_1$ to the set of states $A$ at time $t_2$. It was supposed to be a measurable function with respect to $x$, with expected values of random variables defined as Stieltjes integrals. As we saw, the theory of continuous processes with the Markov property was built directly upon the model of classical physics. Typically the state space of a classical system is a subset of a real space $R^n$, and once the task is to define transition probabilities instead of deterministic evolutions in such a space, it is natural to use Lebesgue measure for a $\sigma$-algebra of subsets of the state space.

A paper preceding the random processes of 1931 by only two years in publication and much less in writing, namely, Kolmogorov (1929b), contains nothing of the physically oriented motivations. Titled 'General theory of measure and the calculus of probability,' it wishes to show the possibility of 'a completely general and purely mathematical theory of probabilities.'[28] Further, 'finding out from the formulation of probability theory those elements which condition its inner logical structure and do not relate at all to its concrete meaning, is sufficient for such a theory.' The theory is consequently wider in its range than a calculus of probability which is only meant to deal with chance phenomena, for the former extends to the realm of pure mathematics.

[28] This rare publication is now available in the collection of selected papers on probability and mathematical statistics, Kolmogorov (1986).

Kolmogorov mentions as an example the distribution of digits of a decimal expansion, a result found with the help of the formulas of the calculus of probability, but not involving any concrete notion of chance. On the relation between measure theory and probability Kolmogorov says that 'the general concept of measure of a set contains the concept of probability as a special case.' Therefore the results of probability theory concerning random variables are special cases of results on measurable functions. The concept of *independence* is central in the application of probability theory to pure mathematics. Kolmogorov says this concept was never before formulated purely mathematically. Obviously, one need for such a definition is the independence of the digits of a decimal expansion. The thought being that arithmetic sequences follow some law or other, their independence property has to be saved from the domain of chance by a purely formal mathematical definition. The conditions for a finitely additive probability are laid down as axioms, and denumerably additive measures are signaled out by the term 'normal.' In a note added to the reprinting, Kolmogorov (1986, p. 472) mentions that this early work did not yet contain the set theoretic notion of conditional probability. One could speak of a set theoretic foundation of the whole of probability theory only after conditional probabilities as well as distributions in infinite product spaces had been incorporated, he says. These are the two mathematical novelties of the *Grundbegriffe der Wahrscheinlichkeitsrechnung*, the book that established those set theoretic foundations for all these decades to follow:

### *7.4(b) Axioms for finitary probability theory*

The book proper starts with the axiom system for a finite set of events (G.W., I. §1). In view of Kolmogorov's position on foundations of mathematics, this is very natural. The famous axioms go as follows (p. 2):

There is a set $E$ of *elementary events* $x, y, z, \dots$. There is a family of subsets $\mathscr{F}$ of $E$. Its members are called *chance events*.

II. $\mathscr{F}$ is a field of sets (that is, closed with respect to unions, intersections and complements).
II. $\mathscr{F}$ contains the set $E$.
III. To each set $A$ of $\mathscr{F}$, a nonnegative real number $P(A)$ is attached. This number $P(A)$ is called the probability of the event $A$.
IV. $P(E) = 1$.
V. If $A$ and $B$ are disjoint, $P(A \cup B) = P(A) + P(B)$.

I shall now proceed to discuss the place of foundational studies in Kolmogorov's book.[29] First of all, he does not offer a formalization of probability in the strict sense of the word. Instead, his is an *informal* axiomatization within intuitive set theory. It is of course straightforward to give a strict formalization by giving the axioms in a formalized system of set theory. Set theory and the measure theoretic way of building up the theory of real functions were the kind of mathematics in which Kolmogorov was educated. His first work, in fact, was on descriptive set theory. The reference for set theory in (G.W.) is the short version (1927) of Hausdorff's treatise.[30]

Kolmogorov shows first that his axiom system is *consistent*. In logical terms, he gives an *interpretation* for the formal axioms. An interpretation, or a *model* in logical terminology, consists of a *domain* $D$, the set of objects the interpretation talks about, and a set of *relations* $F$. These latter specify the functions and relations of the domain that correspond to the functions and relations of the formal axioms. This correspondence has to be such that the relations which interpret the formal notions are fulfilled in the domain. Corresponding assertions about the relations are *true in the model*. Specifically, the axioms correspond to relations which hold in the model. A contradictory axiom system is one that has no models. Conversely, if an axiom system has at least one model, it is noncontradictory. The model Kolmogorov puts up is very simple: For $E$ take any set $\{x\}$, so the domain interpreting $\mathscr{F}$ is the set $\{\emptyset, E\}$. Defining the function $P$ by $P(E) = 1, P(\emptyset) = 0$, it is easy to see that this interpretation fulfils the axioms.

Next Kolmogorov notes that the axiom system is, as he says, *incomplete* (unvollständig). Some caution is in order here. At the time of the writing of Kolmogorov's book, work on foundations of mathematics was in full progress. The great inaugurator of foundational studies was of course Hilbert. His 'metamathematics' had as its objects the formalization, proof of consistency, and creation of a decision method for mathematics. These questions receive a precise formulation for the case of ordinary first-order predicate logic in the book Hilbert and Ackermann (1928). The further problem of the completeness of an

[29] The following remarks are obvious to those trained in mathematical logic. But I hope they are useful to others not previously acquainted with foundational studies. The admirable book *From Frege to Gödel*, edited by the late Jean van Heijenoort, contains all the relevant papers on logic and foundations of mathematics from the times under discussion, including Kolmogorov (1925), a selection of Hilbert's and Brouwer's works, and the two papers by Gödel.

[30] The discussion of measure theoretic probability in the first edition (1914), for which see Sec. 2.1(b), was left out in the shorter 1927 version.

axiomatic system can be explained as follows. An assertion is *logically true* or valid if it holds in all possible models. (There is no counter-example.) An axiom system is *complete* if all true assertions can be formally proved in it. The completeness of predicate logic was proved by Gödel (1930). Next, an axiom system is *incomplete* if there is a true assertion for which there is no formal proof within the system. Gödel's sensational incompleteness theorem (1931) shows that there are true assertions of formalized arithmetic for which there is no such proof. The notion that Kolmogorov is trying to capture in the passage under discussion is another one: the notion of *categoricity* of an axiomatization. An axiomatization is categorical if all its models are isomorphic. Kolmogorov (G.W., p. 3) says that his axiomatization of probability is incomplete because in different problems of probability theory one considers different fields of probability. Obviously, the intention is that there are nonisomorphic fields so that the axioms of probability do not characterize their possible interpretations in a categorical way.[31] A further metamathematical question concerns the mutual *independence* of the axioms. If there is an interpretation that makes all but one of the axioms of some system true, there cannot be any logically correct deduction of that axiom from the others. Kolmogorov seems to take the independence of his axioms for obvious. Indeed, simple considerations show the independence[32]: Not all measures all normalized, so that there cannot be any deduction of axiom IV. There are genuine subadditive measures, so axiom V is independent, and so on. In the treatment of infinite fields of probability, an additional axiom VI is posed. Kolmogorov shows that it is independent of the other axioms. However, under the additional assumption of a finite field of probability, it becomes derivable (G.W., p. 14).

### *7.4(c) The application of probability*

The sense of probability that Kolmogorov endorses is addressed, characteristically, in the chapter dealing with the theory of probability for a finite set of events. The strictly infinitary parts of the theory are purely mathematical, and do not correspond to anything in the empirical world. Kolmogorov borrows von Mises' title, 'the relation

[31] Kolmogorov's terminology is the same as that of Weyl's 1926 book on foundations of mathematics, see Weyl's p. 22.

[32] Arguments of this type can be found in Steinhaus (1923), a paper mentioned by Kolmogorov. Steinhaus refutes Broggi's (1907) suggested derivation of $\sigma$-additivity from finite additivity.

to the world of experience,' for the title of his I.2§, and follows von Mises' presentation of the conditions of the applicability of the theory to the world of experience.[33] The application of probability takes place according to the following scheme (G.W., p. 3):

1. A certain complex *S* of unlimitedly repeatable conditions is assumed.
2. One investigates certain events which may appear in the realization of the conditions *S*. In individual cases of the conditions, the events appear in general in different ways. Let *E* be the set of possible variants $x_1, x_2, \ldots$ of how the events appear. The set *E* contains all variants we hold a priori as possible.
3. If the variant appearing after the realization of conditions *S* belongs to the set *A*, we say the event *A* appeared.
4. Under certain conditions ['into which we do not wish to go here,' as Kolmogorov says] one can assume that to the event *A* a real number *P*(*A*) is attached such that:
   A. If the conditions *S* are repeated a great number of times *n*, one can be *practically certain* that the relative frequency $m/n$ of occurrence of *A* differs only little from *P*(*A*).
   B. If *P*(*A*) is very small, one can be practically certain that *A* does not appear in a single realization of the conditions *S*.

The first requirement limits the application of probability to *repeatable events*. Probability does not concern singular, unrepeatable events. The repetitions have to take place under the same conditions, which further limits the applicability of probability, for many random processes in the empirical world occur under constantly varying conditions. The social events that Kolmogorov mentions in (A.M., p. 415) would be good examples. Strictly speaking only the most basic phenomena are exactly repeatable, typically those that can be reproduced in controlled laboratory conditions. Why repeatability of the same conditions is required by Kolmogorov, becomes understandable when we come to discuss point 4. The second point postulates, rather modestly, only *variability* of the events under realization of the conditions *S*. It does not postulate that the events occur *by chance*. But we saw already that Kolmogorov thinks of probabilistic schemes as models, not 'reality itself.' Then chance in nature itself, in the specific empirical conditions *S* studied, is not a precondition for the applicability of probability. Instead, the applicability of frequentist probability rests on the fulfilment of the requirements under point 4.

[33] See (G.W., p. 3), also note 1 therein. The reference to von Mises is (1931, p. 21).

What, then, are the conditions in 4 Kolmogorov does not want to discuss? I think they are conditions for the empirical applicability of laws of large numbers. The *mathematical* conditions had been studied intensively: Kolmogorov himself, in his (1928a) gave both necessary and sufficient conditions for the convergence of relative frequencies. In 1932 Khintchine had shown that the probabilistic generalization of Birkhoff's ergodic theorem to stationary probabilities gives the most general necessary and sufficient conditions.[34] As noted, Kolmogorov saw probabilistic independence, and weakened but analogous conditions, as the notions which distinguish probability theory from measure theory in general. The application of probability calls for a justification of independence. And indeed, we find Kolmogorov writing: 'After the philosophy of natural science has explained the much debated question concerning the character of the concept of probability itself, one of its most important tasks is the following: to make precise the conditions under which any given real phenomena can be held mutually independent' (G.W., pp. 8–9).[35] He adds that 'this question falls outside the scope of our book.' Much later he said he did not answer the problem of application of probability in 1933 because he did not know what the answer should be.

The requirement 4.A is a conclusion from two things: a law of large numbers and what in philosophical literature is known as the *rule of high probability*. If the probability of an event is close to 1, the occurrence of the event is 'practically certain.' This notion, sometimes appearing under the name of 'Cournot's lemma,' has naturally been debated very much. The subjectivist de Finetti has nothing positive to say about it. In Borel (1924), a representative of the somewhat eclectic French tradition on foundations of probability, on the other hand, a 'grading' of degrees of practical certainty is suggested, as we saw in Section 2.2(a).

The problem with the notion of 'practical certainty' points directly at one of the basic weaknesses of frequentist interpretations of probability. Typically, a law of large numbers would conclude that the limit of relative frequency of an event coincides with the probability of the event, with probability 1. The nature of this theorem of Borel's was not generally understood at first, as the terminology 'Borel's paradox' shows (Section 2.3). Where does the latter probability 1 in Borel's law

[34] See Khintchine (1932c,d). Kolmogorov does not refer to these papers in his book, but naturally knew them while writing it.

[35] In my opinion, Kolmogorov's request for the philosophy of natural science has been best met by studies of probability in dynamical systems. The problem of the real (empirical, physical) meaning of independence is shifted to the level of systems of mathematical physics in Hopf's works, especially Hopf (1934), as discussed in Chapter 5.

come from? It is derived from the former: Each sequence of results is an infinite binary sequence $x = (x_1, x_2, \ldots) \in \{0, 1\}^N$. In the simplest case of an event with probability 1/2 and independent repetitions, the system of finite dimensional probabilities gives the probability $1/2^n$ for each sequence of length $n$. From this probabilistic assumption, a probability measure $P$ is uniquely defined for the space $\{0, 1\}^N$ in which the limits of relative frequencies reside. That space can be thought of as the unit interval [0, 1], which makes $P$ become Lebesgue measure of [0, 1]. The Lebesgue measure of the set of binary reals with 1/2 as the limit of 1's, is itself 1. However, one cannot allow a frequentist reading to this measure, for it takes an infinite time to perform one 'experiment' of choosing a sequence from $\{0, 1\}^N$. This is strictly infinitistic and, as says Kolmogorov, such a situation has nothing to do with the concrete sense of probability. The suggested remedy for these problems is to take a finite version of the law of large numbers. We choose a probability value $1 - \delta$ and an $\varepsilon > 0$. There is an $n$ such that the probability of having a relative frequency within $\varepsilon$ of the probability $P(A)$ of the event, is at least $1 - \delta$. Now we have a single finite event: that of obtaining a relative frequency $y_n = \sum_1^n x_i/n$ within $\varepsilon$ of $P(A)$ in $n$ repetitions. (Incidentally, this construction gives us a way of concluding Kolmogorov's point 4.A from his point 4.B). Consider now the sequence of random variables $y_n$. It is obvious that the average of the $y_n$'s converges to $P(A)$ in the sense that with probability 1, $\lim y_n = P(A)$. To stop this regress of probabilities, something else has to be offered. This something else is in Kolmogorov the notion of 'practical certainty.' Practical certainty is partly a descriptive notion: By following what people actually do, we find out what risks they ignore in their lives. On the other hand, a computation of a sufficiently low probability can lead us into accepting a risk.

According to point 4.B, the interpretation of a sufficiently low probability is contained in the practical certainty of the nonoccurrence of the event, and consequently the practical certainty of the occurrence of the complementary event. A traditional problem with frequentist probability has been, what its meaning is for 'the single case.' Following Kolmogorov, only sufficiently low and sufficiently high probabilities apply directly to the single case. Intermediate values of probability are led back to these by the step we illustrated: that of looking at $n$ repetitions as a new single event.[36]

Kolmogorov's principles 4.A and 4.B are *principles of inference*. The general problem of statistical or inductive inference contains at least

[36]I follow here the illuminating discussion of Martin-Löf (1970), a text unfortunately available only in Swedish.

the following two cases: 1. inferences from frequencies to probabilities, or from data to a probabilistic model more generally, and 2. inferences from probabilities to frequencies, or predictions from a statistical model to properties of observable events more generally. In more modern terms, the first kind of inferences corresponds to problems of *estimation*. The second leads almost immediately to problems of *hypothesis testing*: If a prediction from the model fails to agree (in some suitable sense) with observations, the model is rejected. Coming back to Kolmogorov, his principles 4.A and B purport to give empirical content to the mathematical concepts, not the other way around. In the former, it is assumed that a real number $P(A)$ is adjoined to the event of interest $A$. In 4.B an inference is made from the low probability value to the nonoccurrence of an event. It would seem that 4.A and B are rules of inference of the second kind.

Although Kolmogorov says (G.W., p. 3, note 1) that his 'exposition of the necessary conditions for the applicability of probability to the world of real events follows to a high degree von Mises,' it is in fact different in important respects. In the theory of von Mises, probability is a defined notion, the limit of relative frequency in a collective. A collective is a mathematical model for an unlimited repetition of a chance event. The application of probability is based on showing that the requirements for a collective are met: that the limit of relative frequency exists, and that the events occur in a random order. These are of course idealized requirements and need to be placed with practically applicable criteria when the theory is related to the empirical world. Anyway, it follows that the concept of probability does not apply to predictable sequences, for these allow a way of choosing a subsequence with a differing limit, which contradicts the requirement of randomness. In von Mises this requirement was an expression of an indeterminist philosophy. Kolmogorov, instead, said in 1931, with physical systems in mind, that the applicability of probabilistic schemes stands in no relation to questions of chance in nature. A related problem arises in arithmetical applications of the theory of probability, with binary expansions and with continued fractions, for example. The objects of the theory are in this case sequences following mathematical laws. Random sequences, on the other hand, are thought of as 'lawless.' In the preface to the *Grundbegriffe* Kolmogorov says that there are applications which 'do not have anything to do with the concrete sense of the notions of chance and probability.' This same attitude appears already in the early paper (1929b) on measure theoretic probability. It therefore seems that Kolmogorov allowed a broader applicability for probability theory than von Mises originally did, in that a requirement of randomness would not always be necessary. At large, though, he undoubtedly shared the frequentist idea of probability of von Mises. Also, von Mises in his (1933) says that one can restrict the class of admissible choices of subsequences, which would result in a less stringent criterion of randomness. Those formulas of probability theory that hold also under the wider notion, could become applicable to number theoretic problems, say. I shall say more about the relation of Kolmogorov's measure theoretic probabilities and von Mises' ideas in Section 7.5.

Kolmogorov thinks of the asymptotic formulas of probability theory, limiting distributions, for example, as approximations to probability laws for finite situations with a great number of events (A.M., p. 441). The former are often in parametric form or in some other way that is easier to handle than the corresponding finite versions. In another sense, the probability laws for finite situations are approximations for empirical distributions. This latter sense lies at the basis of the application of probability. It is not an approximation in a strict mathematical sense, in contrast to the former. Otherwise one would not have to say that it is 'practically certain' that an empirical frequency is close to the theoretical probability. But the application of probability is not a mathematical question, either.

Thirty years after the *Grundbegriffe*, Kolmogorov turns again to the theory of von Mises in an attempt at clarifying the application of the theory (1963b). There he mentions one thing whose omission in his book may somewhat puzzle his readers, namely, that in von Mises probability was defined as a *limit* of finite relative frequency. Kolmogorov says he thought in the 1930s that a finite relative frequency notion of probability would not be amenable to a sufficiently satisfying mathematical development. But neither would a limiting frequency notion be acceptable. Such a definition of probability could only be a 'mathematical fiction' which, as Kolmogorov says in his (1963a, p. 254), 'cannot have any objective meaning.' Such were the reasons for resorting to a notion of 'practical certainty.'

It is a fair requirement that an interpretation of probability should also allow of a justification of its properties. This would be achieved if the interpretation justifies the axioms. Kolmogorov, in his 'empirical deduction of the axioms' (G.W., pp. 4–5) tries to offer such a justification. He says that 'one can normally assume' that the system of sets $\mathscr{F}$ to which probabilities are attached forms a field of sets including $E$. This, he maintains, takes care of axioms I and II and the first part of III, requiring the attachment of nonnegative numbers to the events. If the set $E$ is finite, not many would object to having a Boolean algebra of events. On the other hand, no reason is given for the first part of axiom III. The second part of axiom III is fulfilled because relative frequencies are numbers between 0 and 1. But the identification offered between relative frequencies and probabilities is rather debatable: Why should some property valid for relative frequencies hold for probabilities in general? On the contrary, relative frequencies vary from trial to trial, whereas probabilities would remain constant. Now, to proceed to axiom IV, Kolmogorov notes that since the frequency of $E$ is 1, $P(E) = 1$ is justified. Further, the frequencies of two mutually exclusive events add, so that finally axiom V is also justified.

Kolmogorov calls the preceding an 'empirical deduction of the axioms' (G.W., p. 4). He adds two remarks: 1. If two events are separately practically certain, their conjunction is also practically certain, although the 'degree of certainty' is somewhat less. With a great number of assertions, one cannot conclude the practical certainty of their conjunction from the practical certainty of the conjuncts. 2. From $P(A) = 0$ it only follows that the nonoccurrence of $A$ is practically certain in one trial. It does not follow that the occurrence of $A$ is *impossible*. Similarly, for a long series it also only follows that the relative frequency of $A$ is *close to* 0. These remarks indicate that conditions 4.A and B for the application of probability we met above, are principles of inference from probabilities to relative frequencies. They connect the theoretical probability numbers with the empirical world, or at least to our views or expectations of the empirical world. But they were so openly formulated in this respect that one can as well suppose the frequencies as given in them, and make an inference of the first kind: one from frequencies to values and properties of probabilities. And this is what Kolmogorov's 'empirical deduction' appears to do.

Finite relative frequencies, if assumed to be the probability numbers of events, fulfil the axioms of finite probability. But if we use Kolmogorov's principles 4.A and B in the other direction, we can, starting from empirical frequencies, conclude only that, for sufficiently large $n$, the theoretical probabilities are with practical certainty *close* in value to the observed frequencies, not *the same*. For this reason it would only follow that probabilities are approximately, not strictly, between 0 and 1, that they are approximately additive, and so on. On the other hand, setting $P(E) = 1$ and $P(A \cup B) = P(A) + P(B)$ in case $A \cap B = \emptyset$ might be taken as the establishment of two conventions or simplifications. (Kolmogorov's phrase is, 'es scheint angebracht zu setzen...,' G.W., p. 4). One thing he does not mention at all is that finite relative frequencies must, of arithmetic necessity, vary as $n$ obtains successive values. By the same token, finite relative frequencies are rational numbers. But it would be arbitrary to restrict the values of probabilities in this way: Let $P(A \cap B) = P(A)P(B) = 1/2$ and $P(A) = P(B)$. Then $P(A) = \sqrt{2}/2$ which cannot be a finite relative frequency. As it stands, Kolmogorov's 'empirical deduction of the axioms' does not do very much in the direction of justifying the formal properties of probability. It gives *one* method for adjoining real numbers to events so that the axioms become fulfilled: a particular, restricted way of constructing models for the axioms of finite probability. But one can be almost sure that the true probabilities of events would not be rational numbers, unless by sheer accidental coincidence. To start with, a true justification of the axioms would have to show why probabilities must be unique real numbers. An argument to precisely that effect was offered by de Finetti (1931c). He showed directly that without uniqueness, a system of probabilities is not coherent.[37]

### *7.4(d) Experiments*

In (G.W., 1.3§) Kolmogorov gives the now conventional probabilistic readings of set theoretic notions: Disjointness of sets is read as (so many) incompatible events, intersections as the simultaneous realization of the member events, and complements as nonoccurrence of the event (set) complemented. The empty set is an impossible event, the whole space a necessary event. The subset relation $A \subset B$ says that from the occurrence of $A$ the occurrence of $B$ follows. Kolmogorov's list contains one more notion: a finite partition $A_1 + \cdots + A_n = E$ of the space. In probabilistic terms: 'An *experiment* (Versuch) $\mathscr{A}$ consists in the determination of which of the events $A_1$ to $A_n$ occurs. $A_1$ to $A_n$ are the possible results of the experiment' (G.W., p. 6). This intuitive meaning of the notion of a partition is not common in today's axiomatizations of probability theory. In Kolmogorov the treatment of probabilistic independence and of Markov chains is in terms of an experiment.

As mentioned, Kolmogorov thought that the notion of probabilistic independence distinguishes probability theory from measure theory in general. He defines independence as follows: Let $n$ experiments $\mathscr{A}_1, \ldots, \mathscr{A}_n$ be given. If experiment number $i$ has $r_i$ results, the smallest partition $\mathscr{A} = \bigcap_{i=1}^{n} \mathscr{A}_i$ containing all of the $\mathscr{A}_i$ has $r = r_1 \times \cdots \times r_n$ members.[38]

[37] See Sec. 8.4. It is interesting to note that Kolmogorov knew de Finetti's early works on probability. Still, it seems that he never commented on the subjective interpretation. The same remark holds for Khintchine.

[38] Kolmogorov's discussion is not in exactly these terms.

Each member of the smallest partition is of form $\bigcap_i A_{j_i}$, where $A_{j_i}$ belongs to experiment $\mathscr{A}_i$. Experiments $\mathscr{A}_1,\ldots,\mathscr{A}_n$ are *mutually independent* if

$$P\left(\bigcap_i A_{j_i}\right) = P(A_{j_1})\cdots P(A_{j_n}), \qquad 1 \leqslant j_i \leqslant r_i, \quad 1 \leqslant i \leqslant n. \quad (7.11)$$

It follows that any $m < n$ experiments are also mutually independent. The *events* $A_1,\ldots,A_n$ are defined as mutually independent if the $n$ experiments $\{A_i, -A_i\}$ are. Kolmogorov's way of handling independence leads directly to considering conditional probability a random variable: Let $\mathscr{A} = \{A_i\}$ be an experiment. If $P(A_i) > 0$, the conditional probability $P_{A_i}(B)$ of $B$ under the condition $A_i$ is $P(A_i \cap B)/P(A_i)$. Given an event $B$, the *conditional expectation of B after the experiment* $\mathscr{A}$ is the random variable $P_{\mathscr{A}}(B)$ over the results $A_i$ of the experiment $\mathscr{A}$ (G.W., p. 12).

As an example of experiments, Kolmogorov discusses Markov chains. His treatment is not as clear as it could be, because of some confusions. In the ordinary elementary way of handling Markov chains, one assumes a finite set of states $\{1,\ldots,r\}$, with (stationary) transition probabilities $P_{mn}$ for going next to state $n$ if the system now is at state $m$, given as an $r \times r$ matrix. In Kolmogorov, the consecutive states (results in his terminology) are allowed to belong to different sets of states (experiments in his terminology), but the notation does not take care of this.[39] In a reconstructed notation, his definition for a sequence of experiments $\mathscr{A}_1, \mathscr{A}_2, \ldots$ to be a *Markov chain* is

$$P_{\mathscr{A}_1\cdots\mathscr{A}_{n-1}}(A_{j_i}) = P_{\mathscr{A}_{n-1}}(A_{j_i}), \qquad \mathscr{A}_1\cdots\mathscr{A}_{n-1} = \bigcap_{i=1}^{n-1} \mathscr{A}_i,$$
$$1 \leqslant j_i \leqslant r_i, \qquad 1 \leqslant i \leqslant n. \qquad (7.1.2)$$

Next Kolmogorov wants to recover the system of transition probabilities of the usual approach. But we should remember that there is one matrix for each pair of experiments $\mathscr{A}_m, \mathscr{A}_n$, and therefore we must write

$$p_{j_m j_n}(m,n) \equiv P_{A_{j_m}}(A_{j_n}), \qquad 1 \leqslant j_i \leqslant r_i, \quad i = m, n. \qquad (7.13)$$

For given experiments $\mathscr{A}_m, \mathscr{A}_n$, we have an $r_m \times r_n$ matrix. Kolmogorov's notation is $p(m,n)$, which might lead to the assumption that one has an $m \times n$ matrix. Since he allows each experiment $\mathscr{A}_i$ to have a different number $r_i$ of results, an infinity of matrices would be needed.

[39] Specifically, the notation for 'products' on p. 12, line 17, should have italic Latin capital latters.

### *7.4(e) Infinite fields of probability*

Kolmogorov's Chapter II is devoted to infinite fields of probability. The two mathematical novelties by which his book differs from previous formulations of measure theoretic probability concern such fields. These are the theory of conditional probabilities and the construction of a random process as a probability measure over an infinite-dimensional product space.

The presentation of infinite fields of probability begins with the *axiom of continuity* (G.W., p. 13). Let $\bigcap_i A_i$ and $\bigcup_i A_i$ be the finite or denumerable intersections and unions of $A_1, A_2, A_3, \ldots$ . Axiom VI reads:

VI. For a descending sequence (1) $A_1 \supset A_2 \supset \cdots$ of events from $\mathscr{F}$ with (2) $\bigcap_i A_i = \varnothing$, it holds that (3) $\lim P(A_i) = 0$ as $i \to \infty$.

If a field of probability is finite, let $A_k$ be the smallest set in (1). Then, since $\bigcap_i A_i = A_1 \cap \cdots \cap A_k = \varnothing$ by (2), $A_k = \varnothing \in \mathscr{F}$. Therefore $P(A_k) = 0$, so (3) follows (G.W., pp. 13–14). This also proves that the system is consistent and noncategorical. The continuity axiom is, by an easy argument (G.W., p. 14), equivalent to *denumerable additivity*, or $\sigma$-additivity, as it is also called. Assume $A_1, A_2, A_3, \ldots$ form a disjoint sequence of events:

$$P\left(\bigcup_i A_i\right) = \sum_i P(A_i). \tag{7.14}$$

Denumerable additivity is not a universally accepted property of probability measures.[40] Kolmogorov sees it as a mathematical convention, a view based on his finitism (G.W., p. 14):

> Since the new axiom is essential only for infinite fields of probability, it would hardly be possible to explain its empirical meaning in the way sketched for axioms I–V in §2 of the first chapter. In the description of any really observable random process, one can obtain only finite fields of probability. Infinite fields of probability appear only as idealized schemes of real random processes. *We delimit ourselves arbitrarily to schemes which fulfil the continuity axiom VI.*

Axiom VI would not work if the field of events were not closed with respect to denumerable unions and intersections. Kolmogorov refers to Hausdorff (1927, p. 79) the construction of the smallest $\sigma$-field $B\mathscr{F}$ over a given field of sets $\mathscr{F}$. Then he goes on to the extension of a denumerably additive probability $P$ over a field $\mathscr{F}$ into a $\sigma$-field $B\mathscr{F}$

[40] Some of the best known opponents to the unrestricted use of denumerable additivity are de Finetti, Savage and Dubins.

(G.W., p. 16). This is followed by a remark to the effect that even if the events $A$ from $\mathscr{F}$ can be taken as (possibly only approximately) observable real events, it does not follow that this would be the case for the sets of $B\mathscr{F}$. The extended field of probability $(B\mathscr{F}, P)$ remains a purely mathematical construction (p. 16). As we noted in Section 7.2, the infinitary events from $B\mathscr{F}$ are only 'ideal events' in Kolmogorov, and their status is the one Hilbert gave for ideal elements in mathematics in general. 'If the use of probabilities of these ideal events leads to a determination of the probability of a real event in $\mathscr{F}$, it is obviously automatically acceptable from an empirical point of view' (G.W., p. 16).

In chapter III random variables are defined as measurable functions. Then the consistency conditions for a system of finite-dimensional distributions are laid down as pertaining to $n$ random variables (p. 24). These conditions require that the distributions for any $k$ variables, with $k < n$, coincide with marginals of $n$-dimensional distributions. With the systems of finite-dimensional distributions all the prerequisites have been laid for the following paragraph III.4, where the elementary events are points in an infinite-dimensional space. Such product spaces had been considered in measure theory earlier, and even their probabilistic significance had been under some attention.[41] In Kolmogorov's treatise, the product space construction is made for the purpose of a measure theoretic treatment of stochastic processes as follows:

Let $M$ be any set. Then the probability space to be considered is the set $R^M = \{x_\mu\}$ where $\mu \in M$. To given $n$ indices $\mu_1, \ldots, \mu_n$ there corresponds an $n$-dimensional subspace $R^n$. A set $A$ is a *cylinder set* if it is the inverse of the projection from $R^M$ to $R^n$ of a set $A' \subset R^n$ for some $n$. If $A'$ is a Borel set, $A$ also is by definition. Let $\mathscr{F}^M$ be the Borel sets of $R^M$ thus obtained. Its Borel extension is $B\mathscr{F}^M$. If a probability $P$ is given over $\mathscr{F}^M$, $A \in \mathscr{F}^M$ is a cylinder set and its probability $P(A)$ is obtained as follows. There is a set $A'$ of $R^n$ to which $A$ projects such that $P(A) = P_{\mu_1 \cdots \mu_n}(A')$. Here the latter probability is well determined since it is the probability for the $n$ random variables $x_{\mu_1}, \ldots, x_{\mu_n}$. A system of finite-dimensional distributions determines in this way the probabilities for all Borel sets. Therefore it determines a probability $P$ on $\mathscr{F}^M$. By the extension theorem the same holds for $B\mathscr{F}^M$. Kolmogorov's *Hauptsatz*, or what is often called his extension theorem, now shows that a consistent system of finite-dimensional distributions determines a probability $P$ on $\mathscr{F}^M$ and $B\mathscr{F}^M$ fulfilling axioms I–VI (G.W., p. 27).

[41] General product measures are studied in Daniell (1919). In a probabilistic context I have found them prior to 1933 in Tornier and Feller (1930), Tornier (1933) and Ulam (1932), at least.

The Kolmogorov extension theorem allows for two things: First, the discussion of the strong limit theorems of probability in a systematic setting. These theorems typically state that the limit of a denumerable sequence has with probability 1 a certain property. That probability 1 is, after Kolmogorov, the same as the measure in an infinite-dimensional space of all possible sequences. The systematic reason for the connection between strong limit laws and measure theoretic probability is brought into clear light. Second, the extension theorem allows of the construction of a probability law of a stochastic process with an arbitrary index set $M$, starting from the finite-dimensional distributions. (In keeping with the synthetic mode of presentation, stochastic processes are only later mentioned in the book, on p. 39.) The probability is not defined on *all* subsets of $R^M$, however, if the index set $M$ is continuous. Kolmogorov's example is the set defined by requiring $x_\mu$ to be below a given bound for each $\mu$ (G.W., p. 26). There are relatively simple sounding events whose probability is not well defined by Kolmogorov's procedure. Some, however, think they really only sound simple, but are not in fact intrinsically well defined.[42]

In chapter V Kolmogorov develops the second of the two essential novelties of his book, the theory of conditional probabilities for infinite sets of elementary events. It applies to cases where the condition has zero probability, such as one encounters in the theory of Brownian motion, for example. Conditional probabilities are defined as random variables, the case of a zero probability condition being handled with what is today called the Radon–Nikodym theorem.

### *7.4(f) The Impact of the 'Grundbegriffe'*

Before entering into the immediate reception of the *Grundbegriffe*, I add some remarks on the rest of its contents. The last chapter VI, is a treatment of laws of large numbers, a topic to which Kolmogorov had contributed continuously since his first joint paper with Khintchine in 1925. An appendix of the book contains a purely infinitistic theorem, namely, what is called a 0–1 law. As was mentioned, it escapes the concrete sense of chance and probability according to Kolmogorov, whereas some other infinitistic results allow a finitistic reformulation. The 0–1 law gives rather general conditions under which the probability

[42] Doob (1937, 1953) addresses the problems of defining more general measures that would answer questions of probability of boundedness, continuity, and so on, of a continuous time process. Criticisms of such procedures of extending probabilities will be met in the next chapter in connection with de Finetti's views.

of convergence of a sequence can only obtain the values 0 or 1. The *Grundbegriffe* ends with a bibliography on previous works on probability of foundational interest.

The mathematical novelties of Kolmogorov's book, besides the organization of the axiomatization, were the construction of stochastic processes and the general theory of conditional expectations, conditional probabilities in particular. He says himself in the preface that 'these new questions arose out of necessity from certain very concrete physical questions.' As we have seen, conditional probabilities where the condition is drawn from a continuous set (thus having in general zero probability), appear at once in the theory of stochastic processes.

Now to the reception of the Kolmogorovian measure theoretic probabilities. The new approach has certainly been seen as a revolution that made earlier theories obsolete. Some later descriptions by contemporaries or near contemporaries of Kolmogorov are Doob (1989), Cramér (1976), and Loève (1978). The last mentioned takes Kolmogorov's theory as 'the definitive formulation,' and in Doob one finds bewonderment as to why the approach was 'not immediately universally accepted at once' on the ground of 'the uncontroversial nature of the measure theoretic approach' (p. 820). We also read (p. 818) that Kolmogorov's book 'transformed the character of the calculus of probabilities, moving it into mathematics from its previous state as a collection of calculations inspired by a vague nonmathematical context.' Cramér in 1976 puts his words more carefully: 'Looking back towards the beginning of a new era in mathematical probability theory, it seems evident that a real breakthrough came with the publication of Kolmogorov's book' (p. 519). Already in 1938, in his review of 'Lines of development in the calculus of probability,' Cramér had emphasized the continuity, instead of an opposition, between classical and modern probability. He explains briefly the measure theoretic basis and goes on to show how the classical problems appear as special cases in the new theory. 'Here, too, the case turns out of so many revolutionary ideas: The development does not take place as spontaneously as may seem at a first look. The central new ideas partly are only a consequent, necessary redevelopment of a common property of thoughts that one can follow a long way back in time' (Cramér 1939, p. 67). In Cramér's 1937 *Random Variables and Probability Distributions* a vague frequentist idea of probability is first introduced. Different axiomatizations are always possible, he says. Then he explains a bit the frequentist theory of von Mises. Its difficulties in defining the irregularity of collectives justify the choice of Kolmogorov's axiomatic measure theoretic approach, 'at least for the time being' (1937, p. 4). Thus measure theoretic probability is not seen as any necessity, logical, mathematical, historical, or what have you. Nor was it a novelty of Kolmogorov's. Now if we are to wonder why measure theoretic probabilities were not immediately accepted in 1933, then why not equally well already in, say, 1923?[43] And why are there even today alternative approaches to probability?

We have witnessed a long development, which culminated in Kolmogorov's monograph. It is also undeniable that its appearance meant a remarkable advancement in the mathematics of probability. This was mainly felt in the theory of stochastic processes, where Kolmogorov's use of infinite product spaces met with immediate approval, but measure theoretic probability found other uses, too. One of the first to join Kolmogorov was Khintchine, who at the time was developing a probabilistic approach to ergodic theory and the theory of stationary processes (1934). Eberhard Hopf had been using measure theory in his studies of dynamical systems (cf. his 1932a for example). His great paper on probabilistic aspects of dynamical system (1934) immediately took advantage of Kolmogorov's measure theoretic probabilities. J. L. Doob started developing the theory of stochastic processes, in his (1934a,b). Later he gave measure theoretic analogies to von Mises' requirements for the impossibility of a gambling system (1936).[44] His systematic papers (1938, 1937) are devoted to the study of discrete and continuous time stochastic processes. Feller (1936) continued the study of Markov processes initiated by Kolmogorov (1931). Measure theoretic probabilities were also taken into use by Cramér and Lévy in their books that both appeared in 1937. Even on the basis of this very partial list, one can conclude that many of the leading researchers in mathematical probability soon absorbed Kolmogorov's measure theoretic probabilities. Their 'universal acceptance,' on the other hand, took time. This was partly due to resistance from other, competing approaches to probability, notably the theory of von Mises. De Finetti also systematically refused to think that a measure theoretic scheme would be more than a useful way of finding examples. Instead he offered an alternative, stemming from his thought that probability theory must have a form immediately appealing to the 'everyday sense' of probability. That matter shall be discussed in the final chapter of this book. Anyway, as concerns the reception of measure theoretic probabilities, it is a fact that textbook expositions did not start appearing until after the war,

[43] That specific year the journal *Fundamenta Mathematicae* contained two extensive papers on measure theoretic probability.

[44] See Sec. 6.3; Halmos (1938) continues this line of research.

with the exception of Cramér (1937). His 1946 *Mathematical Methods of Statistics* makes systematic use of measure theory, as does Doob's 1953 *Stochastic Processes*. Halmos' 1950 *Measure Theory*, the standard treatise on its topic for a long time, devotes one chapter to probability measures.

Kolmogorov himself surveyed the status of probability theory in a talk in 1934, published as (1935). The essential difference between the classical calculus of probability and the newer research is, he says, that the former only had a finite number of possible results. The modern studies Kolmogorov divided into three main fields, giving in each the most important research directions together with the principal names. It reads like a *Who's Who* of probability theory of the time:

1. Analogy to the measure theory of real variables, with a) general axiomatization of probability: Borel, Fréchet, Kolmogorov, Hopf; and b) stong laws of large numbers: Borel, Cantelli, Slutsky, Fréchet, Khintchine, Kolmogorov, Glivenko, Lévy.
2. New schemes in physical and other applications, with a) the theory of stochastic processes: de Finetti, Hostinsky, Hadamard, von Mises, Kolmogorov, Fréchet, Khintchine, Lévy; and b) the theory of random functions: Wiener, Slutsky, Lévy.
3. New analytical apparatus, with a) equations for stochastic processes: Kolmogorov, Hostinsky, Fréchet, Bernstein, Pontryagin; b) characteristic functions and moments in infinite dimensional and functional spaces: Khintchine; and c) new methods for proving limit theorems: Kolmogorov, Petrovski, Bernstein, Khintchine, Bavli.[45]

In the paper Kolmogorov explains what all these researches are about. Some of them already made use of the brand-new Kolmogorovian measure theoretic probabilities.

## 7.5 THE CURIOUS REAPPRAISAL OF VON MISES' THEORY

The most notable of Kolmogorov's later works, from our point of view, is his reopening of the von Mises style of probability theory in the early 1960s. Other works were, to be sure, mathematically much more important. These would contain at least the entropy theory of dynamical systems and the stability theorem for the three-body problem. But the

[45] All I know about G. M. Bavli is that he translated Kolmogorov's book into Russian in 1936.

matter under discussion in this section continues on a central problem left open in the *Grundbegriffe*, the application of probability.

The *Grundbegriffe* makes it obvious that Kolmogorov shared a frequentist basic conception of probability. It is not the idea of probability as a limit of relative frequency, but the somewhat vague connection between finite relative frequency and probability, associated to the notion of 'practical certainty.' The limiting frequency notion of probability on the other hand does not, according to Kolmogorov (1963b, p. 369), 'contribute anything to substantiate the applicability of the results of probability theory to real practical problems where we always have to deal with a finite number of trials.'

The intuitive idea of a random population (or a sequence obtained from such a population or similar random trial, to be more precise) was the basis of the theory of von Mises (1919b). In his collectives, the limit of relative frequency is postulated to exist. Second, it is required that the limit be the same for subsequences of the original sequence. A similar idea appears in Fisher (1956): Statistical probability is based on the notion of a homogeneous population. Such a population must not admit of 'recognizable subpopulations,' that is, those where the relative proportion of the cases of interest would be different. Fisher thought that his criterion would ultimately be based on a subjective judgment.

Many others besides von Mises and Fisher have felt the need to define what it means for a property to be randomly distributed in a class. Very often the claim, or conclusion, has been that a proper definition would lead only to trivial cases: the property holds universally, or the class has only one member. Otherwise the indeterminism of quantum mechanics would be called for help, as in the so-called propensity approaches to probability.[46] In Kolmogorov (1963b) a new approach to the random distribution of a property in a finite class is proposed. It is suggested from Church (1940), one of the creators of the theory of recursive functions. He had suggested that the methods of choosing a subsequence in a collective should be restricted to those we can actually perform, the algorithmic ones. Church's definition of random sequences applies to infinite sequences, whereas Kolmogorov works with finite repetitions. His basic idea is that 'there cannot be a very large number of simple algorithms' (1963b, p. 369). He puts up a class of selection algorithms and requires invariance of relative frequency with respect to this class. As a consequence of the 'strictly finite nature of the entire conception,' as Kolmogorov (1984, p. 3) puts it, one can

[46] See the various papers of Popper.

only require approximate invariance. Two later papers of his (1965, 1968) lay the basis for the form of the theory which is now rather commonly known and used. Kolmogorov defines the *complexity* of a finite object, typically a (binary) sequence. Sequences that are maximally complex are considered random, and the conclusions of probability theory apply to them. Probability theory itself retains its established measure theoretic form, complemented by the theory of random sequences that gives it a direct way of application.[47] Kolmogorov's initiative was developed by Per Martin-Löf, who studied under him at the time, which resulted in the remarkable papers Martin-Löf (1966, 1971). I shall now review these developments.

In the ordinary formulation of probability theory, the probabilities describe a hypothetical collection, or *ensemble*, of systems. Somehow real samples imitate the ideal properties of the ensemble. The modern theory of random sequences wants to describe a 'typical sequence,' one sharing as much as possible of the probabilistic properties of the ensemble. For this description, a measure of *algorithmic complexity* of individual sequences is proposed. There are variants of the definition, but the basic results are similar in each case. An algorithm (Turing machine) takes binary sequences as arguments and values. The complexity of a sequence $x = x_1 \cdots x_n$, relative to algorithm $A$, is defined as $K_A(x|n) = \min l(p)$, where $l(p)$ is the length of the input $p$ for which $A(p, n) = x$. We use a version of the complexity measure where the length of the sequence is given, so the algorithm does not have to determine separately when to stop. It can be shown, as in Kolmogorov (1965), that there is at least one 'asymptotically optimal' universal algorithm $U$ such that $K_U(x|n) \leqslant K_A(x|n) + c$ for any algorithm $A$. The universal algorithm $U$ computes $x$ in general with the shortest input, save for the constant $c$ that is used for the binary coding of algorithm $A$. From now on we drop the index $U$ and write $K(x|n)$ for $K_U(x|n)$. Here are some of the elementary theorems:

1. The length of a sequence gives an upper bound to its complexity, up to the added constant $c$: Each sequence can be produced one by one, by listing its elements. Define an algorithm by setting $B(p, n) = p$. Then $K_B(x|n) = l(x) = n$. By the inequality above, $K(x|n) \leqslant K_B(x|n) + c = n + c$.
2. There cannot be many sequences with a simple description (that is, low complexity): A counting argument shows that there are at most $2^a$ sequences of length $n$ with $K(x|n) < a$. It follows that there are

[47] I shall soon present some reasons for caution concerning this idea.

sequences with $K(x|n) \geqslant n-1$, so there are sequences of arbitrarily high complexity.

3. The complexity measure is not effectively computable: Assume it is. Start computing the complexity of sequences in lexicographical order. Then there is a one-place algorithm $F$ which, for any $n$, finds the first sequence $x$ such that $K(x|n) \geqslant n-1$. Define $B(p,n) = F(n)$, whence $K_B(x|n) = 0$ and by the fundamental inequality, $K(x|n) \leqslant 0 + c$. By choosing $n > c + 1$ you get a contradiction.
4. The set $I$ of inputs $p$ for which $U(p,n) = x$, with $U$ a universal algorithm satisfying the basic inequality, is not recursive: Assume it is. Find the first one in the lexicographical ordering belonging to $I$. This produces effectively min $l(p)$ so $K(x|n)$ would be computable.[48]

A sequence is defined as being random if it is maximally complex, or $K(x|n) \approx n$. A certain relativity is unavoidable here. One way to put this definition is to say that those sequences are random whose information content cannot be compressed. Kolmogorov (1965) calls in fact his notion a measure of quantity of information. He thought (see his 1968, p. 663) that this algorithmic notion, pertaining to individual objects, would be more fundamental than the usual probabilistic measure of information introduced by Shannon. He also believed that the new theory would give a new way of applying probability theory. One retains the measure theoretic formalism: Its conclusions are applicable once it has been shown that the results or sequences observed are random in the above sense (Kolmogorov 1983, pp. 34–35).[49]

I have not said anything about infinite random sequences yet. Von Mises' original collectives were infinite. On the other hand, we saw that Kolmogorov had been critical of the limiting frequency conception of probability in general. He thought that infinite random sequences are mainly of mathematical interest. The definition of these sequences, carried through by Martin-Löf (1966), proved a bit surprising. It would seem an obvious idea to define an infinite sequence as random if its finite initial segments are finite random sequences. But Martin-Löf (1971) shows that that is not the right way, for there are no such sequences. The complexity of the segments of any infinite binary

[48] These results come mainly from Kolmogorov and Martin-Löf. I have here tried to state and prove the ineffectiveness of the measure $K$ as simply as possible. For further elaborations see van Lambalgen (1989).

[49] As mentioned, the usefulness of this idea is not absolutely clear. The reason is precisely that the measure of complexity is not effectively determinable, so that one could not prove any sequence to be maximally complex.

sequence oscillates below the maximum as given by the length of the segment: Let $f$ be a recursive function of natural numbers such that $\sum 2^{f(i)} = \infty$. Then $K(x|n) < n - f(n)$ holds for infinitely many $n$. For example, setting $f(n) = [\log n]$ fulfils the condition so that the complexity is infinitely often below $n - \log n$ for all binary sequences.

The proper definition of infinite random sequences goes beyond the present discussion. The idea is to locate properties which in the measure theoretic formulation hold with probability 1. However, one cannot require random sequences to fulfil all these properties, for the complement of any one-sequence set would in general have measure 1, whereas the intersection of these sets of measure 1 is empty. Instead, one takes those sets of measure 1 which are relatively simply definable. If a sequence belongs to the complement of such a set, this can be equally simply found out. The sequence then lacks a 'property of randomness.'[50] There would exist other properties of randomness, but one would not be in a position to recognize their failure because of the strongly non-constructive character of the property. This line relates to the complexity measure approach via the following result (Martin-Löf 1971, theorem 5): If there is a constant $c$ such that $K(x|n) \geqslant n - c$ infinitely often, the infinite sequence $x$ is random.

It was shown by Martin-Löf that the limit of relative frequency exists for random sequences. A sufficiently powerful formulation of the second of von Mises' axioms, the 'Regellosigkeitsaxiom,' makes the first axiom on the existence of limit of relative frequency derivable. This has been thought surprising.[51] But infinite random sequences are defined as those which fulfil certain simply definable properties holding with probability 1 in the measure theoretic approach. A law of large numbers gives one property of that kind so that the result is no more unexpected than such a law.

[50] See Martin-Löf (1970).

[51] See for example, Fine (1970, p. 254).

# 8

# *De Finetti's subjective probabilities*

## 8.1 'PROBABILITY DOES NOT EXIST'

The two main interpretive ideas about probability for the times under discussion are the frequentist and the subjectivist. Frequentist probability has had a remarkable role in the development of modern probability. It was the object of von Mises' theory, and it was more or less tacitly assumed as the interpretation of probability by the main proponents of mathematical probability in the 1920s. No attention was paid to the idea of subjective probabilities. The most one can find are Borel's philosophical essay (1924), which is a review of Keynes' *Treatise on Probability* of 1921, Borel's papers on game theory, and Lévy's discussion of subjective probability in his book of 1925.[1] Borel's main contribution to probability (1909a), though, was from much earlier times.

The idea of subjective probabilities was further undermined by the developments in physics: Classical statistical physics already contained a commitment to statistical probability at least,[2] and quantum mechanics brought a new kind of fundamental indeterminism into the description of nature's basic processes in 1926. An epistemic notion of probability must at that time have seemed like a thing from the past, from the shadows of the Laplacian doctrine of mechanical determinism.

In the 1920s philosophical thinking, or at least the part of it sensitive to scientific developments, was transformed through the rise of logical empiricism. Heisenberg's quantum mechanics of 1925 is one example of a scientific theory that shows the mutual effect of philosophical ideas and theory construction. A concept such as the trajectory of an electron

[1] In England the Keynesian logical view of probability had an influence on Frank Ramsey, for which see Sec. 8.4. Harold Jeffreys' somewhat later *Theory of Probability* (1939) also contains a subjectivist philosophy. The effect of these works on mathematical probability is hardly noticeable, except for Ramsey since the 1950s.

[2] See the views of Einstein and others in Chapter 3. However, the epistemic interpretation of probabilities in classical statistical physics, as in Tolman (1938) and later in Jaynes (see his 1983), should also be kept in mind.

is *meaningless* according to strictly empiricist criteria.[3] The concept of probability, understood in the statistical sense, corresponds instead to an observable quantity: The individual transitions of an atom from one energy level to another are governed by a probabilistic law, and with a great number of atoms, a transition probability becomes measurable as the radiation intensity connected to a particular kind of transition through Bohr's frequency condition.

In physics the numbers of individual events are often great, so that the statistical frequencies observed are closely approximated by the theoretical probabilities. The physicist is freed from the worry that the approximation becomes exact only in the limit of an infinite number of events. (He turns the opposite case into a determination of the magnitude of fluctuations.) Physical theory connects probabilistic laws to observable parameters, as in the case of the molecular velocity distribution. It is determined from total energy, that is, basically, from a thermometer reading. Thus von Mises said in 1919 that the application of probability is easiest in physics. Statistics and games of chance are, in this order, progressively more difficult cases (1919b, p. 68).

The identification of probability with a limiting frequency is a theoretical idealization that constitutes a conceptual problem for an empiricist such as von Mises. What is the proper place of such a notion of probability in an empiricist system of knowledge? Second, how is probability connected to the inferential procedures that should somehow prevent it from becoming an 'idle' notion, as the value of a limiting frequency is and remains unknown.[4] Significantly, the other notable logical empiricist involved with probability, Hans Reichenbach, preferred a finite relative frequency view. It was supplemented by a (rather simplistic) rule of inference from data to probability (Reichenbach 1935).

In sum, those who developed mathematical probability in the 1920s had found no place of note for subjective probability. Physical thought had rendered it almost obsolete, it seemed. Those who should have found it congenial to their general philosophical outlooks, found it instead based on a 'mistaken idea.'[5] The scientifically oriented logical empiricists were looking for hard facts rather than individual perceptions of the likelihoods of events. Subjective probability was barely alive around 1930.

[3] As explained in Sec. 4.3.

[4] See Sec. 6.1 for von Mises' notion of 'idle'.

[5] See von Mises (1928, p. 72). He directed his criticism rather directly against arbitrary applications of the principle of indifference.

The merit of revitalizing the theory of subjective probability goes to Bruno de Finetti's youthful works from the late 1920s and early 1930s. New ideas, results and publications kept coming from him at an incredible pace. His achievements add one more significant aspect to a complex story, the development of modern probability. First of all, one has to stress the independence of development in de Finetti. He was following an idea: Its roots were in empiricist and pragmatist philosophy and, more recently, in the operationalization of concepts after the model of simultaneity in special relativity.[6] Probability is a subjective degree of belief of someone, pertaining to anything that someone is uncertain about. De Finetti's probabilism, as he calls it in 1931, is the true heir of the empiricist philosophical tradition in the spirit of David Hume. Probability statements stand on a different level as compared to statements of fact. For they are like reports of sense-data, that is, correct in themselves as accounts of a person's attitudes and beliefs. Once formed, they exist independently of the external circumstances that may have provoked them.

De Finetti was a prodigy who could make his philosophical and conceptual ideas match his mathematical developments. Already in the late 1920s his mathematical results made his name known to the small circle of probability mathematicians. De Finetti had learned the methods for handling probability problems from Castelnuovo's and Lévy's books of 1925–1928. These works taught him the analytical means for arriving at his first results, concerning the class of exchangeable events he had defined in 1928. The next year he was able to derive, with similar means, probabilistic laws for continuous time random processes.

The notion of exchangeability, and especially the representation theorem, forms the centerpiece of those achievements for which de Finetti is best known today. Since the publication in 1964 of the English translation of his 1935 French lectures *La prévision* (1937), the result and its philosophy have been standard themes in foundations of probability. The representation theorem constitutes an almost perfect match between de Finetti's philosophy of probability and his mathematics. I shall address in Section 8.2(b) the historical and philosophical questions of how he arrived at the result, and how his views of its significance related to his overall approach. As he stated it some years later, the task of exchangeability was to eliminate the idea of objective probabilities with unknown constant values, in favor of a subjectively assessed probability that changes according to the rule

[6] The philosophical and ideological background is explained in Jeffrey (1989).

of *Bayesian conditioning*: In the simplest case, one starts with a *prior probability* $P(H)$ of a 'hypothesis' $H$. The latter may be just a guess as to the next result of a trial, or it may be the assignment of a value to a parameter describing the mean, variance, etc., of a distribution, for example. Next, an observation or 'evidence' $E$ is used in the calculation of a *posterior probability*, according to the formula of conditional probability

$$P(H|E) = \frac{P(H \cap E)}{P(E)} \tag{8.1}$$

By Bayes' formula, this can be written as

$$P(H|E) = \frac{P(E|H)}{P(E)} \cdot P(H). \tag{8.2}$$

The first term of the product on the right is the *likelihood* of the observation $E$, that is, the probability of $E$ under the hypothesis $H$, relative to the probability of $E$ without the hypothesis. The rule of Bayesian conditioning can now be put as follows: posterior probability = likelihood × prior probability. According to whether the hypothesis $H$ makes the evidence $E$ obtained more or less probable, the same happens for the posterior probability of the hypothesis itself. The constant probabilities of independent trials, the usual scheme of the objectivistic line of thought, cannot be used to account for probabilistic or statistical inferences in this way. For due to independence, observations have no effect on the posterior probabilities. De Finetti's notion of exchangeability relaxes the independence, but still allows statistically stable frequency phenomena.

De Finetti's work on continuous time random processes, to be treated in Section 8.3, seems to be an equally independent development as his introduction of exchangeability. Here also, profound mathematical results combine with a foundational, programmatic point of view, the 'renunciation of determinism' as he says. The break of the ties between subjective probability and determinism was certainly needed in order for the theory of subjective probability to gain any status of scientific respect. De Finetti saw subjectivism as a way out of an impasse into which he thought the classical epistemic notion of probability had been led. It was brought about through the acceptance of indeterminist ways of thinking in the sciences. Although he appreciated the role of quantum mechanical chance, in his own work he tried to explain how statistical laws prevail already within the framework of classical physics. This was the context in which he developed the theory of continuous time random processes, starting in 1929. He was among the first mathematicians to study such processes, with his notion of a process with independent increments. It is a process in which the changes in non-overlapping time intervals are probabilistically independent. His technical tool here was a new definition of a derivative of a process, which let him write down directly a differential equation for the process. Its solution was the sought for probabilistic law of a continuous time process. In this connection he invented the notion of an infinitely divisible probability distribution also, and considered Markov processes briefly. Very soon he almost stopped working with continuous time processes. The same happened with some others of de Finetti's early research themes.

De Finetti's works up to 1936 were published together in *Scritti (1926–1930)* and *Scritti (1931–1936)* in 1981 and 1991, respectively. The first volume also contains his bibliography.[7] The well-known *La prévision* lectures at the Institut Henri Poincaré in 1935 are de Finetti's own summary of his central ideas and results on foundations of probability. The recently published *Calcolo delle probabilità* (1987) contains de Finetti's lectures at the University of Padua in 1937–1938. It provides the most detailed source of many of de Finetti's early ideas and results on probability.

De Finetti's very first publication (1926) was a mathematical study of Mendelian heredity. He gives no references, but some later recollections identify a popular article as an impetus. The two-part paper (1927a,b) on the diffusion of Mendelian characteristics was presented to the Accademia dei Lincei by the biologist Carlo Foà.[8] In the first part it is interesting to note how he transforms 'the statistical phenomenon of diffusion of a Mendelian characteristic in a population' into 'a problem posed as one of the motion of a point for which Mendel's laws constitute the kinematical law' (1927a, p. 914). Putting down a differential condition for the diffusion amounts to the study of a statistical law that depends on a continuous time parameter. Fifty years later de Finetti mentions a popular article by Foà, and Enriques' book, as works arousing his interest in the diffusion of Mendelian characteristics: 'This was the way that I became attracted to the calculus of probability, the discussions of the concept of probability and its

[7] Many of the papers are reproduced from offprints with a new pagination which is then followed in the bibliography also. A more accessible bibliography, with correct page numbers, is found in *La logica dell'incerto* (1989), a recent collection of de Finetti's essays. That bibliography is based on the one published by Daboni (1987).

[8] The first part (1927a, p. 914) mentions a book of the year 1924, *L'Eredità nell'Uomo* by Paolo Enriques. See also *Scritti (1926–1930)*, p. xvii.

foundations, and the applications of its mathematical elaborations' (1977, p. 219).

The foundational interest behind de Finetti's mathematical work is first brought out when he turns to the question of extending the additivity property of probability from finite to infinite domains (1928a). First he discusses the question in terms of certain limiting processes. These are suggested from the equality of probability and limiting frequency in Cantelli's strong law of large numbers. De Finetti argues by examples that an inner and an outer measure need not coincide, and that a probability can assume any value between these lower and upper bounds (1928a, p. 824). Specifically, one can have a probability 0 for each of a denumerable set of events, the set itself having outer measure 1. In (1930a, p. 157) the question is put in terms of the denumerable additivity of probability. The precise question is, whether in a denumerable case of mutually exclusive events $E_1, E_2, E_3, \ldots$ one always has, for the logical sum $E_1 + E_2 + E_3 + \cdots$, the equality

$$P(E_1 + E_2 + E_3 + \cdots) = \sum_i P(E_i). \tag{8.3}$$

De Finetti (1930b) calls this the problem of the extension of total probability to infinite cases. In a discussion note on the matter with Maurice Fréchet, he objects to Fréchet's attitude. Following what is the case for Lebesgue measure, the latter says there are events that do not have a probability. They do not fulfil the property of denumerable additivity and are therefore not measurable.[9] De Finetti's answer to Fréchet contains an expression for a general principle one can see in action in his later works: 'If... a certain notion, in our case that of probability, has a well-defined sense, the question is not one of stating a convention. Rather, it is one of proving a theorem' (1930b, p. 902). This sense, defined well enough for de Finetti at the time, was not one that would secure denumerable additivity as a necessary property of probabilities.

De Finetti does not at once reveal what the sense or meaning of the notion of probability is for him, but refers instead to forthcoming works. In these he formulates a foundational program, his 'new approach' based on a theory of subjective probabilities. Its first indication is the short paper (1930d) on the logical foundations of probabilistic reasoning. Remarkably, a condensed version of the paper on characteristic functions, where the representation theorem appears, does not say a

[9]See Fréchet (1930a, p. 900), a comment on de Finetti (1930a). Fréchet's two notes are reprinted in the *Scritti (1926–1930)*.

word about the philosophical basis. This is de Finetti (1928b), a paper that came out only in 1932. A three-page summary of the same theme, de Finetti (1929d), stresses the inferential aspect along with a criticism of assuming constant probabilities for events.[10]

The central works in the program for new foundations of probability, the topic of Section 8.4, were the philosophical essay *Probabilismo* (1931b) and the long paper 'On the subjective significance of probability' (1931c). Probability has an intuitive subjective meaning in finite situations: It is a degree of belief. De Finetti shows in (1931c) how the formal properties of probability can be derived from two assumptions: First, that probabilities are (or rather can be) used in choosing a betting ratio. A betting ratio $a/b$, or 'odds' $a$ against $b$, means that by paying $a$, your gain is $a + b$ or your loss is $a$, according to whether the event you bet for occurs or fails to occur. The sums $a$ and $b$ can be normalized into percentages through division by $a + b$. Often $a$ and $b$ are given directly as percentages, in which case you can choose to bet a sum $Sa$ against $Sb$. If your probability for the occurrence of an event is $p$, the odds to be chosen are $p/(1 - p)$. Second, it is assumed that one abstains from placing bets in a way that would lead to a sure loss. This is the *requirement of coherence.*

The betting approach has an air of artificiality in it. De Finetti thinks choices of acceptable betting ratios would be useful means for measuring a person's subjective probabilities. But they would not be constitutive of those probabilities. He developed instead in (1931c) a theory of *qualitative probability*, and studied conditions that would guarantee a conventional numerical representation of the basic qualitative relation $A \succeq B$, stating that the event $A$ is at least as probable as the event $B$.

De Finetti's new foundations for the theory of probability were to be both mathematical and philosophical. I shall emphasize the former in Section 8.4, so that de Finetti's contribution could be more clearly seen as a distinct solution to the question, how to formalize the mathematical theory of probability. Throughout this formalization, de Finetti remained faithful to the principles of his foundational program and his radically empiricist philosophy, such as the requirement that unverifiable infinitary events cannot in general have well-determined probability values. His first note on foundations of probability (1930d) sets in its opening passage the tone of the investigations to come:

The 'truth' of an assertion, of a proposition, one can intend in two ways: either in the objective sense as conformity to an external reality, conceived as

[10] It is a bit problematic to arrange de Finetti's publications in chronological order. He added cross references at proof stage, which sometimes creates confusion in this respect.

independent of us, or in a subjective sense as conformity to our very opinions, impressions, sensations.

The same philosophical distinction applies also to the concept of probability.

For one who attributes an objective meaning to probability, the calculus of probability ought to have an objective meaning, and its theorems ought to express properties that are satisfied in reality. But it is useless to make such hypotheses. It suffices to limit oneself to the subjectivist conception, and to consider probability as a degree of belief a given individual has in the occurrence of a given event. Then one can show that the known theorems of the calculus of probability are necessary and sufficient conditions in order for the given person's opinions not to be intrinsically contradictory and incoherent.

Toward the end of his long career, permeated by probabilistic studies, he put his philosophy more bluntly: The motto of his *Theory of Probability* reads: '*Probability does not exist.*'

## 8.2 EXCHANGEABILITY AND THE REPRESENTATION THEOREM

### *8.2(a) The notion of exchangeability*

De Finetti's representation theorem and its subjectivist interpretation are well known in foundational discussions of probability, especially after the appearance in 1964 of the English translation of *La prévision*, de Finetti's concise summary of his position from the year 1935. The study of exchangeable probabilities is a lively field of mathematical research today. It is finding applications in statistics and physics as well as in other fields.[11]

Let us first get a simple idea of the notion of exchangeability itself. The idea has certainly occurred to many, of considering sequences of trials where each sequence with the same length $n$ and the same number of successes $r$ has the same probability. One special case of such sequences results from Laplace's *rule of succession*: Let $p$ be the unknown probability of success of a simple event. *A priori* one has, according to Laplace, 'no reason' to think that one value of $p$ would be more probable than another. This ignorance of the value of the parameter $p$ is represented by a uniform distribution over all values. Posterior

[11] See the recent reviews Aldous (1985) and Diaconis (1988) for the mathematics. De Finetti's 75th anniversary conference proceedings, Koch and Spizzichino (1982), also surveys applications.

probabilities are calculated according to the scheme of Bayesian conditioning, which gives the rule of succession: The probability of a further success conditional on $m$ successes in $n$ is the number $(m+1)/(n+2)$. It is an example of an exchangeable probability assignment. Others have suggested directly the assignment of equal probabilities, conditional on a given number of successes in $n$.[12] It has not always been realized that this leads beyond the scheme of independent trials.

The condition of exchangeability can be expressed in many ways: The $n$-dimensional distribution function over events has to be *symmetric* in the arguments, the probability is the same independently of the order of successes and failures, the success count is a *sufficient statistic*, and so on. Let us denote, after de Finetti (1930e), by $\omega_r^{(n)}$ the probability of getting $r$ successes in $n$ exchangeable trials. If the events are independent, with a constant success probability $p$, the probability of any one sequence with $r$ successes in $n$ is $p^r(1-p)^{n-r}$. There are $\binom{n}{r}$ such sequences, so that the probability of the success frequency $r/n$ is $\binom{n}{r}p^r(1-p)^{n-r}$. As can be seen, this expression is independent of the order so that independence plus equal probability is a special case of exchangeability. Further, say we have $m$ urns with red and white balls, urn $i$ having the probability $p_i$ of giving a white ball. Assume the prior probability of choosing urn $i$ is $a_i$. In a repetition, an urn is first drawn according to the probabilities $a_i$, and then a ball is drawn from it $n$ times, with replacement. We get from the formula of total probability for the probability of $r$ white balls in $n$ repetitions,

$$\binom{n}{r}\sum a_i p_i^r(1-p_i)^{n-r}. \tag{8.4}$$

This 'mixture of Bernoullian schemes' gives a probability that is independent of order, that is, exchangeable. Next imagine a mechanism of drawing a number $p\in[0,1]$ according to the probability density $f$. Then the overall probability becomes the integral

$$\binom{n}{r}\int_0^1 p^r(1-p)^{n-r}f(p)\,dp. \tag{8.5}$$

This again is exchangeable. De Finetti's representation theorem says

[12] See Dale (1985) and Zabell (1982) for prehistory, Haag and Johnson in particular. Exchangeability was also studied by Frank Ramsey in unpublished work from the late 1920s, cf. 'Rule of succession,' in Ramsey (1991, pp. 279–281).

the converse is true: For any probability law that is exchangeable for all $n$, there is a unique distribution $F$ such that

$$\omega_r^{(n)} = \binom{n}{r} \int_0^1 p^r(1-p)^{n-r}\,dF(p). \tag{8.6}$$

If $F$ has a density $f$, the integral representation has the same expression as in (8.5). De Finetti's philosophical interpretation of the result was that it shows how to eliminate the 'unknown objective probabilities' $p$ of independent events in favor of the subjective ones. Sometimes the notion of 'conditional independence' is used to replace exchangeability. The unknown probability is thought of as a random parameter such that conditional on its value, the events are independent.

### *8.2(b) The 'characteristic function of a random phenomenon' of 1928*

I shall now go into explaining what de Finetti's mathematical concerns were in the work 'Funzione caratteristica di un fenomeno aleatorio' (characteristic function of a random phenomenon), where he found the representation theorem. The theorem results from the application of the method of characteristic functions to exchangeable probabilities. It first appears as formula [20] in (1930e, p. 96). De Finetti assumes $h$ successes in $n$ trials, and the formula reads in the original notation as

$$\begin{aligned}\omega_h^{(n)} &= \binom{n}{h} \int \xi^h(1-\xi)^{n-h}\,d\Phi \\ &= \binom{n}{h} \int (h-n\xi)\xi^{h-1}(1-\xi)^{n-h-1}\Phi(\xi)\,d\xi.\end{aligned} \tag{8.7}$$

This formula is really a by-product of the central mathematical result of the *Funzione caratteristica*: the determination of the limiting distribution of the relative frequency of an exchangeable sequence of events. The latter comes out as follows: First the characteristic function for the distributions of the relative frequencies $h/n$ is determined, then the limit is taken. All the probabilities $\omega_h^{(n)}$ are determined from the limiting characteristic function: The representation theorem (8.7) appears as an expression of this fact (1930e, p. 96). De Finetti's method of proof via the use of characteristic functions is based on the work of Paul Lévy. His *Calcul des probabilités* (1925) contains what are now called the Lévy formulas for going back and forth between probability distributions and their characteristic functions.[13]

A random phenomenon (fenomeno aleatorio) is one of which you can make, or at least conceive of making, any number of trials. The order in which the successes appear must be 'attributable to chance' (1930e, p. 87). Elsewhere de Finetti says an *event* is a well-defined unique fact. Then, instead of speaking of the repetition of events giving a sequence of trials, one should use *phenomenon* as a general notion, with each *trial* of a phenomenon giving an event (1930f, p. 369). The requirement of chance in the order of successes leads, for any $n$ and $h$, to giving a uniform probability distribution over sequences with $h$ successes in $n$ trials.

In modern parlance, de Finetti's random phenomenon is one whose trials give an exchangeable sequence of events. The idea of a repetitive trial relates the case to that of frequentist probability. De Finetti's concern was to show how, and in what sense, an inference from frequency data to probabilities is possible, according to his subjectivist approach.[14] The elements of such inferences have to be made explicit. They are, first, a judgment about the exchangeability of the events, or in de Finettian terminology, having a random phenomenon. Second, a particular a priori distribution is needed or else no inferences are possible. De Finetti's main interest in the paper under discussion was to calculate the probabilities of frequencies for a given exchangeable probability law: That is what 'foresight' means for him. In his original notation, $\omega_h^{(n)}$ stands for the probability of getting $h$ successes in $n$ trials of a random phenomenon. In other words, it is the probability of the value $h/n$ for the random variable $X_n$ giving the frequency in $n$ trials. This random variable has a limiting distribution as $n$ goes to infinity. (Then the limit of relative frequency exists, which means that a law of large numbers is obtained). In one sentence, the *Funzione caratteristica* is a work where the limiting distribution of relative frequency is determined for the case of exchangeability.

One might think that there is a trace of frequentist thinking left in de Finetti when he describes the invariance conditions for an experiment: 'For a succession of events to be trials of the same phenomenon, they have to take place in identical conditions.' But later in the same essay he says that it suffices to have an unlimited

[13] See Lévy (1925). In common with other French probabilists such as Poincaré and Borel, Lévy allows both subjective and statistical probabilities. Castelnuovo (1925–28) is another of de Finetti's sources for the method of characteristic functions.

[14] He called such an inference 'previsione,' or 'foresight' in later translation.

exchangeable sequence: That its members are trials of the same phenomenon is only one significant interpretation (1930e, p. 122). Exchangeability, then, is a probabilistic judgment with a subjective significance only. Events whose probabilities have this significance have to be singular in time and space. Another term with a poignant contemporary air is 'verifiable events.' Obtaining a specific relative frequency $h/n$ as a value of the random variable $X_n$ is one such singular event. The same does not apply to the limit as $n$ goes to infinity. For this reason de Finetti would not find acceptable the description of exchangeability as 'conditional independence.'

After the determination of the limiting distribution of relative frequency, de Finetti determines in the second chapter the effect of observed frequencies on the a priori probability. This result follows from the recurrence relation for the momenta $\omega_n^{(n)}$ by the calculus of finite differences.[15] De Finetti obviously thinks of a sequential type of inference from data. In the third chapter two favorite topics of older texts on probability are found: a posteriori probabilities and the probabilities of hypotheses. De Finetti finds the distinction between a priori and a posteriori probabilities artificial, not conceptually significant. The evaluation of probabilities cannot, in practical situations, be based on 'equally likely cases'; that is the usual way of arriving at a priori probabilities. Such an evaluation is instead based on previous analogical situations, he says (1930e, p. 115). The question of a posteriori probabilities is a question of inference from frequencies to probabilities, or of the 'inversion of Bernoulli's theorem.' De Finetti gives a limit theorem which shows in what sense the probability law for exchangeable events approaches one with independent events with the observed frequency as a priori success probability. Relative frequency in $n$ trials is a random variable, but it need not be the case that its limit also is a random variable, that is, has a probability distribution. In the special case of independence the limit has with probability 1, as the usual way of stating it is, a constant value. For the general exchangeable case it would not be constant.

The mathematical part of de Finetti's essay does not address the question of interpretation. It is discussed in the beginning of the paper (p. 89): The usual talk about unknown constant probability of success for independent events is meaningless according to the subjectivist view. And taking the unknown probability $p$ as a parameter with a 'second-order' probability distribution $\Phi$ has in de Finetti's view, no positive content. The notion of exchangeability or of a random

[15]Cf. de Finetti (1930e, p. 109, formula [34] et seq.).

phenomenon replaces the old scheme of independent constant probabilities. He says that his representation formula [20] 'shows that, assuming the usual conception made sense, a random phenomenon is exactly the case of a phenomenon "with independent events of a constant but unknown probability $p$," where the limiting distribution is interpreted as "the probability law (distribution) of the unknown probability $p$." This constitutes the *formal*, but *only formal* justification of the usual approach. Conceptually it remains always disputable to say the least' (1930e, p. 89). In the main text the formula is given as one which leads from the characteristic function to the exchangeable probabilities $\omega_h^{(n)}$. But in the end it is the latter probabilities that come first, and from which the limiting distribution is recovered.

The interpretation of the representation theorem received a canonical form in de Finetti's later writings. He often expressed himself as having shown how to replace or reduce the 'nebulous metaphysical talk' about objective unknown probabilities in favor of an immediately meaningful subjectivist language.[16] We have seen that only the exchangeable probabilities on the left side of the integral representation formula (8.7) are genuine meaningful probabilities for him. This he later emphasized repeatedly, saying that the unique distribution $\Phi$ over the parameter $\xi$ is 'only formally' a probability.[17] This same attitude is found already in Section 22 of de Finetti's 1931 philosophical essay *Probabilismo*: The probability that the frequency lies within preassigned bounds tends to a limit. This limit is, in a way of expression de Finetti says he does not accept, the probability of the hypothesis that the constant unknown probability of success lies within the given bounds. That hypothesis does not constitute an event whose occurrence could be verified even in principle.

Let us now turn into a discussion of some examples. These clarify de Finetti's thinking about exchangeability and related topics. The 'Funzione caratteristica' contains (Sec. 24) a discussion of phenomena ascribable to several causes. Let us first note that the integral representation formula (8.7) turns into a finite sum of the form given by (8.4) if there is only a finite number $m$ of possible values for the banned 'unknown probability.' It reads

$$\omega_r^{(n)} = \binom{n}{r} \sum a_i p_i^r (1 - p_i)^{n-r}. \tag{8.8}$$

[16] See *La prévision*, (1937, p. 50, or 1964, p. 142). Further similar passages are easy to find in his later publications.

[17] See de Finetti (1937, pp. 48–50, or 1964, pp. 140–42). Those who suggest that de Finetti interpreted the weights over the parameter in the representation formula (8.7) subjectively, are in error.

Reading from left to right, we get that the exchangeable probabilities determine a unique distribution of weights $a_i$ for the different values of the probability $p$, the weights themselves satisfying the conditions $a_i \geqslant 0$ and $\sum_i a_i = 1$. In the other direction, if each value of $p$ corresponds to the scheme of independence and constant probability of success, the left side is exchangeable. It is a mixture of the independent probabilities as one says. De Finetti discusses a case where the phenomenon depends on $m$ incompatible causes having the probabilities $a_1, \ldots, a_m$ (1930e, p. 114). A concrete classical example is offered by the following. We have $m$ urns chosen according to the probabilities given by the $a$'s. Urn number $i$ has black balls in the proportion $p_i$, and this gives the probability of drawing a black ball from that urn. The overall probability for a black ball is given by total probability, in the way of the above sum (8.4). It is a bit surprising that de Finetti does not address the representation theorem in this context. As long as we do not know which urn was drawn, repeated draws of a ball from that same urn are correctly described probabilistically as a sequence of exchangeable events. The difference is not mathematical but conceptual and factual: There is a concrete realization of what in 'the old language' (to emulate de Finetti's terminology) would be called 'probabilities of probabilities.' Here it is verifiable by assumption which urn was drawn. And it is verifiable by convention what the probability $p$ is in each of these, for the probabilities were agreed to be evaluated as proportions in the first place. In the lectures *La prévision* (1937) the same example is discussed. Now the connection with exchangeability is made explicit. De Finetti criticizes the idea that hypotheses about the true value of unknown probability can themselves be bearers of probabilities. Such hypotheses are not even in principle verifiable. In the case of urns it is different: Hypotheses as to their composition 'express an objective fact that we may verify.' This is essential. If we have, for example, an irregular coin, the 'unknown probability' is not definable, de Finetti maintains (1937, pp. 48–49; 1964, p. 142).

As could be seen from the foregoing, the scheme of a mixture of urns would be a natural way of arriving at the notion of exchangeability. When being posed the question, how did he arrive at his famous theorem, de Finetti answered that it was through thinking of a system of several urns with varying compositions.[18] It is also interesting to note that Castelnuovo in his book, known to de Finetti, had discussed the Laplacian mixture of urns. He determined the 'a posteriori'

[18] I have been told this by Prof. Domenico Costantini, who himself made the question to de Finetti.

probability that $r$ white balls in $n$ drawings come from an urn with a given proportion of white balls.[19]

The last and fourth chapter of de Finetti's essay under discussion deals with what he calls equivalent events (eventi equivalenti) and what now are known as finitely exchangeable events. Simple examples of genuinely finite cases are drawings from a finite urn without replacement. I shall return to such cases in the next section.

De Finetti's later papers on exchangeability did not have the extensive and systematic character of his first paper, except for the (1937) lectures. These papers were rather in the way of notes that extended the original idea, and that provided some basic results and examples for others to follow up.

### *8.2(c) Extensions of exchangeability*

The first response to de Finetti's introduction of exchangeability was Alexander Khintchine's 'Sur les classes d'événements équivalents', a three-page note published in the *Mathematicheski Sbornik* in 1932. He noticed that the condition of exchangeability is equivalent to the following: Writing $\omega_n$ for $\omega_n^{(n)}$, exchangeability is equivalent to having the same probability $\omega_n$ for any $n$ successes (so not just consecutive ones). Then, proving by Chebyshev's inequality that the $\omega_n$ are precisely the momenta of the limiting distribution of relative frequency, Khintchine could conclude in very brief steps de Finetti's representation theorem, as well as the strong law of large numbers for exchangeable events, a result Khintchine considered to be the most important one of de Finetti's paper. Such laws were the topic of a related note Khintchine (1932c). There, a sequence of events is defined to follow a strong law if the relative frequency $h_n$ in $n$ trials obeys the following:

$$P(|h_n - h_{n+q}| > \varepsilon) \to 0 \quad \text{when } n \to \infty, \quad \text{for any } q > 0,\ \varepsilon > 0. \tag{8.9}$$

Beyond the case of independence, he notes one case studied by Bernstein, de Finetti's exchangeable sequences, and his own stationary sequences, as defined the very same year in Khintchine (1932a).

Khintchine's note simplified considerably the proof of the representation theorem and related results. De Finetti published in 1933 three papers in which he generalized on that basis his results to real random variables, or random quantities, as he usually preferred to say. These papers are 'Classes of equivalent random quantities' (1933a), 'The law of large numbers in the case of equivalent random

[19] Cf. Regazzini (1987, p. 10).

quantities' (1933b), and 'On the law of distribution of values in a sequence of equivalent random quantities' (1933c). De Finetti had previously used the term 'random phenomenon' for today's 'exchangeability,' but now he followed Khintchine's use of 'equivalent.' In the last of his brief notes de Finetti says he hopes to make a complete study of exchangeable random variables, but nothing of the kind ever appeared.[20] In the first paper (1933a) the probability of an exchangeable event $E$ is written in a general form as

$$P(E) = \int P_\xi(E)\, d\Phi(\xi) \tag{8.10}$$

where $P_\xi(E)$ is the probability corresponding to the case of independence. In (1933b) a sequence $X_1, X_2, \ldots, X_n, \ldots$ of exchangeable random quantities is assumed. Let $Y_n = 1/n(\sum X_i)$. It is shown that it has a distribution $\Phi_n(\xi)$ that tends to a limit $\Phi(\xi)$ as $n \to \infty$, the existence of a limit being an expression of a law of large numbers. In the last paper (1933c) it is noted that for exchangeable random quantities $X_i$, the events $E_i(x)$ stating that $x_i \leqslant x$, are exchangeable. The probability that out of $X_{i_1}, \ldots, X_{i_m}, h$ have value $\leqslant x$ and $m - h$ have value $> x$, is denoted by $\omega_h^{(m)}(x)$. If $F$ is the distribution of $x$, one has specifically $\omega_1^{(1)}(x) = F(x)$.[21]

A short note, given at a conference in 1933 and published as de Finetti (1934), ends this series of studies of exchangeability by a comparison with probabilistic independence. In the objectivist scheme, the latter is often concluded on the basis of 'causal independence.' In order to take into account data, one has to reason in terms of estimates of the values of probabilities. The subjectivist approach is more direct: It lets de Finetti conclude that 'probabilistic independence is not only a lack of causal connection, but a lack of any influence on our judgment of probability' (1934). Exchangeability and its generalizations were meant as manageable schemes where one does have such influence.

The next generalization of exchangeability, *partial exchangeability*, was introduced by de Finetti in 1937. Partial exchangeability is now well known, due to the translation of the original article in Jeffrey (1980) and to other recent publications, such as de Finetti (1972), as well as works of Diaconis and Freedman and others. Once again, his paper (1938a) introduces ideas and gives some basic results, instead of a comprehensive treatment. In place of perfect symmetry, the events are in partial

[20] The closest thing to it would be chapter IV of *La prévision*.

[21] See further *La prévision* (1937, p. 45ff).

exchangeability divided into $k$ classes such that exchangeability prevails within classes. The main result is the representation theorem for a $k$-fold partially exchangeable sequence of events:

$$\omega_{r_1,\ldots,r_k}^{(n_1,\ldots,n_k)} = \binom{n_1}{r_1}\cdots\binom{n_k}{r_k}\int p_1{}^{r_1}(1-p_1)^{n_1-r_1}\cdots$$

$$p_k{}^{r_k}(1-p_k)^{n_k-r_k}dF(p_1,\ldots,p_k). \qquad (8.11)$$

We can also 'read' this equation from right to left, though that is not the direction de Finetti is thinking. We gain the useful insight that partial exchangeability is a mixture of independent but *not* equally probable events. It broadens de Finetti's 'reduction' of objective probabilities in a most natural way. As limiting cases of partial exchangeability, we obtain ordinary exchangeability when $k = 1$, and the case when events are independent between different classes. In that case the distribution $F(p_1,\ldots,p_k)$ reduces to a product $F_1(p_1)\ldots F_k(p_k)$. A discussion of various example cases follows in de Finetti (1938a).

De Finetti devised already in the 1930s his notation for 'predictive probabilities': Assuming that one has obtained the frequency $r/n$ in an exchangeable sequence of trials, the *predictive probability* $p_r^{(n)}$ is the conditional probability of success on the $(n+1)$th trial. Using these probabilities, de Finetti's less known 1952 paper 'Equivalent events and the degenerate case' introduces another case of exchangeable probabilities. He gives a geometric representation of the situation as a two-dimensional random walk in a lattice: Each success is one step right and up, failure a step right and down. The *degenerate case* of exchangeability obtains if some of the predictive probabilities $p_r^{(n)}$ have values 0 or 1.[22]

A further case of exchangeability, one that is now called *finite exchangeability*, is found already in de Finetti's first essay on exchangeability (1930e). He used the term 'equivalent events' (eventi equivalenti) for the finite case, in distinction to random phenomena (fenomeni aleatori), which are infinitely exchangeable events. Simple examples of genuinely finite cases are offered by drawings from a finite urn without replacement. De Finetti characterizes those situations in which a finite sequence of exchangeable events is a random phenomenon, that is, a segment of an infinite exchangeable sequence. The condition is that the function $\Phi$ determined by the limit of characteristic functions is a genuine probability distribution: a nondecreasing function $\Phi$ such that

[22]The case was further studied by Daboni (1953).

$\Phi(x) = 0$ if $x < 0$ and $\Phi(x) = 1$ if $x > 1$. He gives an algorithm for checking whether a finite sequence is extendible to an infinite one.[23]

Besides the remarks in *La Prévision* (1937), I do not know that de Finetti returned to finite exchangeability before the article 'On the continuation of exchangeable random processes' (1969).

In view of the acknowledged role the representation theorem enjoys today, it seems strange that as late as the 1940s it was not necessarily taken as the most profound result about exchangeable probabilities. In Fréchet's book of 1943, exchangeability is addressed at length. Mainly, he is playing with the calculus of finite differences, obtaining all means of equalities and inequalities. The representation theorem is almost casually mentioned as 'une formule obtenue par Khintchine et de Finetti.'[24]

### *8.2(d) Of generalizations and connections and the ability to notice them*

There is a broader notion of a law of large numbers than that of a unique limit up to a set of exceptions of measure 0. Formulated for simple events, as in (8.9), it only requires that there exist a limiting distribution for the limit of relative frequency. Stationarity is a condition that characterizes (is equivalent with) the existence of a limiting frequency. As to the case of unique limits, there is a condition that characterizes it, too. Independence is a special case. Then, calling probabilities fulfilling this latter condition ergodic, we look at their mixtures. It turns out that these are precisely the stationary probabilities, of which exchangeability is a special case. There is a theorem for the more general pair of concepts, ergodicity and stationarity, exactly analogous to de Finetti's representation theorem, namely, the *ergodic decomposition theorem.* It gives a unique decomposition of stationary measures into ergodic parts, and

[23] See (1930e, p. 125), also de Finetti (1937, pp. 36–37, or 1964, pp. 129–30). Questions of the extendibility of finitely exchangeable sequences of events have been studied by Jaynes (1986). He refers to *La Prévision* but does not seem to have checked the details. Following a remark to the effect that Laplace never suspected 'that every exchangeable sequence can be so [by the "Laplace–de Finetti" integral formula] represented,' he allows negative values for the function $\Phi$. Then the integral representation holds also for genuinely finitely exchangeable sequences. Comparing de Finetti (1930e, p. 125), we notice that de Finetti saw this already in 1930. Jaynes' paper is very interesting though, and contains an explicit method for answering marginalizability questions. Such questions extend naturally to the partial case, as in my (1991).

[24] See Fréchet (1943). p. 112. On p. 108 we read: 'nous adopterons plutôt l'expression d'événements échangeables proposée par Polyà.' This should correct the (rather insignificant) misinformation, perpetuated also by de Finetti, that the term exchangeable should have been Fréchet's invention.

each part or component can be taken as a subset of the set of all infinite sequences. Denoting by $\mathscr{D}$ the decomposition, a stationary probability $P_S(E)$ is given as the integral over $\mathscr{D}$ of ergodic probabilities $P_D(E)$ of components $D$, with weights $\mu(D)$:

$$P_S(E) = \int_{\mathscr{D}} P_D(E) d\mu(D) \tag{8.12}$$

Within a component, asymptotic statistical properties are identical. In the case of simple events these properties are characterized by the limits of relative frequencies. The overall probability is a mixture or weighted average of these limits. The weights come from, or give, the relative measures of the components.[25]

The stationary and ergodic probabilities of (8.12) can be interpreted in several ways. A de Finettian way would be to say: I have decided not to change my subjective probabilities unless I observe something relevant; therefore my probabilities are stationary. The ergodic probabilities $P_D(E)$ are mere mathematical fictions. Another interpretation would be that we are faced with an ergodic component, the distribution $\mu$ over the components giving our prior subjective uncertainty as to which component it is. In fact, de Finetti discussed an example of urns that has, quite concretely, only one component materialized. There is another example with a similar character, inspired, it seems, by one of Lévy's examples: de Finetti considers a roulette wheel with a uniform probability distribution for a pointer in it. The circumference is divided into two parts, red and black, with an unknown proportion $p$ for red. This $p$ can be determined by a preceding experiment with a random variable $X$ between 0 and 1 with distribution $F$ (1930e, p. 98). The arrangement for materializing any probability $p$ is similar to one found in Lévy's book (1925, p. 88).

Several other interpretations of stationary probabilities could be cited from probabilistic and physical literature, for which see my (1982). However, this much reconstruction of de Finetti's position will have to do here, especially since it seems that he himself did not realize the potentials of the ergodic theory in relation to his work on foundations of probability. A colloquium of foundations of probability was held in Geneva in 1937, more or less the last get-together of probability theorists before the war. Its proceedings give a unique view of the status of probability theory and the debates over its foundations in the late 1930s.

[25] The ergodic decomposition can also be treated as a simplicial decomposition. The ergodic measures are the extreme points of the simplex of stationary measures.

These proceedings are supplemented by a summary of all the talks and discussions that de Finetti undertook to prepare. Eberhard Hopf gave a talk on the ergodic decomposition of classical dynamical systems, but judging from de Finetti's summary, it does not seem that the probabilistic significance of the result would have become clear to him.[26]

Let us now have a somewhat broader look at the historical and conceptual situation. To me, the time around the year 1932 appears to be a culmination point of the researches leading to modern probability, a time when a sufficiently general view of probability theory as an abstract mathematical discipline was reached. That made it possible to see connections between apparently different fields. Here are some of the reasons for this contention.

Early in 1932 von Neumann published his mean ergodic theorem for classical dynamical systems. The average of time averages of a phase function converges to the phase average of that function. Meanwhile, he had explained it to G. D. Birkhoff, who sharpened the result: Not only does the average of time averages converge, but each individual time average converges to the same limit, up to a set of exceptions of measure 0.[27] Von Neumann's paper also contained an outline for the result according to which a classical dynamical system has a unique decomposition into ergodic parts.

Khintchine and Hopf were both fast in discovering the essential content of Birkhoff's ergodic theorem: that it can be given a purely probabilistic formulation in terms of measure-preserving transformations. Khintchine was led to define the notion of a (strict sense) stationary process on the basis of the notion of stationary motions in dynamical systems.

De Finetti had defined the concept of an exchangeable sequence of events in 1928, and proved a law of large numbers for them. The difference compared to the case of independence was that there is no unique limit of relative frequency. One has instead the kind of mutual convergence, as in Khintchine's formula (8.9). Khintchine showed in 1932 the law of large numbers for the general stationary case and noted that de Finetti's limit theorem is a special case since exchangeability is a special case of stationarity. De Finetti had obtained his representation theorem (8.7) in 1928, and it can be put as saying that exchangeable

[26] See the French monograph de Finetti (1939), pp. 34–37. There is a shorter Italian version de Finetti (1938c), pp. 21–23.

[27] Birkhoff got his paper published in 1931 already, even though it was a later work and the journal was the same as von Neumann's. These developments are detailed in Sec. 3.2.

probabilities have a unique decomposition into independent parts. Here is a diagram of the situation.[28]

| classical ergodic thm | *Kh-32* | probabilistic ergodic thm | *Kh-32* | law of large no.s for exchg. |
|---|---|---|---|---|
| *vN-32* | | ? | | *deF-28* |
| classical erg. dec. | ? | erg. dec. for stationarity | ? | de Finetti representation |

Three 'missing links' are needed to make the diagram complete, though if Khintchine had gotten one of them, the rest would probably have followed. Why did he not see that there would have been a straightforward step from classical ergodic decomposition to a genuine probabilistic one? On the other hand, how come did de Finetti not ever realize the connection between that decomposition and his own representation theorem?

A probabilistic ergodic decomposition seems to surface toward the end of the 1940s. That de Finetti's theorem is a special case was realized at least by Ryll-Nardzewski (1957), although, since the formulation is very abstract, it is not immediately noticeable. A general approach to these themes can be found in Dynkin's 1978 'Sufficient statistics and extreme points'.

## 8.3 STOCHASTIC PROCESSES: THE RENUNCIATION OF DETERMINISM

Old-fashioned subjective probability is based on determinism. The world evolves according to its unchanging laws, and probability arises merely from the lack of precise knowledge. The task of the probabilist is to formulate a set of alternatives that are symmetric relative to his ignorance. That warrants dividing probability uniformly over the alternatives, usually finite in number.

It seems a bit surprising that de Finetti in no way shares the above notion of epistemic probability. On the contrary, right from the start he clearly takes side with the indeterminism inspired by the recent quantum mechanics. More generally speaking, it was of course necessary to free subjective probability from the determinist doctrine, if it were ever to become scientifically respectable. De Finetti's earliest works in this direction are the 'Le leggi differenziali e la rinunzia al determinismo' (Differential laws and the renunciation of determinism, 1931a) and the

[28] The abbreviated names and dates can be read as 'established the proper connection between...'

remarks at the beginning and the end of his philosophical manifesto, *Probabilismo*.

The work 'renouncing' determinism starts with the contrasting of two positions: The first says: 'there are necessary and immutable laws; phenomena in nature are determined from their antecedents with precision and absolute certainty.' The second says: 'there are no proper laws: our previsions cannot be certain, only more or less (or even immensely) truthlike or *probable*; so-called laws of nature are only expressions of statistical regularities' (1931a, p. 64). While finding that the debate between the two is still open, and to be decided by future development of science, there is no doubt what de Finetti expects of the outcome. Once causality is being dismissed with, there enters the question of how a science is possible without it. The alternative de Finetti offers in these quotations, namely, statistical regularities, naturally leads to the question, regularities of what. De Finetti's subjectivism would not permit ascribing statistical laws directly to nature.

De Finetti saw very clearly what his denial of the reality of statistical laws amounts to philosophically. He wrote some years later that there seems to exist an 'insurmountable abyss' between the two types of laws of physics today, the deterministic and the statistical (1937, p. 64; 1964, pp. 154–55):

> To overcome this abyss, the point of view adopted so far leads quite naturally to a solution. It is exactly the opposite of what one usually envisages. Instead of extending the realistic character of classical laws to probabilistic laws, one can try to make these classical laws share the subjective character of statistical laws.

He added that from an operational point of view, there is no sense in demanding laws to exist, or to be verified for the reason that they are true. In boldness, de Finetti can here be compared with Pierre Duhem's well-known interpretation of Galileo. He did not, says Duhem, show that a realistically interpreted physics, dealing with terrestrial matters, in fact extends beyond the 'sublunar sphere.' Instead, what is shown is that the instrumentalistic interpretation, in antiquity and medieval times often given to astronomical theories, should be extended to earthly physics also.

Let us now turn back to the possibility of a science not based on causality. De Finetti is careful enough to make a conditional statement (1931a, p. 64): 'If we refute determinism, we have to wholly accept the second position.... Previsions are no more certain, but only more or less probable.' Then he continues by arguing that even if determinism were not refuted, there still would exist a probabilistic science. Such a science is in fact independent of the principle of causality, for that principle 'becomes useless in practice anyway,' he contends. But de Finetti is not happy with the existing foundation of the emerging probabilistic science (p. 65):

Another doubt may remain. The very principles of the calculus of probability and the statistical laws deduced from them are usually explained and justified in a way which, to say the least, echoes the determinist convictions of the one stating them. Is it possible to establish them and to justify them while remaining within the new conception?

That question was answered in a long paper he published the same year.[29]

The philosophical remarks of the paper on the renunciation of determinism are preliminary to a very interesting topic to which we now turn our attention. Several of de Finetti's early publications from 1929 on were devoted to the creation of a theory of a particular kind of continuous time random processes. As I have indicated repeatedly, the mathematical theory of such processes seems like a remarkable case of 'simultaneous discovery.' De Finetti, for his part, invented what are called *processes with independent increments.* A particular case of such processes is offered by Brownian motion, the theory of which had been developed by Einstein and other physicists, starting in 1905, and by Wiener in the 1920s. Bachelier had made a first attempt in 1900 at a theory, for price fluctuations in the stock market, that has the same mathematical substance. De Finetti gives no references and his work seems a completely independent development. The results he was able to show established his name as a probability mathematician in the early 1930s.[30] In his papers de Finetti was approaching a certain characterization of the class of processes he had defined, without actually achieving it as a general result. That characterization was instead obtained by Kolmogorov in 1932, as reported in the previous chapter: If the value of a random variable with independent increments varies continuously in time, it follows a normal law. If it varies discontinuously, it follows a Poisson law.

A possible origin for the idea of studying continuous time random processes was noted above already. In his studies of the diffusion of genetic characters, de Finetti had noticed the analogy between kinematics

[29] This paper de Finetti (1931c) is the central one in his 'new approach,' his theory of subjective probabilities, to be addressed in the next section. The long paper on exchangeability is not contained in a list of works in which the new approach is reported to be developed. See de Finetti (1931a), p. 65 note 1.

[30] See, for example, Khintchine's book of 1933.

and the time development of a statistical law. His central finding in 1929 was a new definition of the derivative of a random process. The problem was to characterize 'the instantaneous action of aleatory factors' (1929a, p. 165). Following de Finetti's presentation for a while, consider a random variable $X$ for continuous time. The task is to determine the probability law for $X$ for arbitrary times from its law for an initial time $\lambda_0 = 0$. There are three cases, analogous to the situation in physics[31]: 1. The law is known, 2. The law is given through a differential equation, 3. The law is given by an integral equation. Three types of situations for the probabilistic dependence are defined. First, the change $X(\lambda_0) - X(\lambda)$ from time $\lambda_0$ to $\lambda$ is independent of the changes on preceding intervals. Second, it depends only on the value $X(\lambda_0)$ at present. This constitutes the case of a *Markovian* process, as one now says. As in Kolmogorov (1931), the study of this kind of dependence is suggested directly from the physical analogy. The third kind of dependence would be one where the entire history enters. The first of these cases is the one of a process with independent random increments. If the increments depend only on the time difference $\lambda - \lambda_0$, one has a 'fixed' probability law (1929a, p. 164).

The difficulty in creating a mathematical theory of continuous time processes is that 'a variable $X$ subject to accidental variations' does not in general have a derivative. Even if the interval of time goes to zero, the variation of $X$ does not go to zero linearly in the argument, so there is no derivative in the ordinary sense. De Finetti shows as an example that if a variable $X$ is subject to 'accidental variations' following a normal law, no ordinary derivative exists (1929a, p. 168). That problem was resolved as follows: By the method of characteristic functions, he came to think of what are known as infinitely divisible distributions, really one of his inventions at the spot. These are probability laws that can be given, for any $n$, as a sum of $n$ independent increments. By setting for the time $\lambda = 1/n$, the limit as $n \to \infty$ yields the sought for definition of a derivative at time $\lambda = 0$: First let $X(\lambda) = 0$ for $\lambda = 0$, and denote the probability law (cumulative distribution) for $X(\lambda)$ by $\Phi_\lambda(\xi)$. Its characteristic function is

$$\psi_\lambda(t) = \int e^{it\xi}\, d\Phi_\lambda(\xi). \qquad (8.13)$$

The characteristic function of a sum of independent random variables is the product of the individual characteristic functions, which gives the expression $[\psi_\lambda(t)]^n$ for the characteristic function of the sum of $n$ independent increments. Setting $\lambda = 1/n$, de Finetti (1929a, p. 166) obtains

[31] De Finetti refers to Volterra's *Fractions des lignes* for the analogy, in his (1929a, p. 165).

a definition of the derivative of $\psi$ at time $\lambda = 0$,

$$\psi'(t) = \lim_{n \to \infty} [\psi_{1/n}(t)]^n. \tag{8.14}$$

By connecting $\psi'(t)$ to the derivative $\Phi'_\lambda(\xi)$, the general problem of determining the distribution $\Phi_\lambda(\xi)$ when $\Phi'_\lambda(\xi)$ is known was well posed.

One may intuitively motivate de Finetti's procedure for the new definition of a derivative as follows. The greater the number of independent increments, the more statistically stable their average effect is. In the limit fluctuations vanish, and the random impacts are, in a statistical equilibrium, producing exactly their expected effect. The effect can be given a differential condition. For de Finetti, a frequentist-style explanation of the above kind would have been problematic. But arguments of this type are commonly used in statistical mechanics.

At the end of the first short article on continuous time random processes the practical interest of the new approach is emphasized, 'now that one tends to see in physical laws only expressions of statistical regularity' (1929a, p. 168). Examples are discussed briefly in the next paper (1929b), where singularities are allowed in the distribution of $X$. De Finetti calls points of discontinuity of a distribution its 'exceptional values.' At an exceptional value $\xi$, the probability that $X$ is within $\varepsilon$ of $\xi$ does not approach 0 as $\varepsilon \to 0$. The main result of the paper, however, concerns the opposite case of processes with continuous trajectories or sample paths: If $X(\lambda)$ has stationary ('fixed' in de Finetti's terminology) increments and if it is continuous in time $\lambda$, its probability law $\Phi_\lambda(\xi)$ also is continuous in $\xi$. Laws with 'exceptional values' are discussed through examples. An infinity of such values obtains if $X(\lambda)$ takes only integer values, growing by 1 say, at each increment. Examples would be the number of births, or deaths, or goals in soccer (one of de Finetti's passions), and so on. If the law is stationary, there is a constant $p$ such that $pd\lambda$ gives the probability of a discontinuity during the interval $d\lambda$. For $n$ increments one obtains

$$p_n(\lambda) = \frac{p^n \lambda^n}{n!} e^{-p\lambda} \tag{8.15}$$

the Poisson formula, or 'law of rare events' (1929b, pp. 328–29).[32] A further, specific example of purely discontinuous changes is offered by the kinetic theory of gases (1929b, p. 329, note). The velocity of a molecule is a function of time, experiencing only discontinuous random

[32] Now at last the comment has to be made that radioactivity forms a chapter of its own in the prehistory of continuous time processes.

increments, at the collisions. The change, however, depends on the velocity of the molecule, through a differential condition. It corresponds to case 2. of de Finetti's classification above.

The third paper (1929c) of the series assumes $X(\lambda)$ continuous. Its integral exists therefore and is itself a random variable. An elementary example of a differential equation with a stochastic term is treated: the law of free fall. The Galilean (deterministic) solution for the equation of motion gives distance traveled as a function $f(\lambda)$ of time. With a reference to Poincaré, de Finetti says that the concept of a law of nature is probabilistic and approximative. Therefore the increase of $f(\lambda)$ from time $\lambda$ to $d\lambda$ is a random variable with a high probability of being close to the deterministic value $f(\lambda + d\lambda) - f(\lambda)$. It follows that if the successive random increments in the distance traveled are independent, there is no instantaneous velocity (as there is no ordinary derivative). If one tries the equation for velocity as a basis, it follows equally that there is no instantaneous acceleration. De Finetti says, an empirical quantity is not physically exactly determined, or at least it can never be known experimentally with exactness. His operationalism makes it meaningless to ask for instantaneous values (1929c, pp. 552–53).

A somewhat later paper de Finetti (1931d) concludes the series on continuous time random processes, a topic to which he was to return only in passing, in the note (1938b).

It was a philosophical principle of de Finetti's that unverifiable, infinitary events cannot have well-determined probability values. This same empiricist philosophy is emphasized in the paper on differential laws and the renunciation of determinism: The most general kind of experiment consists in the measurement of a random quantity $X(\lambda)$ at a finite number of instants $\lambda_1, \dots, \lambda_n$. A property the occurrence of which can be decided by such an experiment is *empirical* (1931a, p. 69). These are the only properties that have determinate probabilities; an example is the property that $X(\lambda) > 0$ at a given instant $\lambda \in (0, 1)$. A property is *semiempirical* if it is possible that its occurrence or failure to occur could be decided. Thus the claim that $X(\lambda) > 0$ at least at one instant $\lambda \in (0, 1)$ could turn out true, and the claim that $X(\lambda) > 0$ at every instant $\lambda \in (0, 1)$, false. If neither of these can be the case, such as for the claim that $X(\lambda) > 0$ an infinity of times, the property is *transcendent*. Since for a semiempirical property $P$ there are empirical properties $A$ and $B$ such that $P \to A$ or $B \to P$ and since the probability of the consequent in an implication cannot be less than that of the antecedent, the probabilities of semi-empirical properties can have bounds different from 0 and 1. But any passing beyond these bounds, unless one has $P \to A$ and $A \to P$ for $A$ empirical, is based on 'totally illusory methods' (1931a, p. 71). In his brief

1938 review of random functions, de Finetti repeats that from the 'finitist' point of view, a property such as the continuity of the trajectory of a random process has no determinate probability. One can only ask for the probabilities of events depending on a finite number of time coordinates.

## 8.4 FOUNDATIONS FOR THE THEORY OF PROBABILITY

So far we have discussed de Finetti's mathematical works on exchangeability and continuous time random processes. They can be taken as chapters in conventional mathematical probability. We have also encountered some of his philosophical and foundational ideas. Starting in 1930, de Finetti began to develop a new approach to the mathematical foundations of probability, being strictly guided by his subjectivist philosophy. The first indications of this theory are two short papers (1930d, f). A fairly extensive exposition is 'Sul significato soggettivo della probabilità' (On the subjective significance of probability 1931c). This work never seems to have received much attention, probably partly for the language, partly for its style and its unorthodox ideas. It does not look much like a mathematical paper at all, and it is not so immediate to see that the foundations for a remarkable new mathematical approach to probability were being laid down there.[33]

De Finetti's theory of probability can be divided into two parts. The first is a *quantitative* one. Its task is 'to measure subjective probability, that is, to transform our degree of uncertainty into a determination of a number' (1931c, p. 302). The most convenient way is to assume a banker or bookie who has to accept bets at a betting ratio he is free to choose. In the second part of de Finetti's theory, the somewhat artificial and extraneous betting approach, as he describes it, is replaced by an *axiomatic theory of qualitative probability*.

### *8.4(a) Quantitative probability as based on the notion of coherent bets*

Every probability problem has a *logical* and an *empirical part* (1931c, p. 299). In the logical part, the task is to determine all the *admissible*

[33] Why it was published in the Polish *Fundamenta Mathematicae*, a journal devoted to set theory and topology, I do not know. Previously it was announced under the title 'Fondamenti per la teoria della probabilità' (1930d, p. 259), whence the title of this section. Elsewhere de Finetti (1930f, p. 370) refers to what appears to be this paper, as forthcoming in the *Giornale dell'Istituto Italiano degli Attuari*.

probability assignments, relative to a class of events $\mathscr{E}$. Second, assuming that the probabilities of the events in $\mathscr{E}$ are given, the task is to determine the most general class of events whose probabilities themselves are determined by the probability assignments in $\mathscr{E}$. This determination can be strict, or it can consist of lower or upper bounds. The admissible probability assignments are characterized by the requirement of *coherence*: One must not accept betting schemes that lead to a sure loss.

The choice of a particular assignment belongs to the empirical part of a probability problem. The objectivist thinks that in any problem there is only one correct assignment. De Finetti meets their criticisms in advance, so to speak: If one is able to show a result that holds for all admissible probability assignments, it would hold particularly for the one the objectivist thinks is the only correct one (p. 299).

De Finetti's theory of probability contains a primitive notion of *events*. These are identified with logical entities, categorical propositions capable of being verified, but of which we do not yet know whether they are true or false.[34] Given two events $E'$ and $E''$, their *logical sum* is the event $E' + E''$, which is true when $E'$ or $E''$ is true, and false otherwise. The *logical product* $E'E''$ is true when both $E'$ and $E''$ are true, and false otherwise. If $E'E''$ is impossible, $E'$ and $E''$ are *incompatible*. The *negation* $\sim E$ is true exactly when $E$ is false. Two events are *equal*, $E' = E''$, if the verification of $E'$ is equivalent to the verification of $E''$. Finally, $E'$ *implies* $E''$ is defined as $(E' \rightarrow E'') = \sim(E' \sim E'') = \sim E' + E''$. An event $E$ is *logically dependent* on $E_1, \ldots, E_n$ if its truth or falsity is logically determined from that of the $E_i$. In (1930f, p. 370) a random quantity $\lambda$ is defined through $\lambda = 1$ if $E$ is true, $\lambda = 0$ if $E$ is false. Later de Finetti used events directly as 0–1-valued random quantities, writing the indicator functions as $|E|$ (1937–8, p. 25). I shall follow this practice.

The probability assignment $P(E) = p$ is explained as follows. The bookie assumed above has the freedom to choose $p$, but he must accept bets of any sum $S$ for the event $E$. A *bet relative to* $P(E)$ is any random quantity

$$G = (|E| - p)S \tag{8.16}$$

where the bettor is free to propose the stake $S$. Thus if $E$ occurs, we may write $G(E) = (1 - p)S = S - pS$; if not, $G(\sim E) = -pS$. The quantity $G$ is called the *gain* of the bettor. The stakes $S$ can be chosen positive or

[34] In (1930f, pp. 368–9) de Finetti states that the ordinary way of considering events as *sets* of possible elementary events is 'devoid of any intrinsic significance.' It leads to studying probability with concepts of measure theory, this being only 'an illusory external analogy.'

negative. This just reverses $E$ and $\sim E$, so one can in this way choose to bet for or against $E$. For $n$ events $E_1, \ldots, E_n$ and their probabilities $p_1, \ldots, p_n$ the gain is the linear combination

$$G = (|E_1| - p_1)S_1 + \cdots + (|E_n| - p_n)S_n. \tag{8.17}$$

A probability assignment $P(E_1) = p_1, \ldots, P(E_n) = p_n$ is defined to be *coherent* if the gain $G$ cannot be made positive in every case possible, regardless of how the adversary of the bookie chooses the sums $S_1, \ldots, S_n$ (1931c, p. 308). Here are some of the basic consequences of coherence:

1. For one event $E$, the probability $P(E)$ has to be *a unique real number* with

I. $0 \leqslant P(E) \leqslant 1$.

Else one is incoherent. If the bookie accepted first a bet $(|E| - p')S'$ for $E$ with the probability $p'$ and then another bet $(|E| - p'')S''$ with $p''$, the gain would be

$$G = (|E| - p')S' + (|E| - p'')S''. \tag{8.18}$$

Then

$$\begin{aligned} G(E) &= (1 - p')S' + (1 - p'')S'' \\ G(\sim E) &= -p'S' - p''S''. \end{aligned} \tag{8.19}$$

If $p' \neq p''$, there are choices of $S'$ and $S''$ which make these values of $G$ positive under both $E$ and $\sim E$. This is just an elementary result about solutions to linear equations (1931c, p. 309).

2. If $E$ is certain, or if $E$ is impossible, respectively, one has to have

II. $P(E) = 1$ for $E$ certain.
$P(E) = 0$ for $E$ impossible.

In the first case of $E$ certain, $(|E| - p) = 1 - p$, so if $p \neq 1$ there is a choice of $S$ such that $G = (1 - p)S > 0$. On the other hand, from $p = 1$ it follows that $G = 0$ so that coherence is fulfilled. Similar reasonings show that the impossible event has to have probability 0, and that values outside the extremes 0 and 1 are incoherent.

3. The essential property of probabilities is *additivity*: If $E_1, \ldots, E_n$ form a complete set of incompatible events (meaning exactly one of the $E_i$ has to be true),

$$P(E_1 + \cdots + E_n) = P(E_1) + \cdots + P(E_n). \tag{8.20}$$

Let us see why this holds, assuming for simplicity that there are just two events. Let $P(E_1) = p_1$ and $P(E_2) = p_2$. Since $E_1 + E_2$ is certain by the assumption of a complete set, $P(E_1 + E_2) = 1$. We first show that

$p_1 + p_2 = 1$ is necessary and sufficient for coherence. Now

$$G = (|E_1| - p_1)S_1 + (|E_2| - p_2)S_2. \tag{8.21}$$

Then

$$G(E_1) = (1 - p_1)S_1 - p_2S_2$$
$$G(E_2) = -p_1S_1 + (1 - p_2)S_2. \tag{8.22}$$

For any values of $G(E_1)$ and $G(E_2)$ there are solutions in $S_1$ and $S_2$ to these two linear equations, except when $p_1 + p_2 = 1$. Then, if $p_1 + p_2 \neq 1$, the gains could be made always positive. It follows that coherence is equivalent to $P(E_1 + E_2) = P(E_1) + P(E_2)$ for $E_1$, $E_2$ complete and incompatible. A simple consequence is that $P(\sim E) = 1 - P(E)$. Additivity for a complete finite class of $n$ incompatible events follows by generalization of the above argument to $n$ unknowns (cf. 1931c, p. 310). Next, assuming $E_1$ and $E_2$ are any two incompatible events, the sum $E_1 + E_2 + \sim(E_1 + E_2)$ is complete and $E_1 + E_2$ is incompatible with its negation. Then $1 = P(E_1 + E_2 + \sim(E_1 + E_2)) = P(E_1) + P(E_2) + P(\sim(E_1 + E_2)) = P(E_1) + P(E_2) + 1 - P(E_1 + E_2)$, so that

III. $P(E_1 + E_2) = P(E_1) + P(E_2)$

for arbitrary incompatible events.

The requirement of coherence of a probability assessment is summarized in the preceding postulates I–III which can be taken as an axiomatization of probability (1931c, pp. 312–13).

Given $n$ events $E_1, \ldots, E_n$, let $X = x_0 + x_1|E_1| + \cdots + x_n|E_n|$. Then the 'mathematical expectation' of $X$ is

$$M(X) = x_0 + x_1P(E_1) + \cdots + x_nP(E_n). \tag{8.23}$$

Expectation is a linear operator: $M(X + Y) = M(X) + M(Y)$. Putting $X = |E|$, we get the result $P(E) = M(|E|)$. This relation 'expresses the interpretation of probability as a betting ratio' (1937–8, p. 32). How is this? There is an intuitive notion of expectation behind the quantitative mathematical one, and at the heart of the former lies the principle de Finetti once formulated as follows: 'To be able to enjoy a certain advantage $V$, whatever its nature, in case the event $E$ verifies, we are disposed to sustain the more grave sacrifices the more probable $E$ appears' (1931c, p. 324). A more down-to-earth wording would be that we are prepared to pay the more for a bet the higher the corresponding probability is: Let a bet $G = (|E| - p)$ be proposed where the stake is a unit sum. The betting ratio determined by $p$ is $p{:}1 - p$. The gain $G$ is a random variable, and by taking expectations we get $M(G) = M(|E| - p) = M(|E|) - p = P(E) - p$. A *fair* bet has zero expectation of gain, so that we get $P(E) - p = 0$, or $P(E) = p$, or probability interpreted by a betting ratio. De Finetti wants to keep separate the judgment that a bet is fair and the actual acceptance of a bet, for one may accept unfair bets in single cases, through the desire of risk taking, for example (1931c, pp. 303–304). We could, of course, say that the 'advantages' are not the same as the direct monetary values, but some other quantities, reflecting the desirabilities of the different end result. That move could bring us back the equality $M(G) = 0$. De Finetti though, in keeping separate considerations of fairness and the actual behavior taken, is obviously indirectly suggesting that considerations of utility should not be mixed with the problem of assessing probability.

De Finetti invented in his (1930f) a geometric representation of probabilities. Let $\mathscr{E}$ contain the events $E_1, \ldots, E_n$. The $n$-fold logical products of the $E_i$ or their negations are called *constituents*.[35] Say, for $n = 3$, we have the $2^3$ constituents $E_1E_2E_3$, $\sim E_1E_2E_3, \ldots, \sim E_1 \sim E_2 \sim E_3$. Any two constituents are incompatible, and all the constituents together form a complete class of events. Let the constituents of $\mathscr{E}$ be $C_1, \ldots, C_s$. An arbitrary logically dependent event $E$ is then a logical sum of the form $C_{h_1} + \cdots + C_{h_k}$ of $k \leqslant s$ constituents. If $P(E) = p$ and $P(C_{h_i}) = c_{h_i}$,

$$p = c_{h_1} + \cdots + c_{h_k}. \tag{8.24}$$

For each of the $E_i$ there is a linear equation of this same form. Total probability gives one more equation, $\sum c_i = 1$, so there are altogether $n + 1$ linear equations for the probabilities of the $E_i$. If the equation (8.23) is linearly dependent on these equations, one says $E$ is linearly dependent on $E_1, \ldots, E_n$. It is 'semidependent' if there is a constituent $C_i$ such that $E \rightarrow C_i$ or $C_i \rightarrow E$. This gives an upper or lower bound for the probability of $E$, through the general rule stating that if $A \rightarrow B$, $P(A) \leqslant P(B)$.

Here is another way of looking at the above. The event $E$ is linearly dependent on $E_1, \ldots, E_n$ if and only if its indicator function $|E|$ is a linear function of the indicators of the $E_i$,

$$|E| = x_0 + x_1|E_1| + \cdots + x_n|E_n|. \tag{8.25}$$

Specifically, if $E$ is logically dependent, with $E = C_{h_1} + \cdots + C_{h_k}$,

$$|E| = |C_{h_1}| + \cdots + |C_{h_k}|. \tag{8.26}$$

Taking expectations on both sides we obtain the familiar formula $P(E) = p = c_{h_1} + \cdots + c_{h_k}$.

With the help of constituents, the geometry of the set of coherent or admissible probability assignments has the following structure. With $s$ constituents, the probability assignments are points of an

[35] From Boole, through Medolaghi (1907).

$(s-1)$-dimensional simplex. To each of the $s$ extremal points one associates a constituent. Let $P_i$ be the probability assignment that gives probability 1 to the constituent $C_i$ (and 0 to the rest). The admissible probability assignments are linear combinations of these, with $c_i \geqslant 0$ for $1 \leqslant i \leqslant s$ and $\sum_i c_i = 1$:

$$P = c_1 P_1 + \cdots + c_s P_s. \tag{8.27}$$

If $P'$ and $P''$ are two admissible probabilities, their linear combinations $P = \lambda P' + \mu P''$, with $\lambda$, $\mu \geqslant 0$, $\lambda + \mu = 1$, also are admissible. Therefore the set of probabilities over a given class of events is *convex*. It is also *closed*: If $P_1, P_2, \ldots$ are probability laws for $\mathscr{E}$, if for every event $E$ the limit of $P_n(E)$ as $n \to \infty$ exists, that limit

$$P(E) = \lim_{n \to \infty} P_n(E) \tag{8.28}$$

also is an admissible probability assignment (1930f, p. 372; 1937, p. 12).

This geometric point of view may look more familiar if we think of the constituents as binary sequences of length $n$ (for $n$ events). Each such sequence is associated with one of the $2^n$ extreme points of a simplex in $2^n - 1$ dimensions. The probabilities of constituents become the $n$-dimensional probabilities $p_n(x_1, \ldots, x_n)$ with $x_i = 0$, 1, $1 \leqslant i \leqslant n$. Any probability assignment over the sequences is represented by a point of a $(2^n - 1)$-dimensional simplex. The representation $P = c_i P_1 + \cdots + c_s P_s$ means that a given point $P$ is a unique weighted average of the extreme points, with the $c_i$ as weights. On the other hand, any weighted average of the extreme points is, by convexity, a point of the simplex. The terminology from statics can be taken in quite a concrete sense. To any distribution of unit mass over the extreme points of a simplex there corresponds a unique point as barycenter, and the other way around. Only the two lowest dimensions can be visualized: If $n = 1$, we have $P(1) = p_1$ and $P(0) = p_2$ with $p_1 + p_2 = 1$. If as the simplex we take the unit interval $[0, 1]$ with the result 0 represented by 0 of $[0, 1]$, 1 by 1, the barycenter is simply $0 \cdot p_2 + 1 \cdot p_1 = p_1$, the success probability $p_1$. If $n = 2$, we have $P(1,1) = p_1$, $P(1,0) = p_2$, $P(0,1) = p_3$, and $P(0,0) = p_4$. Any probability over two simple events is given by a point $(p_1, p_2, p_3)$, with $p_4 = 1 - (p_1 + p_2 + p_3)$, and so on. It is rather interesting to study the set of all possible probability assignments through this geometric representation. For example, exchangeability for two simple events reduces the dimension by 1 since it gives the equation $P(1,0) = P(0,1)$. The admissible probabilities can in this case be given as points in the triangle whose vertices are $(1,0)$, $(0,0)$, and $(0,1)$.[36] De Finetti's geometric

[36] The simplex structure of exchangeable probabilities is a chapter of its own, for which see Dynkin (1978).

point of view reduces concern about the algebraic structure of the class of events to a minimum, giving instead a representation of the relations between the probability numbers themselves.

In his (1931c) de Finetti promised to give treatments of conditional probabilities and of frequencies soon. These can be found developed only in his (1937) and more fully in the Padua lectures (1937–8). An event $E'$ *conditional* on $E$, to be written as $E'|E$, is true if $E'E$ is true, false if $E$ is true and $E'$ false, and *indeterminate* if $E$ is false. Then $P(E'|E) = p$ is the probability one would choose for a bet for $E'$ in case $E$ is verified, but which bet is canceled if $E$ fails to be verified (1937, p. 13; 1937–8, p. 42). From coherence, one can show directly the *theorem of composite probabilities* (1937, p. 14; 1937–8, pp. 42–43):

$$P(EE') = P(E) \cdot P(E'|E). \tag{8.28}$$

Note that this does not require that $P(E) > 0$. If that inequality holds, one obtains the usual formula of conditional probability, $P(E'|E) = P(EE')/P(E)$. Note also that this is not a definition, but a theorem derived from the requirement of coherence. The proof goes as follows (1937–8, pp. 42–43): Let $P(E) = p$, $P(EE') = p'$, and $P(E'|E) = p''$. Let three bets be accepted, following these numbers, for arbitrary sums $S$, $S'$, and $S''$: 1. $(|E| - p)S$, 2. $(|EE'| - p')S'$, and 3. $|E|(|E'| - p'')S''$. If $\sim E$ is true, the first and second bets are lost, the third is canceled. If $E \sim E'$ is true, the first bet is won, the second and third are lost. If $EE'$ is true, all three are won. Accepting all three bets, the gain is

$$G = (|E| - p)S + (|EE'| - p')S' + |E|(|E'| - p'')S''. \tag{8.30}$$

We get for the values of gains

$$\begin{aligned} G(\sim E) &= -pS - p'S' \qquad (8.31)\\ G(E \sim E') &= -pS - p'S' - p''S'' + S \\ G(EE') &= -pS - p'S' - p''S'' + S + S' + S''. \end{aligned}$$

Choosing $S = p''S''$ and $S' = -S''$, as one may since the stakes are arbitrary, the gain will reduce in all three cases to $(p' - pp'')S''$. Unless $p' - pp'' = 0$, this gain can be made always positive. Therefore coherence requires that $P(EE') = P(E) \cdot P(E'|E)$. If $P(E) > 0$, the condition $p' - pp'' = 0$ is also sufficient (1937–8, p. 43). Another very simple argument for the formula of composite probabilities is as follows. A unit bet relative to $P(E) = p$ is given by $G = |E| - p$. Taking expectations, we get $M(G) = M(E) - p = p - p = 0$, in other words, a fair bet. Now, starting from the definition of the event $E'|E$, and putting $P(E'|E) = p$, the gain is

$$G = |E|(|E'| - p). \tag{8.32}$$

From the requirement of fairness we get

$$M(G) = M(|E||E'| - |E|p) = P(EE') - P(E)\cdot p = 0, \quad \text{so } P(EE') = P(E)\cdot p \tag{8.33}$$

These are de Finetti's two arguments for the rule of Bayesian conditioning.[37]

Let us proceed to the second of the topics de Finetti (1931c) promised to handle soon, namely, frequency: The number of successes in $n$ events $E_1, \ldots, E_n$ is the random quantity

$$X_n = |E_1| + \cdots + |E_n|. \tag{8.34}$$

One gets, by linearity of expectation,

$$M\left(\frac{X_n}{n}\right) = \frac{1}{n}\sum M(E_i) = \frac{1}{n}\sum P(E_i) \tag{8.35}$$

Specifically, if $P(E_i)$ is a constant $p$ for every $E_i$, $M(X_n/n) = p$, that is to say, *probability equals the expectation of frequency* (1937–8, p. 37). Exchangeability would of course form a most interesting and tractable special case to which the preceding dictum applies. De Finetti suggests that it can also be used to explain why probabilities so often are assessed on the basis of actually observed frequencies (1937–8, pp. 36–37).

De Finetti refers in his essay (1931c, p. 302) to a passage in Bertrand's *Calcul des probabilités* that contains some subjectivistic elements. There it is said that the probability of an event for a person is $p$ if that person were prepared to exchange the consequences attached to the occurrence of the event to identical consequences attached to a drawing from an urn with a probability equal to $p$ (Bertrand 1888, p. 27). If I make the 'probability evaluations' 9/10 for the doctor coming upon being called and 1/3 for a successful treatment, the probability of cure through treatment is, '*for me*' as Bertrand emphasizes, $9/10 \times 1/3 = 3/10$. He says that one's probability evaluations must follow the rules of total and composite probability, but nothing more is made of these ideas. De Finetti wrote later that Bertrand introduced subjective probabilities 'for the sole purpose of opposing them with other "objective probabilities"' (1937, p. 6, note).

There are some clear resemblances between de Finetti's theory of

[37] Recent discussions on Bayesian conditioning have centered on what is called temporal coherence. If the events $E$ and $E'$ refer to different times, would a Bayesian still have to compute his posterior probabilities according to the rule of conditioning, or would other ways of revising one's probability assessment be allowed? For a review, see Diaconis and Zabell (1982).

subjective probability and Emile Borel's ideas from 1924, explained in Section 2.2(b). De Finetti commented these ideas in very positive terms, in his essay on Borel.[38] Borel had also been trying to formulate a theory of games in the later 1920s, a theory that subsequently proved a very natural field of application for subjective probability. Borel's 1924 paper was written in the form of reflections on Keynes' *Treatise on Probability* that had appeared in 1921. Keynes' main idea was that probability is a *logical* relation between a hypothesis $h$ and evidence $e$. The idea has led to the study of 'inductive logics.' In these one tries to capture principles of logical symmetry for a language in which $h$ and $e$ are expressed, in order to obtain a unique 'confirmational probability' $c(h|e)$ for $h$ given $e$. The influence of inductive logics on probability mathematics has been marginal at most.

Keynes' book and person instead had an influence on Frank Ramsey, who in 1926 gave a talk on 'Truth and probability,' published posthumously in Ramsey (1931). Subjective probabilities are *degrees of belief* rendered operational through a measurement procedure (1931, p. 166ff.): 'The old-established way of measuring a person's belief is to propose a bet, and to see what are the lowest odds which he will accept' (p. 172). But this is a very restricted idea. Ramsey gives instead an axiomatization of degrees of belief that establishes a preference ordering between 'worlds' denoted by $\alpha, \beta, \tau, \ldots$. Equivalence classes of equally preferred worlds are 'values.' Ramsey's axioms lead to representing values by real numbers, denoted by the same symbols as the worlds. A person's degree of belief in the truth of a proposition $p$ can be defined as follows. Assume he is indifferent between the alternatives of having $\alpha$ for certain, or $\beta$ if $p$ is true and $\tau$ if $p$ is false. Then degree of belief in $p$ is defined as $(\alpha - \tau)/(\beta - \tau)$. Behind this definition we find the principle of mathematical expectation: $\alpha = \text{prob}(p)\beta + (1 - \text{prob}(p))\tau$. Ramsey says his definition 'amounts roughly to defining the degree of belief in $p$ by the odds at which the subject would bet on $p$' (pp. 179–80). He goes on to outline how to derive the laws of probability from his axioms. Further on he states that if someone's preferences violated the axioms, a book could be made against him (p. 182).

The result that coherence is, for a field of events, equivalent to the rules of finitely additive probability, is sometimes called the Ramsey–de Finetti theorem. Whereas de Finetti very definitively proved this result, known also under the later name of 'Dutch book theorem,' Ramsey's notion of 'consistency' relates in the first place to his axioms for

[38] This is de Finetti (1939a). I have not found evidence of early influence of Borel on de Finetti.

preferences. He does not show in any detail how a book could be made if one were inconsistent in his sense. That is understandable also, since Ramsey's paper was not at all technical in nature, but rather introduced new ideas.[39] He only gave it as a talk; the publication was from a posthumous manuscript. He said that working out the details would be 'rather like working out to seven places of decimals a result only valid for two' (p. 180). Ramsey's work does not seem to have had much influence before the 1950s. Today it is seen as the first axiomatization of Bayesian decision theory.

### *8.4(b) The axiomatization of qualitative probability*

Let us now turn back to de Finetti. The approach to subjective probability through betting is in de Finetti's words somewhat 'artificial and extraneous.' Its advantage is that it permits a straightforward operationalization of subjective probability. De Finetti's more fundamental theory in his 'Sul significato soggettivo' (1931c) is based on purely qualitative notions, these being closer to the 'everyday sense' of probability he thought is the primary one. A first motivation for a qualitative theory of probability is as follows. The *classical definition* of probability is that it is the number of favorable cases divided by the number of all possible cases. These cases can always be given as results of drawings of balls from urns. In subjectivist terms, the classical definition requires that we judge equally probable the drawing of any ball from an urn. With $n$ balls the probability of a given ball to be drawn is $1/n$. If we put $m$ white and $n-m$ red balls in the urn, the probability of drawing a white one is $m/n$. The subjective probability $P(E)$ of an arbitrary event $E$ can be determined if we are able to compare $P(E)$ to the probabilities $m/n$ in drawings from urns, for any choices of $m, n$. For this means that $P(E)$ can be compared to any rational number, a procedure that defines a unique real number (1931c, p. 315).

The *axiomatic theory of qualitative probability* is based on a primitive concept $\succeq$, read as 'is not less probable than' (or put it as 'at least as probable as' if you want a formulation in positive terms). As before, there is a class of events $\mathscr{E}$ closed with respect to logical sums, products, and negations. There are four axioms for the qualitative probability ordering (1931c, pp. 321–22):

I. $\succeq$ is connected: for any two events $E'$ and $E''$, $E' \succeq E''$ or $E'' \succeq E'$.

If $E' \succeq E''$ and $E'' \succeq E'$, $E'$ and $E''$ are defined as *identically probable*,

[39] A comparison of the philosophical views of Ramsey and de Finetti is given in Galavotti (1991).

$E' \cong E''$. If $E' \succeq E''$ but not $E'' \succeq E'$, $E'$ is more probable than $E''$, $E' \succ E''$.

II. If $A$ is certain and $B$ impossible and if $E$ is neither, $A \succ E \succ B$.

III. $\succeq$ is transitive: $E' \succeq E$ and $E \succeq E''$ imply $E' \succeq E''$.

IV. If $E_1$ and $E_2$ are both incompatible with $E$, $E + E_1 \succeq E + E_2$ if and only if $E_1 \succeq E_2$. Specifically, if $E_1 \cong E_2$, $E + E_1 \cong E + E_2$.

The first three postulates are 'as banal as not to merit discussion' (1931c, p. 322). The fourth postulate is the essential one. An event $E_1$ can be substituted in a logical sum with another identically probable one if logical relations are not altered. The probability of a logical sum of incompatible events 'is a "function" of the probabilities of the events themselves,' as de Finetti says (1931c, p. 322).

Axioms I to IV lead to a conventional quantitative probability $P$ through comparison with the probabilities of drawings from urns. The matter is somewhat tricky because of axiom II. For observe that any possible event is qualitatively strictly more probable than an impossible event $B$. If one has a nonimpossible event $E$ of zero probability, $P(E) = P(B)$. Consequently $P(B) \geqslant P(E)$, but still $\sim(B \succeq E)$ since $E \succ B$. The qualitative probability ordering is not simply representable through equating $P(E) \geqslant P(E')$ with $E \succeq E'$. Therefore de Finetti's qualitative notion of *identical* probability, $E \cong E'$, is not the same as *equal* probability, $P(E) = P(E')$. This is a bit problematic since the qualitative probability ordering is supposed connected. Axiom I requires that we can decide whether $E \succeq E'$ or $E' \succeq E$, but we cannot put simply $P(E) \geqslant P(E')$ or $P(E) \leqslant P(E')$, even if $E'$ is drawing a white ball from an urn with a probability $P(E') = m/n$. Instead de Finetti proceeds as follows: If $E \cong E'$ and $P(E') = m/n$ as above, set $P(E) = m/n$. If $E$ is an arbitrary event and if there is a real number $x$ such that for any event $E'$ with a *rational* probability $P(E') = y$, $x > y$ implies $E \succ E'$ and $y > x$ implies $E' \succ E$, the probability of $E$ is defined as $P(E) = x$. De Finetti's axiom II renders the structure of the probability ordering non-Archimedean, as he observes (1931c, p. 324). A probability is 'infinitely small' if it is less than the probability of an event from a complete class of $n$ incompatible identically probable events, for any $n$ (ibid.).

The axioms for qualitative probability are reproduced in *La Prévision* (pp. 4–5). A fifth axiom is added for handling qualitative conditional probabilities of conditional events $E'|E$ as defined above (p. 14):

V: If $E' \rightarrow E$ and $E'' \rightarrow E$, $E' \succeq E''$ if and only if $E'|E \succeq E''|E$.

This axiom also appears in the 1937–8 lectures (p. 48).

Serge Bernstein's long paper on the axiomatic foundations of the calculus of probability, published in 1917 in the proceedings of the mathematical society of Kharkov, Ukraine, antedates de Finetti's qualitative approach by more than ten years. Bernstein's axioms for qualitative probability contain the following, with $\Omega$ certain and $O$ impossible events (1917, p. 227):

I. If $A \neq \Omega, \Omega \succ A$. It follows that if $A \neq O, A \succ O$.
II. If $A \cong A_1$, $B \cong B_1$, and $(A \text{ and } B) = (A_1 \text{ and } B_1) = O$, it follows that $(A \text{ or } B) \cong (A_1 \text{ or } B_1)$.

Hardly anyone outside Russia and the Soviet Union can be expected to have read the article. Kolmogorov in his *Grundbegriffe* mentions it as giving an axiomatization with a set of basic concepts different from his. The axioms are of course quite similar to those of de Finetti, but whether Bernstein's idea of qualitative probability influenced de Finetti, remains undecided.

The main task of de Finetti's theory of qualitative probability was to show the connection between the 'everyday sense' and the mathematical sense of probability. The former has a value 'absolutely superior' to the latter (1931c, p. 301). To meet the consequences of such hard choices, de Finetti was prepared to introduce infinitesimal probabilities. Although he repeated the suggestion for creating a theory of zero probabilities (see, for example, his 1936a), nothing definitive came out of it. Non-standard probability was only invented in the 1970s.[40] More important for de Finetti's probability mathematics proved his constant denial of the necessity of denumerable additivity, on the basis that no justification for it has been offered from the 'everyday sense' of probability as codified in the principle of coherence and its consequences. I shall discuss that matter in the final section.

The paper on the subjective significance of probability (1931c) is de Finetti's main work on his foundational system. It took a long time for his ideas to be well noticed. A proof of de Finetti's failure to communicate his theory is its omission from the bibliography on foundations of probability contained in Kolmogorov's 1933 book. In 1935 Fréchet had invited de Finetti to give a series of lectures at the Institut Henri Poincaré. These lectures, published in 1937, made de Finetti's ideas more accessible.[41] A sympathetic attitude toward de Finetti's ideas on subjective probability is found in Polya's article (1941) on 'plausible

[40] See Nelson (1987) for an introduction.

[41] Though it must be noted that the *Annales de l'Institut Henri Poincaré* are not very easy to find.

reasoning,' and in Ville's monograph (1939). De Finetti's foundational system only started becoming better known in the 1950s, especially through the influence of Savage's formulation of Bayesian statistics in 1954. The emergence of game theory and decision theory provided a natural context for de Finetti's subjective probabilities.

De Finetti had laid the foundations of a general theory of subjective probabilities. One of his particular motivations had been to meet the challenge provided by indeterminism. It seems fair to say that despite his physical motivations, he did not really carry through the program of subjective probabilities in physics, say, with adequate motivations for interpreting subjectively probability in statistical mechanics and quantum theory.

## 8.5 THE PROBLEM OF DENUMERABLE ADDITIVITY

In this book modern probability has been defined as the kind of probability theory in which infinity plays an essential role. Its ways of handling the infinite modern probability borrowed mainly from measure theory. The most typical kind of infinitary event in probability theory derives from the limiting behavior of a denumerable sequence of random quantities. The first mathematical result here was Borel's strong law of large numbers of 1909, stating that the limit of the relative frequency of 1's, in the binary expansion of an arbitrary real number $x \in [0, 1]$, has with probability 1 the value 1/2. The sense of 'probability 1' in the theorem is that exceptions have measure 0.[42] Such a 'strong law' is, also typically, based on the assumption of denumerable additivity. What kind of stand do the three main approaches to probability, the frequentist, the measure theoretic, and the subjectivist, take with respect to denumerable additivity? In the second approach it is usually accepted without question, as part of measure theoretic probability. It is seen as a 'mathematical convenience.' For example, if we only assume finite additivity, we can usually derive only upper and lower bounds for the limit of relative frequency. Denumerable additivity implies a strong law of large numbers in which these bounds coincide, and that may be a remarkable mathematical convenience. But how should one legitimate the infinitary computation rule for probabilities that is used in deriving a unique bound? That is, put in a somewhat more general formulation, what does a probability of a strictly infinitary event mean?

[42] Borel had his reservations about this 'geometric point of view' and preferred a reading in terms of his theory of 'denumerable probabilities.' Cf. Sec. 2.2(c).

Kolmogorov's reason for assuming denumerable additivity was that it would be a harmless idealization: If the probability $P(E)$ of a finitary event $E$ can be derived with the help of probabilities of infinitary events, involving computations by the rule of denumerable additivity, there would also be a derivation without them. Following Hilbert, Kolmogorov thought the infinite is an extension of properly finitist mathematics, such that the above property of derivations of results with a finitary meaning holds. Infinite events would work as 'ideal elements' in probability theory, and their use would not extend the class of provable truths about the finite. That, of course, turned out to be a mistaken idea after Gödel showed that Hilbert's program for saving the consistency of mathematics cannot be carried through with finitary means.

Others have approached denumerable additivity directly. The interpretation of probability as a limit of relative frequency is easily seen to be sufficient for the properties of finitely additive probability. But instead of justifying also denumerable additivity, the frequentist interpretation contradicts it. The third main contemporary approach to probability, in addition to frequentist and measure theoretic probabilities, is best represented by de Finetti's subjectivism. He would not accept a crucial question such as denumerable additivity to be decided through a convention, as we saw in connection with his dispute with Fréchet. The answer, if any, should instead follow from the meaning of probability. De Finetti's contention was that the requirement of coherence captures exactly those principles of probability that are generally valid: Probabilities are unique real numbers between 0 and 1 that have to be added according to the rule of finite additivity. This was in 1930–1931. Much later he mentions in passing the following idea: Instead of requiring that the gain must not be positive in every possible case, one would require that it must not be positive or zero. A sure gain means a sure loss for the other party in a bet. The weaker condition would prohibit betting if one could only lose or turn out even.[43] Why should one ever accept such a situation? De Finetti notes in his extensive 1949 study of the axiomatization of probability that requiring merely nonnegative gains would mean that all possible events must have positive probability—a circumstance that would seriously restrict the scope of probability (1949, Sec. 20). In conclusion, no generally applicable way of extending coherence has been proposed. Coherence does not require denumerable additivity, but neither does it contradict it. One has to look for other kinds of arguments.

De Finetti's first concerns about denumerable additivity derive from

[43] This suggestion, known today as strict coherence, was first studied by Shimony (1955).

examples in his 1928 paper 'On denumerable and geometric probabilities', indicating the possibility of a denumerable class of equiprobable events. An example would be the 'random choice' of a natural number. Such a choice cannot be represented in a system of probability including the principle of denumerable additivity. Still, it seems sensible, at least in judgments such as: The probability that a randomly chosen integer is pair, is one-half. In (1930g) de Finetti introduced the related *paradox of nonconglomerability*: Let $N$ and $M$ be natural numbers 'chosen at random independently of each other' (p. 6). For a given value $n$, the probability of $M \leqslant n$ is 0, that is, the probability of the inequality $M \leqslant N$ conditional on any of the possible hypotheses $N = 1, N = 2, N = 3, \ldots$ is 0. Therefore the probability of $M \leqslant N$ is 0. A similar reasoning shows that the probability of $N \leqslant M$ also is 0, which is impossible. There is a hidden assumption behind the paradox: Let the hypotheses $H_1$, $H_2$, $H_3, \ldots$ be mutually exclusive and jointly exhaustive. The *conglomerative property* of conditional probabilities says that if $p' \leqslant P(E|H_i) \leqslant p''$ holds for each $H_i$, $p' \leqslant P(E) \leqslant p''$. But this assumption is a special case of denumerable additivity.

De Finetti seems to think that denumerable additivity would be an optional property. It would then be a question of the accuracy of our probabilistic vision to see whether denumerable additivity holds in a given case. Is de Finetti's attitude eclectic, or is there some principle, so far not found, that will go deeper into infinitary events and their probabilities, so as to decide in which cases measure theory's simple-minded extrapolation from the finite to the infinite is justified? At present it seems that the foundations of the topic remain as open as ever.

SUPPLEMENT

# *Nicole Oresme and the ergodicity of rotations*

Nicole Oresme lived from approximately 1325 to 1382. He was a philosopher, mathematician and churchman.[1] We shall here be interested in a very particular aspect of his work: the incommensurability of celestial motions. In many ways, though, it was at the center of his achievements. As background for the discussion of Oresme's mathematical results, let us review the elements of Ptolemaic astronomical models. Ptolemy, the greatest of the applied scientists of antiquity, in his astronomy assumed the Earth to be immobile, with the planets, the Sun and Moon orbiting around it in a motion consisting of several (up to three) uniform circular motions. There is a great circle, the *epicycle*, on which is attached the center of another circle, the *deferent*. On this circle, finally, is located the mobile object. Spatial coordinates are determined against the 'sphere of the fixed stars.' It of course rotates once a day around the Earth. Different combinations of sense and speed of rotation of the circles are able to account for phenomena such as the retrograde motion of a planet.

## 1 THE QUESTION OF THE PERIODICITY OF THE UNIVERSE

Starting with the Greeks, who are said to have invented the geometrical representation of the motions of celestial bodies, there has been a debate about the character of such geometric models. The crucial issue was, whether the models pertained directly to reality, or whether they were to be taken just as instruments for prediction. In this question, Oresme held a realist position, taking the pointlike accuracy of the geometric presentation as literally true of the universe. According to an Aristotelian doctrine that prevailed until the sixteenth century, the heavenly motions

[1] The Harvard *Source Book on Mediaeval Science*, edited by Edward Grant, gives a general orientation on the subject matter. It also includes excerpts of the original texts of Oresme that concern us here.

were thought 'immutable,' remaining the same for all eternity. Each celestial body was believed to have, with good justice if the geometric model is taken for real, a proper period after which it returns to exactly the same point relative to the fixed stars. Oresme's great problem was, how do the circulation times of the several bodies relate to each other. That is, is there a common integral measure of revolutions after which all the bodies will have returned to their previous positions? This is known as the question of the *commensurability versus incommensurability* of celestial motions. As a simplification, it is assumed throughout that the motions take place in the plane of the ecliptic.

In other terms Oresme's question is: do the motions of the starry heavens have an exact period, or is the motion aperiodic? There is a long philosophical tradition to this problem, going back to antiquity. In ancient times it was common to believe in exact periodicity, a belief following from the assumption that the times of circulation of the celestial bodies bear a relation to each other, expressible in (rational) numbers. According to the Stoic doctrine of 'eternal return,' the heavens determine earthly events as well, so that here, too, all things would repeat themselves.[2] In Platonic philosophy, the assumed least common period of motions in the universe is known as the Great Year. An idea of such a year can already be found in Babylonian astronomy. There it is based, however, on the use of purely arithmetic calculations of the least common multiple for all the known heavenly revolutions. Naturally, some common lower bound on the accuracy of measurement had to be adopted, although we do not know more about this than the numbers delivered down to us on the Babylonian clay tablets. Only with the geometric way of imagining the heavenly motions, and together with the early discovery of incommensurable magnitudes, does it become possible to ask whether the astronomical periods are commensurable or incommensurable. The first known person to have suggested the possibility of incommensurability was Theodosius of Bithynia (about 180 B.C.). In a tract known as *De diebus et noctibus*, he remarks that if the length of the year consists of 365 days and a rational fraction, the Sun's motion will be reproduced exactly after a definite period. But if the fraction is incommensurable to the length of the day, the solar events

[2] This doctrine was repeated in modern times by Friedrich Nietzsche. He tried to prove eternal return by assuming an atomistic, discrete and finite universe. It would somehow have to sample all of its finitely many possible states. On a more scientific level, eternal return reappeared in speculations about the foundations of statistical mechanics, as recurrence of the state of a stationary system. See Brush (1976, 1978) for illuminating discussions of these topics.

would not recur the same way.[3] At least two other passages on the incommensurability of celestial motions are known from antiquity.[4]

Observations are discrete and of finite accuracy by their very nature. This is reflected in the notation for astronomical measurements, our sexagesimal one stemming from the Babylonians. In antiquity and medieval times, a lower bound to the accuracy of measurement was set by the resolution of the bare eye. It is said that such a best possible bound was reached by Tycho Brahe. The question of commensurability versus incommensurability of celestial motions depends on a distinction that goes beyond the discrete. It is therefore of no interest for the practical astronomer, but remains a mathematical and philosophical question.

According to the Platonic doctrine of *saving the appearances*, astronomical theories only have to represent faithfully observed phenomena. The first principles guiding the setting up of astronomical models were, according to a well-known treatise by Geminus, 'that the movements of the stars are regular, uniform and constant.'[5] In other respects, the astronomer had all the freedom to pose his models to account for the phenomena. The first principles, stemming from Aristotle's philosophy, combined with the methodology of choosing the simplest hypotheses. When a geometric model was proposed, it went through an *analytic procedure* where one writes down, as we would say, equations that combine the unknown parameters of the model to directly observable quantities, such as the lengths of seasons. The procedure is beautiful theoretically seen, but absolute determinations of the dimensions of a model turned out badly. Already the closest thing to determine, the lunar parallax, led Ptolemy to a lunar distance too small by a couple of orders of magnitude. What was left were proportions, the question of their commensurability or incommensurability being independent of absolute scale.

## 2 THE DENSITY OF ROTATIONS OF A CIRCLE BY AN IRRATIONAL ANGLE

Oresme was clearly aware of the fact that observations, because of their finite character, would be unable to determine the answer to his question. He compared the alternatives instead by deriving mathematical consequences from both, and devised an empirically oriented application

[3] The relevant part of Theodosius' book are translated in Grant (1966, pp. 79–84).

[4] These have been located by Neugebauer (1975, p. 749).

[5] See Cohen and Drabkin (1948).

of frequentist probability for showing that the 'odds' are for incommensurability. Some of the mathematical results Oresme achieved comparing commensurability and incommensurability, are among the most remarkable ones between ancient and modern times, particularly one concerning the density of rotations of a circle by an irrational angle.

In previous chapters the nature of the motion we now call ergodic has come up in various forms. Oresme gives a beautiful description of such motion for the case of the Sun's annual and diurnal (daily) rotations. It was the same case that had concerned Theodosius. If the relation of the two rotations is incommensurable, the Sun's center *B* will describe a path between the two tropics we have no difficulty in calling *ergodic*[6]:

> It follows from all this that *B* describes daily a new spiral—imagined as motionless in space—which it has never before traversed. And so, by its track, or imagined flow, *B* seems to extend a spiral line that has already become infinite from the infinite spirals that were described in the past. This infinite spiral line is constituted from the spirals that were described between the two tropics, and from the intersections between the earlier and later spirals. Consequently, through an eternal time, *B* appears twice in any point of intersection—and only once, or never, in any of the other points.
>
> In accordance with what has been imagined here, the whole celestial space between the two tropics is traversed by *B*, leaving behind a web- or net-like figure expanded through the whole of this space. The structure of this figure was already infinitely dense through the course of an infinite past time, and yet, nonetheless, it will be made continually more dense, since it produces a new spiral every day.

Oresme further concludes that it is 'verisimile and probable' that a suggested proportion is incommensurable. Therefore it is probable that the universe is aperiodic, so that nothing ever recurs in exactly the same way. He is thus performing the step—in some way at least—from individual ergodic motions to the ergodicity of all possible motions; so to the ergodicity of a system of motion as a whole. Let us now see how these extraordinary conclusions were arrived at.

The source of interest here is Oresme's tract on the commensurability versus incommensurability of celestial motions. In this work properties of one or several uniform circular motions are studied.[7] In the first part, Oresme studies the situation of rotations with commensurable circulation times, in the second that of incommensurable motions. There the

[6] See Oresme's *De proportionibus proportionum* and *Ad pauca respicientes* (pp. 275–9). These two treatises are abbreviated as PP and AP, respectively.

[7] The text, dating back to around 1360, is edited with an introduction, English translation and commentary by Grant. Page references are to this edition, abbreviated as CI.

following proposition appears: 'No sector of a circle is so small that two such mobiles could not conjunct in it at some future time, and could not have conjuncted in it sometime' (CI p. 253). Imagining the pointlike bodies to rotate in the same sense in the same plane, points of conjunction are those points of the circle in which they are at the same time. The arcs between two successive points of conjunction are equal in length, as are the angles (CI p. 257). Therefore Oresme is actually studying what are now called rotations of a circle by a given angle. If by 'sector' we understand any closed interval of the circle, Oresme's proposition says rotations with an irrational angle give a dense set of points, a result usually seen as a special case of Kronecker's theorem.[8] From Dirichlet we learn that the special case was 'known since a long time,' this statement being as of 1842.[9]

Oresme proves in the first part of his treatise that two bodies in a commensurable motion have a finite number of conjunction points, and it follows that their motion is periodic (CI p. 199). In the second part he proves that in the case of incommensurability, there is an infinity of conjunction points: Any arc between two points of conjunction is cut by a further conjunction. At several places Oresme seems to take it for granted that one may infer from this what his proposition asserts. What is missing from the argument is that the length of all the arcs produced by the conjunction points diminishes without limit.[10] But by following Oresme's (rather verbal) proof and ensuing remarks, it is possible to account for the omissions in his argumentation. The reasoning can be brought very close to Oresme in all respects.

Here is an outline of Oresme's proof: Starting from a point of conjunction *d*, the next one is *e*, and after a finite number of applications of the angle corresponding to the arc *de*, you cut *de* in two, after another circle one of the parts, and so on. Oresme says 'this process can be carried into infinity by always dividing the circle into smaller parts ad infinitum. Thus no part of the circle will remain, but that it could not, at some time, be imagined as divisible in this way' (CI p. 255). Oresme concludes the proof of his proposition by stating that 'no sector of the circle will be so small but that at some time in the future the mobiles could not conjunct in some point of it, and this is what we have proposed.' (CI pp. 255–57).

Oresme does not explicitly prove that the parts 'diminish into infinity.' However, he provides a further proposition that contains an appropriate

[8] See Kronecker (1884), or his collected works, vol. 3, p. 49.

[9] See Dirichlet (1842), as reprinted in his collected works, vol. 1, p. 635.

[10] This is passed by without comment in Grant (cf. CI p. 45) and in another work Grant (1961, p. 445), as well as in my previous study (1981).

method of proof, one in which an arbitrary part is cut into half, into infinity. This is his proposition 5, saying that mobiles in incommensurable motion 'will conjunct infinitely close to any given point of conjunction, and have already conjuncted infinitely near to it' (CI p. 259). In the proof he assumes another point given, and then takes a third halfway between these two, and so on into infinity. On the basis of his proposition 4, there is a conjunction point in each of the parts that approach the original given point: As he says, the 'conjunctions will approximate infinitely close' (CI pp. 259–61). Here we have a method of producing ever smaller arcs all over the circle. Its proper application gives an elementary argument for the density of rotations of a circle by an irrational angle. In sum, Oresme's proof for the density of rotations of a circle by an irrational angle, while not entirely conclusive, can be completed by elementary arguments used by himself in the very same context.[11]

Oresme approached the density theorem by geometric means. This was a necessity since he had no number concept available for formulating the property; a number for him meant a rational number. Incommensurable quantities were represented geometrically, and they were unequal to any such numbers. A comparison with Dirichlet (1842, p. 635), and some of the developments reported in Section 2.4(a), is instructive in this respect. By definition, $\alpha$ is irrational if $m\alpha - n \neq 0$ for every integer $m$, $n$. What the density theorem says in these terms is that there still are infinitely many pairs $m$, $n$ such that $m\alpha - n < 1/n$. In the geometric Oresmian reading this means that for an incommensurable angle, corresponding to an irrational $\alpha$, the $m$-fold application exceeds the starting point by an arc length of less than $1/n$, assuming the circumference to be of unit length.

In reading Oresme (for example, CI p. 257), one gains the impression that he thought that the incommensurable conjunctions that so fascinated his mind would be equally distributed in all directions. He says, for example, that 'by means of the greatest inequality, which departs from every equality, the most just and established order is preserved' (CI p. 257). But it would be too much to expect any results in the direction of an equidistribution in Oresme.[12] Instead, he moves in a different direction, in trying to prove that the case of incommensurability and aperiodicity of celestial motions is the *probable* one.

[11] I give a detailed reconsideration of Oresme's proof in a forthcoming paper in *Historia Mathematica* [vol. 20, 1993].

[12] See Sec. 2.4(a) for such results.

In connection with his studies of laws of motion, Oresme considered a certain function proposed by Thomas Bradwardine. This led him into a further study resulting in the *De Proportionibus Proportionum.* We may describe as follows some of its results relevant here, allowing ourselves the freedom of referring to incommensurables as irrational numbers. Two rational numbers $m/n$ and $r/s$ are given at will. An exponent has to be found such that one of these, when raised to that exponent, gives as value the other. By computing long series of cases in which the numbers succeed each other in a systematic way, Oresme realized that rational numbers are in most cases related by an irrational exponent. With 100 numbers he had 4950 cases to be investigated, and only 25 proved to be commensurable (PP p. 259). The longer the series, the rarer the rational cases seem to be. Oresme says that if instead of 100 numbers we take 200 or 300, the ratio of irrational to rational cases would be much greater (PP p. 249). In the tract on the incommensurability of celestial motions he concludes that 'it is verisimile and probable' that any two proposed numbers are related incommensurably (CI p. 321). In PP the same conclusion is drawn for celestial magnitudes in general (pp. 303–305).

Oresme does not say that probability is the *same* as a frequency in a class from which the case of interest is chosen. He made instead a statistical inference of sorts from his sample of size 4950 that had given the relative frequency of 1/198 for the case of commensurability. With 200 or 300 numbers this size grows also, but the relative frequency of rational exponents goes down from the above value. This, naturally, is a consequence of his happy choice of the function whose values he studied: If in the 'Oresme equation' $(m/n)^x = r/s$, we take logarithms and solve for $x$, $x = (\log r - \log s)/(\log m - \log n)$. Oresme was led to study a function which rarely has rational values. We must keep in mind that he made his conclusions some three hundred years before any systematic development of a calculus of probability, and another two hundred and fifty years before the application of probabilistic arguments to value distribution questions of real functions. Oresme draws a connection to gambling, though: Taking natural numbers in series, 'the number of perfect or cube numbers is much less than other numbers, and as more numbers are taken in the series, the greater is the ratio of noncube to cube numbers or nonperfect to perfect numbers.' Assuming now a number 'wholly unknown... it will be likely that such an unknown number would not be a cube number. A similar situation is found in games where, if one should inquire whether a hidden number is a cube number, it is safer to reply in the negative since this seems more probable and likely' (PP pp. 249–51). This is the conclusion one has to take with respect to the incommensurability of the exponent in Oresme's equation also. For 'if one reflected carefully he would find that between ratios of rational ratios [that is, exponents relating rational numbers] those which are rational are fewer than cube numbers in an aggregate of numbers' (PP p. 251).

Because of 'the defects of the senses' (AP p. 427), the question of incommensurability cannot be decided. 'Thus it is sufficient for a good astronomer to judge motions and aspects near a point' (AP p. 429). Arguments for incommensurability have to be of a different kind. The first one is Oresme's 'proof' that incommensurability is the 'probable and likely' case. Second, deriving different properties from incommensurability, he argued that it is probable that the celestial system shares those properties: 'If the antecedent is probable, the consequent will be probable' (AP p. 423). Oresme had also several philosophical and aesthetic arguments for incommensurability. The latter is obvious from his vivid description of the crisscross path described by the Sun's center as quoted above. One philosophical argument is an application of what is known as the principle of plenitude that we can put as follows: That which is possible will eventually take place. Assuming commensurability, Oresme notes that parts of the ecliptic would be 'deprived of a conjunction between the sun and moon' (CI p. 319). Such consequences of commensurability are 'improbable as well as inappropriate for the beauty of this world..., one should be able to say that there is no part of the ecliptic so small that the sun and moon could not conjunct there sometime, or have not already conjuncted there' (CI p. 319). Amidst this rhetoric, Oresme seems to forget his general proof that any three planets can only conjunct either not at all, or else once, or an infinite number of times (CI p. 267).

Another philosophical merit of incommensurability is that it provides arguments against certain Aristotelian doctrines. In an aperiodic universe there is an instant at which the Sun in its motion combining two uniform circular components is, in some direction, at its farthest from the Earth. By the probable incommensurability of the two rotations, that point at that direction will only be reached once in the infinite history and future of the universe. The cone of the Earth's shadow will then be at its longest, and a point in space will be in shadow that never in the infinite past was 'deprived' of the Sun's light, nor will ever be after that instant. There is thus something in the universe that had no beginning but comes to an end, namely, that the point in question was lit by the Sun. And there is something that begins but never ends, namely, that the point will be lit by the Sun. Oresme gives this as an argument against the Aristotelian doctrine of generation and corruption. It can be found together with many similar ones in the *Livre du ciel et du monde* (for example, p. 199). These arguments show the seriousness with which Oresme took the exact geometrical configurations of the celestial system of epicycles. That attitude can be compared with the contents of his proposition 10 (CI p. 269), saying that 'if three or more mobiles are moved with incommensurable velocities, they could never be so close but that, at some other time, they could not be even closer; and so into infinity.[13] Similarly, the shadowy cone cast by the Earth would come an infinite number of times infinitely close to the point only once deprived of the Sun's light. This type of motion was studied in modern times by Poincaré in his 1890 work on the three-body problem. Oresme's results are easy consequences of Poincaré's recurrence theorem, stating the quasi-periodicity of the motions of any isolated mechanical system, excepting a set of possible motions of zero probability.

[13] The condition of the theorem is of course superfluous out of context; two or more would do as well here.

# *Bibliography*

*Abbreviations*: *CP* = Collected Papers, *CW* = Collected Works, *GA* = Gesammelte Abhandlungen, *OB* = Oeuvres de Emile Borel, *WA* = Wissenschaftliche Abhandlungen. *Comptes rendus* are those of the Paris Academy.


Abraham, R. (1967) *Foundations of Mechanics*, W.A. Benjamin, New York.

Aldous, D. J. (1985) Exchangeability and related topics, *Lecture Notes in Mathematics* **1117**, 1–198.

Aleksandrov, P., N. Akhiezer, B. Gnedenko and A. Kolmogorov (1969) Sergei Natanovich Bernstein, *Russian Mathematical Surveys* **4**, 169–176.

Amaldi, E. (1979) Radioactivity, a pragmatic pillar of probabilistic conceptions, in G. Toraldo di Francia ed. *Problems in the Foundations of Physics*, 1–28 (= Proceedings of the International School of Physics «Enrico Fermi», Course LXXII), North-Holland, Amsterdam.

Antretter, G. (1989) *Von der Ergodenhypothese zu stochastischen Prozessen: Die Entfaltung der Theorie der Markov-Ketten und -Prozessen vor dem Hintergrund statistisch-mechanischer Probleme*, Schriftliche Hausarbeit aus dem Fach Mathematik, Universität München.

Bachelier, L. (1900) *Théorie de la spéculation*, (in the series 'Thèses présentées a la faculté des sciences de Paris pour obtenir le grade de docteur ès sciences mathématiques') Gauthier-Villars, Paris. Also published in *Annales Scientifiques de l'Ecole Normale Supérieure* **17**, 21–86.

Bachelier, L. (1912) *Calcul des probabilités*, Gauthier-Villars, Paris.

Barnes, R. B. and S. Silverman (1934) Brownian motion as a natural limit to all measuring processes, *Reviews of Modern Physics* **6**, 162–192.

Barone, J. and A. Novikoff (1978) A history of the axiomatic formulation of probability from Borel to Kolmogorov: Part I, *Archive for History of Exact Sciences* **18**, 123–190.

Bell, J. S. (1987) *Speakable and Unspeakable in Quantum Mechanics*, Cambridge University Press.

Bernays, P. (1922) Zur mathematischen Grundlegung der kinetischen Gastheorie, *Mathematische Annalen* **85**, 242–255.

Bernstein, F. (1912a) Ueber eine Anwendung der Mengenlehre auf ein aus der

Theorie der säkularen Störungen herrührendes Problem, *Mathematische Annalen* **71**, 417–439.

Bernstein, F. (1912b) Ueber geometrische Wahrscheinlichkeit und über das Axiom der beschränkten Arithmetisierbarkeit der Beobachtungen, *Mathematische Annalen* **72**, 585–587.

Bernstein, S. (1917) An essay on the axiomatic foundations of probability theory (in Russian), *Communications de la Société mathématique de Kharkow* **15**, 209–274.

Bertrand, J. (1888) *Calcul des probabilités*, Gauthier-Villars, Paris.

Birkhoff, G. and J. von Neumann (1936) The logic of quantum mechanics, *Annals of Mathematics* **37**, 823–843.

Birkhoff, G. D. (1927) *Dynamical Systems*, American Mathematical Society, New York.

Birkhoff, G. D. (1931) Proof of the ergodic theorem, *Proceedings of the National Academy of Sciences* **17**, 656–660.

Birkhoff, G. D. (1932) Probability and physical systems, *Bulletin of the American Mathematical Society* **28**, 361–379.

Birkhoff, G. D. and B. O. Koopman (1932) Recent contributions to the ergodic theory, *Proceedings of the National Academy of Sciences* **18**, 279–282.

Birkhoff, G. D. and P. Smith (1928) Structure analysis of surface transformations, *Journal de mathématiques* **7**, 345–365.

Bishop, E. (1967) *Foundations of Constructive Analysis*, McGraw-Hill, New York.

Bohl, P. (1909) Ueber ein in der Theorie der säkularen Störungen vorkommendes Problem, *Journal für Mathematik* **135**, 189–283.

Bohlin, K. (1897) Hugo Gyldén. Ein biographischer Umriss nebst einigen Bemerkungen über seine wissenschaflichen Methoden, *Acta mathematica* **20**, 397–404.

Bohlmann, G. (1901) Lebensversicherungs-Mathematik, in *Encyclopädie der mathematischen Wissenschaften*, vol. 1, part 2, 852–917, Teubner, Leipzig.

Bohr, N. (1913) On the constitution of atoms and molecules, *Philosophical Magazine* **26**, 1–25, Part II, ibid., **26**, 476–502, Part III, ibid., **26**, 857–875.

Bohr, N. (1928) The quantum postulate and the recent development of atomic theory, *Nature* **121**, 580–590.

Boltzmann, L. (1866) Ueber die mechanische Bedeutung des zweiten Hauptsatzes der Wärmetheorie, *WA 1*, 9–33.

Boltzmann, L. (1868a) Ueber die Integrale linearer Differentialgleichungen mit periodischen Koeffizienten, *WA 1*, 43–48.

Boltzmann, L. (1868b) Studien über das Gleichgewicht der lebendigen Kraft zwischen bewegten materiellen Punkten, *WA 1*, 49–96.

Boltzmann, L. (1868c) Lösung eines mechanischen Problems, *WA 1*, 97–105.

Boltzmann, L. (1871a) Ueber das Wärmegleichgewicht zwischen mehratomigen Gasmolekülen, *WA 1*, 237–258.

Boltzmann, L. (1871b) Einige allgemeine Sätze über Wärmegleichgewicht, *WA 1*, 259–287.

Boltzmann, L. (1871c) Analytischer Beweis des zweiten Hauptsatzes der mechanischen Wärmetheorie aus den sätzen über das Gleichgewicht der lebendigen Kraft, *WA 1*, 288–308.

Boltzmann, L. (1872) Weitere Studien über das wärmegleichgewicht unter Gasmolekülen, *WA 1*, 316–402. English: Brush (1966).

Boltzmann, L. (1875) Ueber das Wärmegleichgewicht von Gasen, auf welche äussere Kräfte wirken, *WA 2*, 1–30.

Boltzmann, L. (1877a) Bemerkungen über einige Probleme der mechanischen Wärmetheorie, *WA 2*, 112–148.

Boltzmann, L. (1877b) Ueber die Beziehung zwischen dem zweiten Hauptsatze der mechanischen Wärmetheorie und der Wahrscheinlichkeitsrechnung respektive den Sätzen über das Wärmegleichgewicht, *WA 2*, 164–223.

Boltzmann, L. (1878) Weitere Bemerkungen über einige Probleme der mechanischen Wärmetheorie, *WA 2*, 250–288.

Boltzmann, L. (1881) Referat über die Abhandlung von J. C. Maxwell "Ueber Boltzmanns Theorie betreffend die mittlere Verteilung der lebendigen Kraft in einem System materieller Punkte.", *WA 2*, 582–595.

Boltzmann, L. (1884) Ueber die Eigenschaften monozyklischer und damit verwandter Systeme, *WA 3*, 122–152.

Boltzmann, L. (1887) Ueber die mechanischen Analogien des zweiten Hauptsatzes der Thermodynamik, *WA 3*, 258–271.

Boltzmann, L. (1892) III. Teil der Studien über Gleichgewicht der lebendigen Kraft, *WA 3*, 428–453.

Boltzmann, L. (1895a) On certain questions of the theory of gases, *WA 3*, 535–544.

Boltzmann, L. (1895b) On the minimum theorem in the theory of gases, *WA 3*, p. 546.

Boltzmann, L. (1896–98) *Vorlesungen über Gastheorie*, vols. 1 and 2, Barth, Leipzig.

Boltzmann, L. (1896) Entgegnung auf die wärmetheoretischen Betrachtungen des Hrn. E. Zermelo, *WA 3*, 567–578.

Boltzmann, L. (1897a) Zu Hrn. Zermelos Abhandlung "Ueber die mechanische Erklärung irreversibler Vorgänge", *WA 3*, 579–586.

Boltzmann, L. (1897b) Ueber einen mechanischen Satz Poincaré's, *WA 3*, 587–595.

Boltzmann, L. (1898) Ueber die sogenannte *H*-Kurve, *WA 3*, 629–637.

Boltzmann, L. (1909) *Wissenschaftliche Abhandlungen*, (ed. Hasenöhrl), vols. 1–3, Barth, Leipzig.

Boltzmann, L. and J. Nabl (1907) Kinetische Theorie der Materie, *Encyklopädie der mathematischen Wissenschaften*, vol. V, part 1, 493–557, Teubner, Leipzig.

Borel, E. (1898) *Leçons sur la théorie des fonctions*, Gauthier-Villars, Paris. 2nd ed. 1914.

Borel, E. (1903) Contribution à l'analyse arithmétique du continu, *OB 3*, 1439–1485.

Borel, E. (1905a) Remarques sur les principes de la théorie des ensembles, *OB 3*, 1251–1252.

Borel, E. (1905b) Remarques sur certaines questions de probabilité, *OB 2*, 985–990.

Borel, E. (1905c) Cinq lettres sur la théorie des ensembles, *OB 3*, 1253–1265.

Borel, E. (1906a) Sur les principes de la théorie cinétique des gaz, *OB 3*, 1669–1692.

Borel, E. (1906b) La valeur pratique du calcul des probabilités, *OB 2*, 991–1004.

Borel, E. (1908a) Sur les principes de la théorie des ensembles, *OB 3*, 1267–1269.

Borel, E. (1908b) Les «paradoxes» de la théorie des ensembles, *OB 3*, 1271–1276.

Borel, E. (1909) *Eléments de la théorie des probabilités*, Hermann, Paris. English translation of 2nd French edition: *Elements of the Theory of Probability*, Prentice Hall, Englewood Cliffs 1965.

Borel, E. (1909a) Les probabilités dénombrables et leurs applications arithmétiques, *OB 2*, 1055–1078. Original: *Rendiconti del Circolo Matematico di Palermo* **27**, 247–270.

Borel, E. (1912) *Notice sur les travaux scientifiques*, Gauthier-Villars, Paris. Also *OB 1*, 119–190.

Borel, E. (1912a) Sur un problème de probabilités relatif aux fractions continues, *OB 2*, 1085–1091.

Borel, E. (1912b) Le calcul des intégrales définies, *OB 2*, 827–878.

Borel, E. (1912c) La philosophie mathématique et l'infini, *OB 4*, 2130.

Borel, E. (1913) La mécanique statistique et l'irréversibilité, *OB 3*, 1697–1704.

Borel, E. (1914) *Le hasard*, Alcan, Paris. New ed. 1948.

Borel, E. (1914a) Exposé français de l'article sur la Mécanique statistique (P. et T. Ehrenfest), Supplement II, *OB 3*, 1713–1733.

Borel, E. (1920) Radioactivité, probabilité, déterminisme, *OB 4*, 2189–2196.

Borel, E. (1924) A propos d'une traité de probabilités, *OB 4*, 2169–2184. Reprinted as note II in Borel (1939).

Borel, E. *et al.* (1925–1939) *Traité du Calcul des Probabilités et de ses applications*, 4 vols. in 19 parts, Gauthier-Villars, Paris.

Borel, E. (1925) *Mécanique statistique classique*, (= Traité, vol. 2, part 3), Gauthier-Villars, Paris.

Borel, E. (1926) *Applications à l'Arithmétique et à la théorie des fonctions* (= Traité, vol. 2 part 1), Gauthier-Villars, Paris.

Borel, E. (1927) Sur le système de formes linéaires à déterminant symétrique et la théorie générale du jeu, *OB 2*, 1123–1124. English: Econometrica **21** (1953).

Borel, E. (1937) Sur l'imitation du hasard, *OB 2*, 1159–1161.

Borel, E. (1939) *Valeur pratique et philosophie des probabilités*, (= Traité, vol. 4, part 3), Gauthier-Villars, Paris.

Borel, E. (1972) *Oeuvres de Emile Borel*, vols. 1–4, Editions du CNRS, Paris.

Born, M. (1926a) Zur Quantenmechanik der Stossvorgänge, *Zeitschrift für Physik* **37**, 863–867.

Born, M. (1926b) Quantenmechanik der Stossvorgänge, *Zeitschrift für Physik* **38**, 803–827.

Born, M. (1927) Quantenmechanik und Statistik, *Die Naturwissenschaften* **15**, 238–242.

Born, M. (1954) Die statistische Deutung der Quantenmechanik, *Les Prix Nobel en 1954*, 79–90, Stockholm 1955. Also in *Ausgewählte Abhandlungen*, vol. II, 430–441.

Born, M. (1961) Bemerkungen zur statistischen Deutung der Quantenmechanik, in *Werner Heisenberg und die Physik unserer Zeit*, 103–118, Vieweg, Braunschweig. Also in *Ausgewählte Abhandlungen*, vol. II, 454–469.

Born, M. and W. Heisenberg (1927) La mécanique des quanta, pp. 143–184 in *Electrons et Photons. Rapports et Discussion du Cinquième Conseil de Physique*, Gauthier-Villars, Paris 1928.

Born, M., W. Heisenberg and P. Jordan (1926) Zur Quantenmechanik II, *Zeitschrift für Physik* **35**, 557–615.

Born, M. and P. Jordan (1925a) Zur Quantentheorie aperiodischer Vorgänge, *Zeitschrif für Physik* **33**, 479–505.

Born, M. and P. Jordan (1925b) Zur Quantenmechanik, *Zeitschrift für Physik* **34**, 858–888.

Bose, S. N. (1924) Plancks Gesetz und Lichtquantenhypothese, *Zeitschrift für Physik* **26**, 178–181.

van Brakel, J. (1985) The possible influence of the discovery of radio-active decay on the concept of physical probability, *Archive for History of Exact Sciences* **31**, 369–385.

Bridges, D. and F. Richman (1987) *Varieties of Constructive Mathematics*, Cambridge University Press.

Brodén, T. (1900) Wahrscheinlichkeitsbestimmungen bei der gewöhnlichen Kettenbruchentwicklung reeller Zahlen, *Öfversikt af Kongliga Vetenskaps-Akademiens Förhandlingar*, 239–266.

Brodén, T. (1901a) *Bemerkungen über Mengenlehre und Wahrscheinlichkeitstheorie, durch einen Schrift des Herrn A. Wiman veranlasst.* Malmö.

Brodén, T. (1901b) *Noch einmal die Gyldén'sche Wahrscheinlichkeitsfrage*, Skånska Lithografiska Aktiebolaget, Malmö.

Broggi, U. (1907) *Die Axiome der Wahrscheinlichkeitsrechnung*, Dissertation Göttingen, as partly reprinted in Schneider (1988), 367–377

de Broglie, L. (1924) *Recherche sur la théorie des quanta*, Masson, Paris.

Brouwer, L. (1921) Besitzt jede reelle Zahl eine Dezimalbruchentwicklung?, *CW 1*, 236–245.

Brouwer, L. (1924a) Intuitionistische Zerlegung mathematischer Grundbegriffe, *CW 1*, 275–280.

Brouwer, L. (1924b) Beweis, dass jede volle Funktion gleichmässig stetig ist, *CW 1*, 286–290.

Brouwer, L. (1975) *Collected Works*, vol. 1, North-Holland, Amsterdam.

Bruns, H. (1906) *Wahrscheinlichkeitsrechnung und Kollektivmasslehre*, Teubner, Leipzig.

Brush, S. ed. (1965, 1966, 1972) *Kinetic Theory*, vols. 1–3, Pergamon Press.

Brush, S. (1971) Proof of the impossibility of ergodic systems: the 1913 papers of Rosenthal and Plancherel, *Transport Theory and Statistical Physics* **1**, 287–311.

Brush, S. (1976) *The Kind of Motion We Call Heat*, 2 vols., North-Holland, Amsterdam.

Brush, S. (1978) *The Temperature of History: Phases of Science and Culture in the Nineteenth Century*, Burt Franklin & Co, New York.

Brush, S. (1983) *Statistical Physics and the Atomic Theory of Matter*, Princeton University Press.

Buchholz, H. (1904) Poincarés Preisarbeit von 1889/1890 und Gyldéns Forschung über das Problem der drei Körper in ihren Ergebnissen für die Astronomie, *Physikalische Zeitschrift*, **5**, 180–186.

Cantelli, F. (1916a) Sulla legge dei grandi numeri, *Atti della R. Accademia dei Lincei, Memorie della classe di scienze fisiche, matematiche, e naturali* **11**, 330–349.

Cantelli, F. (1916b) La tendenza ad un limite nel senso del calcolo delle probabilità, *Rendiconti del Circolo Matematico di Palermo* **41**, 191–201.

Cantelli, F. (1917) Sulla probabilità come limite della frequenza, *Atti della Reale Accademia dei Lincei, Rendiconti* **26**, 39–45.

Carathéodory, C. (1919) Ueber den Wiederkehrsatz von Poincaré, *Sitzungsberichte der Preussischen Akademie der Wissenschaften*, 1919, 579–584.

Castelnuovo, G. (1925–28) *Calcolo delle probabilità*, 2nd ed., Zanichelli, Bologna.

Champernowne, D. G. (1933) The construction of decimals normal in the scale of ten, *Journal of the London Mathematical Society* **8**, 254–260.

Chandrasekhar, S. (1943) Stochastic problems in physics and astronomy, *Reviews of Modern Physics* **15**, 1–89.

Chapman, S. (1928) On the Brownian displacements and thermal diffusion of grains suspended in a non-uniform fluid, *Proceedings of the Royal Society* **A119**, 34–60.

Church, A. (1940) On the concept of a random sequence, *Bulletin of the American Mathematical Society* **46**, 130–135.

Clausius, R. (1857) Ueber die Art der Bewegung, welche wir Wärme nennen, *Annalen der Physik und Chemie* **100**, 353–380.

Clausius, R. (1858) Ueber die mittlere Länge der Wege, welche bei der Molecularbewegung gasförmiger Körper von den einzelnen Molecülen zurückgelegt werden; nebst einigen anderen Bemerkungen über die mechanische Wärmetheorie, *Annalen der Physik* **105**, 239–258.

Clausius, R. (1868) On the second fundamental theorem of the mechanical theory of heat, *Philosophical Magazine* **35**, 405–419.

Cohen, M. and I. Drabkin, eds. (1948) *A Source Book in Greek Science*, Harvard University Press.

Copeland, A. H. (1928) Admissible numbers in the theory of probability, *American Journal of Mathematics* **50**, 153–162.

Copeland, A. H. (1931) Admissible numbers in the theory of geometrical probability, *American Journal of Mathematics* **53**, 153–162.

Copeland, A. H. (1937) Consistency of the conditions determining kollektivs, *Transactions of the American Mathematical Society* **42**, 333–357.

Cover, T. M., P. Gacs and R. M. Gray (1989) Kolmogorov's contributions to information theory and algorithmic complexity, *The Annals of Probability* **17**, 840–865.

Cramér, H. (1937) *Random Variables and Probability Distributions*, Cambridge University Press.

Cramér, H. (1939) Entwicklungslinien der Wahrscheinlichkeitsrechnung, *Neuvième congrès des mathématiciens scandinaves*, pp. 67–86, Mercators Tryckeri, Helsingfors.

Cramér, H. (1946) *Mathematical Methods of Statistics*, Princeton University Press.

Cramér, H. (1976) Fifty years of probability theory: some personal recollections, *The Annals of Probability* **4**, 509–546.

Daboni, L. (1953) Aspetti di una interpretazione geometrica per le probabilità di eventi equivalenti, *Rendiconti di Matematica della Università di Parma* **4**, 145–165.

Daboni, L. (1987) Bruno de Finetti, *Bollettino dell'Unione Matematica Italiana* **7**, 283–308.

Dale, A. I. (1985) A study of some early investigations into exchangeability, *Historia Mathematica* **12**, 323–336.

Daniell, P. J. (1919) Integrals in an infinite number of dimensions, *Annals of Mathematics* **20**, 281–288.

Davis, M. ed. (1965) *The Undecidable*, Raven Press, New York.

Diaconis, P. (1988) Recent progress on de Finetti's notions of exchangeability, in J. M. Bernardo et al. eds. *Bayesian Statistics 3*, 111–125, Oxford University Press.

Diaconis, P. and S. Zabell (1982) Updating subjective probability, *Journal of the American Statistical Association* **77**, 822–830.

Dirac, P. (1925) The fundamental equations of quantum mechanics, *Proceedings of the Royal Society* **A109**, 642–653.

Dirac, P. (1926a) Quantum mechanics and a preliminary investigation of the hydrogen atom, *Proceedings of the Royal Society* **A110**, 561–579.

Dirac, P. (1926b) The physical interpretation of the quantum dynamics, *Proceedings of the Royal Society* **A113**, 621–641.

Dirac, P. (1927) The quantum theory of the emission and absorption of radiation, *Proceedings of the Royal Society* **A114**, 243–265.

Dirichlet, P. (1842) Verallgemeinerung eines Satzes aus der Lehre von den Kettenbrüchen nebst einigen Anwendungen auf die Theorie der Zahlen, *Bericht über die Verhandlungen der Königlichen Preussischen Akademie der Wissenschaften*, Jahrg. 1842, 93–95.

Doob, J. L. (1934a) Stochastic processes and statistics, *Proceedings of the National Academy of Sciences* **20**, 376–379.

Doob, J. L. (1934b) Probability and statistics, *Transactions of the American Mathematical Society* **36**, 759–775.

Doob, J. L. (1936) Note on probability, *Annals of Mathematics* **37**, 363–367.

Doob, J. L. (1937) Stochastic processes depending on a continuous parameter, *Transactions of the American Mathematical Society* **42**, 107–140.

Doob, J. L. (1938) Stochastic processes with an integer-valued parameter, *Transactions of the American Mathematical Society* **44**, 87–150.

Doob, J. L. (1953) *Stochastic Processes*, Wiley, New York.

Doob, J. L. (1989) Kolmogorov's early work on convergence theory and foundations, *The Annals of Probability* **17**, 815–821.

Dutka, J. (1985) On the problem of random flights, *Archive for History of Exact Sciences* **32**, 351–375.

Dynkin, E. B. (1978) Sufficient statistics and extreme points, *The Annals of Probability* **6**, 705–730.

Ehrenfest, P. and T. (1907) Ueber zwei bekannte Einwände gegen das Boltzmannsche *H*-Theorem, *Physikalische Zeitschrift* **8**, 311–314.

Ehrenfest, P. and T. (1911) Begriffliche Grundlagen der statistischen Auffassung in der Mechanik, *Encyclopädie de mathematischen Wissenschaften*, vol. 4, part 32, Teubner, Leipzig. English: *The Conceptual Foundations of the Statistical Approach in Mechanics*, Cornell University Press, Ithaca, 1959.

Einstein, A. (1902) Kinetische Theorie des Wärmegleichgewichtes und des zweiten Hauptsatzes der Thermodynamik, *Annalen der Physik* **9**, 417–433.

Einstein, A. (1903) Eine Theorie der Grundlagen der Thermodynamik, *Annalen der Physik* **11**, 170–187.

Einstein, A. (1904) Zur allgemeinen molekularen Theorie der Wärme, *Annalen der Physik* **14**, 354–362.

Einstein, A. (1905a) Ueber einen die Erzeugung und Verwandlung des Lichtes betreffenden heuristischen Gesichtspunkt, *Annalen der Physik* **17**, 132–148.

Einstein, A. (1905b) Ueber die von der molekularkinetischen Theorie der Wärme geforderte Bewegung von in Flüssigkeiten suspendierten Teilchen, *Annalen der Physik* **17**, 549–560.

Einstein, A. (1906a) Eine neue Bestimmung der Moleküldimensionen, *Annalen der Physik* **19**, 289–306.

Einstein, A. (1906b) Zur Theorie der Brownschen Bewegung, *Annalen der Physik* **19**, 371–381.

Einstein, A. (1907a) Ueber die Gültigkeitsgrenze des Satzes vom thermodynamischen Gleichgewicht und über die Möglichkeit einer neuen Bestimmung der Elementarquanta, *Annalen der Physik* **22**, 569–572.

Einstein, A. (1907b) Theoretische Bemerkungen über die Brownsche Bewegung, *Zeitschrift für Elektrochemie* **13**, 41–42.

Einstein, A. (1908) Elementare Theorie Brownschen Bewegung, *Zeitschrift für Elektrochemie* **14**, 235–239.

Einstein, A. (1909) Zum gegenwärtigen Stand des Strahlungsproblems, *Physikalische Zeitschrift* **10**, 185–193.

Einstein, A. (1910) Theorie der Opaleszenz von homogenen Flüssigkeiten und Flüssigkeitsgemischen in der Nähe des kritischen Zustandes, *Annalen der Physik* **33**, 1275–1298.

Einstein, A. (1911) Bemerkungen zu den P. Hertzschen Arbeiten: mechanische Grundlagen der Thermodynamik, *Annalen der Physik* **34**, 175–176.

Einstein, A. (1913) Statistische Mechanik, Vorlesungsnachschrift von Walter Dällenbach, Sommersemester 1913, to appear in *CP*, vol. 3.

Einstein, A. (1914a) Zum gegenwärtigen Stande des Problems der spezifischen Wärme, *Abhandlungen der Deutschen Bunsen-Gesellschaft*, No. 7, 330–364.

Einstein, A. (1914b) Méthode pour la détermination de valeurs statistiques d'observations concernent des grandeurs soumises à des fluctuations irrégulieres, *Archives des sciences physiques et naturelles* **37**, 254–256.

Einstein, A, (1916) Zur Quantentheorie der Strahlung, *Mitteilungen der Physikalischen Gesellschaft Zürich* **16**, 47–62.

Einstein, A. (1917) Zur Quantentheorie der Strahlung, *Physikalische Zeitschrift* **18**, 121–128.

Einstein, A. (1925) Quantentheorie des einatomigen idealen Gases. 2. Abhandlung, *Sitzungsberichte der Preussischen Akademie der Wissenschaften*, 1925, 3–14.

Einstein, A. (1949) Remarks concerning the essays brought together in this cooperative volume, in P. A. Schilpp ed., *Albert Einstein: Philosopher-Scientist*, 665–688, The Library of Living Philosophers, Evanston, Illinois.

Einstein, A. (1987-) *The Collected Papers of Albert Einstein*, Princeton University Press.

Einstein, A., B. Podolsky and N. Rosen (1935) Can quantum mechanical description of reality be considered complete? *Physical Review* **47**, 777–780.

Engel, E. (1992) *A Road to Randomness in Physical Systems*, Springer (Lecture Notes in Statistics **71**).

Exner, Felix (1900) Notiz zu Brown's Molecularbewegung, *Annalen der Physik* **2**, 843–847.

Exner, Franz (1919) *Vorlesungen über die physikalischen Grundlagen der Naturwissenschaften*, Franz Deuticke, Vienna.

Faber, G. (1910) Ueber stetige Funktionen, *Mathematische Annalen* **69**, 372–443.

Fechner, T. (1987) *Kollektivmasslehre*, Engelmann, Leipzig.

Feller, W. (1936) Zur Theorie der stochastischen Prozesse, *Mathematische Annalen* **113**, 113–160.

Feller, W. (1968, 1971) *An Introduction to Probability Theory and Its Applications*, vol. 1, 3rd ed., vol. 2, 2nd ed, Wiley, New York.

Feynman, R. (1948) Space-time approach to non-relativistic quantum mechanics, *Reviews of Modern Physics* **20**, 367–387.

Feynman, R. and A. Hibbs (1965) *Quantum Mechanics and Path Integrals*, McGraw-Hill, New York.

Fine, T. (1970) On the apparent convergence of relative frequency and its implications, *IEEE Transactions on Information Theory* **16**, 251–257.

de Finetti, B. (1926) Considerazioni matematiche sull'eredità mendeliana, *Metron* **6**, 3–41.

de Finetti, B. (1927a) Conservazione e diffusione dei caratteri mendeliani. Nota I. Caso panmittico. *Rendiconti della R. Accademia Nazionale dei Lincei* **5**, 913–921.

de Finetti, B. (1927b) Conservazione e diffusione dei caratteri mendeliani. Nota II. Caso generale. *Rendiconti della R. Accademia Nazionale dei Lincei* **5**, 1024–1029.

de Finetti, B. (1928a) Sulle probabilità numerabili e geometriche, *Rendiconti del Reale Istituto Lombardo di Scienze e Lettere* **61**, 817–24.

de Finetti, B. (1928b) Funzione caratteristica di un fenomeno aleatorio, *Atti del Congresso Internazionale dei Matematici* vol. 6, 179–90, Zanichelli, Bologna. (Published 1932).

de Finetti, B. (1929a) Sulle funzioni ad incremento aleatorio, *Rendiconti della R. Accademia Nazionale dei Lincei* **10**, 163–168.

de Finetti, B. (1929b) Sulla possibilità di valori eccezionali per una legge di incrementi aleatori, *Rendiconti della R. Accademia Nazionale dei Lincei* **10**, 325–329.

de Finetti, B. (1929c) Integrazione delle funzioni a incremento aleatorio. *Rendiconti della R. Accademia Nazionale dei Lincei* **10**, 548–553.

de Finetti, B. (1929d) Funzione caratteristica di un fenomeno aleatorio, *Bollettino dell'Unione Matematica Italiana*, No. 2, 94–96.

de Finetti, B. (1930a) Sui passaggi al limite nel calcolo delle probabilità, *Rendiconti del Reale Istituto Lombardo di Scienze e Lettere* **63**, 155–166.

de Finetti, B. (1930b) A proposito dell'estensione del teorema delle probabilità totali alle classi numerabili, *Rendiconti del Reale Istituto Lombardo di Scienze e Lettere* **63**, 901–905.

de Finetti, B. (1930c) Ancora sull'estensione alle classi numerabili del teorema delle probabilità totali, *Rendiconti del Reale Istituto Lombardo di Scienze e Lettere* **63**, 1063–1069.

de Finetti, B. (1930d) Fondamenti logici del ragionamento probabilistico, *Bollettino dell'Unione Matematica Italiana*, No. 5, 258–261.

de Finetti, B. (1930e) Funzione caratteristica di un fenomeno aleatorio, *Memorie della R. Accademia dei Lincei* **4**, 86–133.

de Finetti, B. (1930f) Problemi determinati e indeterminati nel calcolo delle probabilità, *Rendiconti della R. Accademia Nazionale dei Lincei* **12**, 367–73.

de Finetti, B. (1930g) Sulla proprietà conglomerativa delle probabilità subordinate, *Rendiconti del Reale Istituto Lombardo di Scienze e Lettere* **63**, 414–418.

de Finetti, B. (1931a) Le leggi differenziali e la rinunzia al determinismo, *Rendiconti del Seminario Matematico della R. Università di Roma* **7**, 63–74.

de Finetti, B. (1931b) Probabilismo, *Logos* (Napoli), 163–219. English: *Erkenntnis* **31** (1989), 169–223.

de Finetti, B. (1931c) Sul significato soggettivo della probabilità, *Fundamenta Mathematicae* **17**, 298–329.

de Finetti, B. (1931d) Le funzioni caratteristiche di legge istantanea dotate di valori eccezionali, *Rendiconti della R. Accademia Nazionale dei Lincei* **14**, 259–265.

de Finetti, B. (1933a) Classi di numeri aleatori equivalenti, *Rendiconti della R. Accademia Nazionale dei Lincei* **18**, 107–110.

de Finetti, B. (1933b) La legge dei grandi numeri nel caso dei numeri aleatori equivalenti, *Rendiconti della R. Accademia Nazionale dei Lincei* **18**, 203–207.

de Finetti, B. (1933c) Sulla legge di distribuzione dei valori in una successione di numeri aleatori equivalenti, *Rendiconti della R. Accademia Nazionale dei Lincei* **18**, 279–284.

de Finetti, B. (1934) Indipendenza stocastica ed equivalenza stocastica, *Atti della Società Italiana per il Progresso delle Scienze*, (XXII Riunione), vol. 2, 199–202.

de Finetti, B. (1936a) Les probabilités nulles, *Bulletin des Sciences mathématiques* **60**, 275–288.

de Finetti, B. (1936b) Statistica e probabilità nella concezione di R. von Mises, *Supplemento Statistico ai Nuovi Problemi di Politica, Storia ed Economia* **2**, 9–19.

de Finetti, B. (1937) La prévision: ses lois logiques, ses sources subjectives, *Annales de l'Institut Henri Poincaré* **7**, 1–68.

de Finetti, B. (1937–8) *Calcolo delle probabilità*, typescript for the academic year 1937–8, University of Padua. Published as *Atti dell'XI Convegno, Torino-Aosta, 9–11 Settembre 1987*, Associazione per la Mathematica Applicata alle Scienze Economiche Sociali.

de Finetti, B. (1938a) Sur la condition de «equivalence partielle», *Actualités scientifiques et industrielles* **739**, 5–18.

de Finetti, B. (1938b) Funzioni aleatorie, *Atti del I Congresso dell'Unione Matematica Italiana*, 413–416, Zanichelli, Bologna.

de Finetti, B. (1938c) Resoconto critico del colloquio di Ginevra intorno alla teoria delle probabilità, *Giornale dell'Istituto Italiano degli Attuari* **9**, 3–42.

de Finetti, B. (1939) *Compte rendu critique du colloque du Genève sur la théorie des probabilités*, Hermann, Paris (Actualités scientifiques et industrielles **766**).

de Finetti, B. (1939a) Punti di vista: Emile Borel, *Supplemento Statistico ai nuovi Problemi di Politica, Storia ed Economia* **5**, 61–71.

de Finetti, B. (1949) Sull'impostazione assiomatica del calcolo delle probabilità, *Annali Triestini* **19**, 29–81. English: de Finetti (1972).

de Finetti, B. (1952) Gli eventi equivalenti e il caso degenere, *Giornale dell'Istituto Italiano degli Attuari* **15**, 40–64.

de Finetti, B. (1964) Foresight: its logical laws, its subjective sources, in H. Kyburg and H. Smokler eds., *Studies in Subjective Probability*, 97–158, Wiley, New York.

de Finetti, B. (1969) Sulla proseguibilità di processi aleatori scambiabili, *Rendiconti dell'Istituto di Matematica dell'Università di Trieste* **1**, 53–67.

de Finetti, B. (1972) *Probability, Induction and Statistics*, Wiley, New York.

de Finetti, B. (1974–75) *Theory of Probability*, 2 vols., Wiley, New York.

de Finetti, B. (1977) Discussion note in *Rapporti tra biologia e statistica*, p. 219, (Contributi del Centro Linceo interdisciplinare di scienze matematiche e loro applicazioni N. 37), Accademia Nazionale dei Lincei, Rome.

de Finetti, B. (1981) *Scritti (1926–1930)*, Cedam, Padua.

de Finetti, B. (1989) *La logica dell'incerto*, Saggiatore, Milano.

de Finetti, B. (1991) *Scritti (1931–1936)*, Pitagora Editrice, Bologna.

Fisher, R. (1956) *Statistical Methods and Scientific Inference*, Oliver and Boyd, Edinburgh.

Fokker, A.D. (1914) Die mittlere Energie rotierender Dipole im Strahlungsfeld, *Annalen der Physik* **43**, 810–820.

Forman, P. (1971) Weimar culture, causality, and quantum theory, 1918–1927: Adaption by German physicists and mathematicians to a hostile intellectual environment, *Historical Studies in the Physical Sciences* **3**, 1–115.

Fréchet, M. (1915) Sur l'intégrale d'une fonctionelle étendue à un ensemble abstrait, *Bulletin de la Société mathématique de France* **43**, 248–265.

Fréchet, M. (1930a) Sur l'extension du théorème des probabilités totales au cas d'une suite infinie d'événements, *Rendiconti del Reale Istituto Lombardo di Scienze e Lettere* **63**, 899–900.

Fréchet, M. (1930b) Sur l'extension du théorème des probabilités totales au cas d'une suite infinie d'événements, *Ibid.* **63**, 1059–1062.

Fréchet, M. (1943) *Les probabilités associées a un systéme d'événements compatibles et dependants, cas particuliers et applications*, Hermann, Paris. (Actualités scientifiques et industrielles **942**).

Fürth, R. (1920) *Schwankungserscheinungen in der Physik*, Vieweg, Braunschwig.

Galavotti, M. (1991) The notion of subjective probability in the work of Ramsey and de Finetti, *Theoria* **57**, 239–259.

Gamow, G. (1928) Zur Quantentheorie des Atomkernes, *Zeitschrift für Physik* **51**, 204–212.

Garber, E., S. Brush and C. Everitt, eds. (1986) *Maxwell on Molecules and Gases*, MIT Press, Cambridge.

Gardner, W. (1987) *Statistical Spectral Analysis: A Nonprobabilistic Theory*, Prentice-Hall, Englewood Cliffs.

Gentzen, G. (1933) On the relation between intuitionist and classical arithmetic, in *The Collected Papers of Gerhard Gentzen*, 53–67, North-Holland, Amsterdam 1969. (Translation of unpublished German original.)

Gibbs, J. W. (1902) *Elementary Principles in Statistical Mechanics*, as reprinted by Dover, New York 1962.

Glivenko, V. (1929) Sur quelques points de la logique de M. Brouwer, *Académie Royale de Belgique, Bulletin de la Classe des Sciences* **15**, 183–188.

Gödel, K. (1930) Die Vollständigkeit der Axiome des logischen Funktionenkalküls, *Monatshefte für Mathematik und Physik* **37**, 349–360. English: van Heijenoort 1967.

Gödel, K. (1931) Ueber formal unentscheidbare Sätze der Principia mathematica und verwandter Systeme I, *Monatshefte für Mathematik und Physik* **38**, 173–198. English: van Heijenoort 1967.

Gödel, K. (1933) Zur intuitionistischen Arithmetik und Zahlentheorie, *Ergebnisse eines mathematischen Kolloquiums* **4**, 35–38.

Gouy, L. (1888) Note sur le mouvement brownien, *Journal de physique théorique et appliquée* **7**, 561–564.

Grad, H. (1952) Statistical mechanics, thermodynamics and fluid dynamics of systems with an arbitrary number of integrals, *Communications in Pure and Applied Mathematics* **5**, 455–494.

Grant, E. (1961) Nicole Oresme and the commensurability or incommensurability of celestial motions, *Archive for History of Exact Sciences* **1**, 420–458.

Grant, E. ed. (1974) *A Source Book in Medieval Science*, Harvard University Press.

See also under *Nicole* Oresme

Gurney, R. W. and E. U. Condon (1929) Quantum mechanics and radioactive disintegration, *The Physical Review* **33**, 127–140.

Gyldén, H. (1888a) Om sannolikheten af inträdande divergens vid användning af de hittils brukliga methoderna att analytiskt framställa planetariska störingar, *Öfversikt af Kongliga Vetenskaps-Akademiens Förhandlingar* **45**, 77–87.

Gyldén, H. (1888b) Quelques remarques relativement à la représentation de nombres irrationnels au moyen des fractions continues, *Comptes rendus* **107**, 1584–1587.

Gyldén, H. (1888c) Om sannolikheten att påträffa stora tal vid utvecklingen af irrationella decimalbråk i kedjebråk, *Öfversikt af Kongliga Vetenskaps-Akademiens Förhandlingar* **45**, 349–358.

Gyldén, H. (1888d) Quelques remarques relatives à la représentation de nombres irrationanels au moyen des fractions continues, *Comptes rendus* **107**, 1777–1781.

ter Haar, D. (1967) *The Old Quantum Theory*, Pergamon Press, Oxford.

de Haas-Lorentz, G. L. (1913) *Die Brownsche Bewegung und einige verwandte Erscheinungen*, Vieweg, Braunschweig.

Hadamard, J. (1927) Sur le battage des cartes, *Comptes rendus* **185**, 5–9.

Hadamard, J. (1928) Sur le battage des cartes et ses relations avec la mécanique statistique, *Atti del Congresso Internazionale dei Matematici* Bologna 1928, vol. 5, 133–139. (Published in 1932.)

Halmos, P. (1938) Invariants of certain stochastic transformations: the mathematical theory of gambling systems, *Duke Mathematical Journal* **5**, 461–478.

Halmos, P. (1950) *Measure Theory*, Van Nostrand, New York.

Hanle, P. (1977) The coming of age of Erwin Schrödinger: His quantum theory of ideal gases, *Archive for History of Exact Sciences* **17**, 165–192.

Hardy, G. H. and J. E. Littlewood (1914) Some problems of Diophantine approximation, *Acta mathematica* **37**, 155–190.

Hardy, G. H. and E. M. Wright (1960) *An Introduction to the Theory of Numbers*, 4th ed., Oxford University Press.

Hausdorff, F. (1914) *Grundzüge der Mengenlehre*, De Gruyter, Leipzig.

Hausdorff, F. (1927) *Mengenlehre*, De Gruyter, Berlin and Leipzig.

Hawkins, T. (1970) *Lebesgue's Theory of Integration. Its Origins and Development*. The University of Wisconsin Press, Madison.

Heidelberger, M. (1987) Fechner's indeterminism: From freedom to laws of chance, in Krüger et al. (1987), vol. 1, 117–156.

van Heijenoort, J. ed. (1967) *From Frege to Gödel. A Source Book in Mathematical Logic, 1879–1931*, Harvard University Press.

Heisenberg, W. (1925) Ueber quantenmechanische Umdeutung kinematischer und mechanischer Beziehungen, *Zeitschrift für Physik* **33**, 879–893. English: van der Waerden (1967), 261–276.

Heisenberg, W. (1926) Schwankungserscheinungen und Quantenmechanik, *Zeitschrift für Physik* **40**, 501–506.

Heisenberg, W. (1927) Ueber den anschaulichen Inhalt der quantentheoretischen Kinematik und Mechanik, *Zeitschrift für Physik* **43**, 172–198.

Heisenberg, W. (1930) *The Physical Principles of the Quantum Theory*, University of Chicago Press.

Heisenberg, W. (1969) Die Quantenmechanik und ein Gespräch mit Einstein, in Heisenberg's *Der Teil und das Ganze*, 90–100, Piper Verlag, Munich.

Herschel, J. (1850) Quetelet on probabilities, *Edinburgh Review* **92**, 1–57.

Heyting, A. (1930) Die formalen Regeln der intuitionistischen Logik, *Sitzungsberichte der Preussischen Akademie der Wissenschaften 1930*, 42–56.

Heyting, A. (1934) *Mathematische Grundlagenforschung. Intuitionismus. Beweistheorie*, Springer, Berlin.

Hilbert, D. (1899) *Grundlagen der Geometrie*, 2nd ed. Teubner, Leipzig 1903.

Hilbert, D. (1900) Mathematische Probleme, *GA 3*, 290–329.

Hilbert, D. (1912) *Grundzüge einer allgemeinen Theorie der linearen Integralgleichungen*, Teubner, Leipzig.

Hilbert, D. and W. Ackermann (1928) *Grundzüge der theoretischen Logik*, Springer, Berlin.

Hlawka, E. and C. Binder (1986) Ueber die Entwicklung der Gleichverteilung in den Jahren 1909 bis 1916, *Archive for History of Exact Sciences* **36**, 197–249.

Hopf, E. (1932a) Theory of measure and invariant integrals, *Transactions of the American Mathematical Society* **34**, 373–393.

Hopf, E. (1932b) On the time average theorem in dynamics, *Proceedings of the National Academy of Sciences* **18**, 93–100.

Hopf, E. (1934) On causality, statistics and probability, *Journal of Mathematics and Physics* (MIT), **13**, 51–102.

Hopf, E. (1935) Remarks on causality and probability, *Journal of Mathematics and Physics* (MIT), **14**, 5–9.

Hopf, E. (1936) Ueber die Bedeutung der willkürlichen Funktionen für die

Wahrscheinlichkeitstheorie, *Jahresbericht der Deutschen Mathematiker-Vereinigung* **46**, 179–195.

Hopf, E. (1937) *Ergodentheorie*, Springer, Berlin.

Hopf, E. (1937a) Ein Verteilungsproblem bei dissipativen dynamischen Systemen, *Mathematische Annalen* **114**, 161–186.

Hopf, E. (1938) Statistische Probleme und Ergebnisse in der klassischen Mechanik, *Actualités scientifiques et industrielles* **737**, 5–16.

Hostinsky, B. (1920) Sur une nouvelle solution du problème de l'aiguille, *Bulletin des Sciences mathématiques* **44**, 126–136.

Hostinsky, B. (1926) Sur la méthode des fonctions arbitraires dans le calcul des probabilités, *Acta mathematica* **49**, pp. 95–113.

Hostinsky, B. (1928) Sur les probabilités relatives aux transformations répétées, *Comptes rendus* **186**, 59–61.

Hostinsky, B. (1929a) Sur la théorie générale des phénomènes de la diffusion, in F. Leja ed., *Comptes rendus du 1 Congrès des mathématiciens des pays slaves*, Warszava 1929, 341–347. (Published Warsaw 1930).

Hostinsky, B. (1929b) Ein allgemeiner Satz über die Brown'sche Bewegung, *Physikalische Zeitschrift* **30**, 894–895.

Hostinsky, B. (1930) Diskussion über Wahrscheinlichkeit, *Erkenntnis* **1**, 284–285.

Hostinsky, B. (1931) *Méthodes générales du Calcul des Probabilités*, Gauthier-Villars, Paris.

Hughes, R. I. G. (1989) *The Structure and Interpretation of Quantum Mechanics*, Harvard University Press.

Ising, G. (1926) A natural limit for the sensibility of galvanometers, *Philosophical Magazine* **1**, 827–834.

Jaynes, E. T. (1983) *Papers on Probability, Statistics and Statistical Physics*, Reidel, Dordrecht.

Jaynes, E. T. (1986) Some applications and extensions of the de Finetti representation theorem, in P. K. Goel and A. Zellner, eds., *Bayesian Inference and Decision Techniques with Applications*, 31–42, North-Holland, Amsterdam.

Jeffrey, R. C. (1989) Reading *Probabilismo, Erkenntnis* **31**, 225–237.

Jeffreys, H. (1939) *Theory of Probability*, Oxford University Press.

Jordan, K. (1972) *Chapters on the Classical Calculus of Probability*, Akadémiai Kiado, Budapest. Hungarian original 1956.

Jordan, P. (1926a) Bemerkung über einen Zusammenhang zwischen Duanes Quantentheorie der Interferenz und den de Brogliesche Wellen, *Zeitschrift für Physik* **37**, 376–382.

Jordan, P. (1926b) Ueber quantenmechanische Darstellung von Quantensprüngen, *Zeitschrift für Physik* **40**, 661–666.

Jordan, P. (1926c) Ueber eine neue Begründung der Quantenmechanik, *Zeitschrift für Physik* **40**, 809–838.

Jordan, P. (1927) Kausalität und Statistik in der modernen Physik, *Die Naturwissenschaften* **15**, 105–110.

Jorland, G. (1987) The Saint Petersburg paradox 1713–1937, in L. Krüger ed. (1987), vol. 1., 157–190.

Kappler, E. (1938) Ueber Geschwindigkeitsmessungen bei der Brownschen Bewegung einer Drehwaage, *Annalen der Physik* **31**, 377–397.

Kennard, E. (1927) Zur Quantenmechanik einfacher Bewegungstypen, *Zeitschrift für Physik* **44**, 326–352.

Keynes, J. M. (1921) *A Treatise on Probability*, Macmillan, London.

Khintchine, A. (1923) Ueber dyadische Brüche, *Mathematische Zeitschrift* **18**, 109–116.

Khintchine, A. (1924) Ueber einen Satz der Wahrscheinlichkeitsrechnung, *Fundamenta Mathematicae* **6**, 9–20.

Khintchine, A. (1926) Ideas of intuitionism and the struggle for content in contemporary mathematics (in Russian), *Vestnik Kommunisticheskaya Akademiya* **16**, 184–192.

Khintchine, A. (1929) Von Mises' theory of probability and the principles of physical statistics (in Russian), *Uspekhi Fizicheski Nauk* **9**, 141–166.

Khintchine, A. (1932a) Sulle successioni stazionarie di eventi, *Giornale dell'Istituto Italiano degli Attuari* **3**, 267–272.

Khintchine, A. (1932b) Sur les classes d'événements équivalents, *Matematicheski Sbornik* **39**, 40–42.

Khintchine, A. (1932c) Remarques sur les suites d'événements obeissant à la loi des grands nombres, *Matematicheski Sbornik* **39**, 115–119.

Khintchine, A. (1932d) Zu Birkhoffs Lösung des Ergodenproblems, *Mathematische Annalen* **107**, 485–488.

Khintchine, A. (1933) *Asymptotische Gesetze der Wahrscheinlichkeitsrechnung*, Springer, Berlin.

Khintchine, A. (1933a) Zur mathematischen Begründung der statistischen Mechanik, *Zeitschrift für Angewandte Mathematik und Mechanik* **13**, 101–103.

Khintchine, A. (1934) Korrelationstheorie der stationären zufälligen Prozesse, *Mathematische Annalen* 109, 604–615.

Khintchine, A. (1938) Zur Methode der willkürlichen Funktionen, *Matematicheski Sbornik* **45**, 585–588.

Khintchine, A. (1954) Die Methode der willkürlichen Funktionen und der Kampf gegen den Idealismus in der Wahrscheinlichkeitsrechnung, *Sowjetwissenschaft-Naturwissenschaftliche Abteilung* **7**, 261–273.

Khintchine, A. (1961) R. Mises' frequentist theory and contemporary ideas in probability theory (in Russian), *Voprosy Filosofii* **15**, No. 1, 92–102, No. 2, 77–89.

Khintchine, A. and A. Kolmogorov (1925) Ueber Konvergenz von Reihen, deren Glieder durch den Zufall bestimmt werden, *Matematicheski Sbornik* **32**, 668–676.

Klein, M. (1970) *Paul Ehrenfest*, North-Holland, Amsterdam.

Klein, M. (1973) The development of Boltzmann's statistical ideas, in *Acta Physica Austriaca*, Supplementum 10, 53–106.

Knobloch, E. (1987) Emile Borel as a probabilist, in Krüger et al. (1987), vol. 1, 215–233.

Koch, G. and F. Spizzichino, eds. (1982) *Exchangeability in Probability and Statistics*, North-Holland, Amsterdam.

Kohlrausch, F. (1906) Ueber Schwankungen der radioaktiven Umwandlung, *Wiener Berichte* **115**, 673–678.

Kohlrausch, F. (1926) Der experimentelle Beweis für den statistischen Character des radioaktiven Zerfallgesetzes, *Ergebnisse der Exakten Naturwissenschaften* **5**, 192–212.

Koksma, J. (1936) *Diophantinsche Approximationen*, Springer, Berlin.

Koksma, J. (1962) The theory of asymptotic distribution modulo one, in J. Koksma and L. Kuipers (1962) 1–21.

Koksma, J. and L. Kuipers, eds. (1962) *Asymptotic Distribution Modulo 1*, Nordhoff, Groningen.

Kolmogorov, A. (1923) Une série de Fourier-Lebesgue divergente presque partout, *Fundamenta Mathematicae* **4**, 324–328.

Kolmogorov, A. (1925) On the principle of excluded middle (in Russian), *Matematicheski Sbornik* **32**, 646–667. English: van Heijenoort (1967).

Kolmogorov, A. (1926) Une série de Fourier-Lebesgue divergente partout, *Comptes rendus* **183**, 1327–1328.

Kolmogorov, A. (1928a) Ueber die Summen durch den Zufall bestimmter unabhängiger Grössen, *Mathematische Annalen* **99**, 309–319.

Kolmogorov, A. (1928b) On operations on sets (in Russian), *Matematicheski Sbornik* **35**, 414–422.

Kolmogorov, A. (1928c) Sur une formule limite de M. A. Khintchine, *Comptes rendus* **186**, 824–825.

Kolmogorov, A. (1929a) Ueber das Gesetz des iterierten Logarithmus, *Mathematische Annalen* **101**, 126–135.

Kolmogorov, A. (1929b) General theory of measure and the calculus of probability (in Russian), as reprinted in Kolmogorov (1986), 48–58.

Kolmogorov, A. (1929c) Bemerkungen zu meiner Arbeit «Ueber die Summen zufälliger Grössen», *Mathematische Annalen* **102**, 484–488.

Kolmogorov, A. (1931) Ueber die analytischen Methoden in der Wahrscheinlichkeitsrechnung, *Mathematische Annalen* **104**, 415–458.

Kolmogorov, A. (1932a) Sulla forma generale di un processo stocastico omogeneo (Un problema di Bruno de Finetti), *Atti della R. Accademia dei Lincei, Rendiconti* **15**, 805–808.

Kolmogorov, A. (1932b) Ancora sulla forma generale di un processo stocastico omogeneo, *Atti della R. Accademia dei Lincei, Rendiconti* **15**, 866–869.

Kolmogorov, A. (1932c) Zur Deutung der intuitionistischen Logik, *Mathematische Zeitschrift* **35**, 58–65.

Kolmogorov, A. (1933) *Grundbegriffe der Wahrscheinlichkeitsrechnung*, Springer, Berlin.

Kolmogorov, A. (1933a) Zur Theorie der stetigen zufälligen Prozesse, *Mathematische Annalen* **108**, 149–160.

Kolmogorov, A. (1934) Zufällige Bewegungen (Zur Theorie der Brownschen Bewegung), *Annals of Mathematics* **35**, 116–117.

Kolmogorov, A. (1935) On some contemporary trends in probability theory (in Russian), in *Trudy II Vsesojuznogo Matematicheskogo Sbesda. Leningrad, 24–30 Yunya 1934 G.*, vol. 1, 349–358.

Kolmogorov, A. (1936a) Anfangsgründe der Theorie der Markoffschen Ketten mit unendlich vielen möglichen Zuständen, *Matematicheski Sbornik* **1**, 607–610.

Kolmogorov, A. (1936b) Zur Theorie der Markoffschen Ketten, *Mathematische Annalen* **112**, 155–160.

Kolmogorov, A. (1937a) Ein vereinfachter Beweis des Birkhoff-Khintchineschen Ergodensatzes, *Matematicheski Sbornik* **2**, 367–368.

Kolmogorov, A. (1937b) Zur Umkehrbarkeit der statistischen Naturgesetze, *Mathematische Annalen* **113**, 766–772.

Kolmogorov, A. (1940) On a new confirmation of Mendel's laws, *Comptes rendus de l'Academie des Sciences URSS* **27**, 37–41.

Kolmogorov, A. (1941) Valeri Ivanovich Glivenko (1897–1940). Obituary (in Russian), *Uspekhi Matematicheski Nauk* **8**, 379–383.

Kolmogorov, A. (1948) Evgeni Evgenevich Slutsky. Obituary (in Russian), *Uspekhi Matematicheski Nauk* **3**, Pt. 4, 143–151.

Kolmogorov, A. (1956) (See next item.)

Kolmogorov, A. (1963a) The theory of probability, in A. D. Aleksandrov, A. Kolmogorov and M. Lavrent'ev (eds.) *Mathematics, Its Contents, Methods, and Meaning*, vol. 2, 229–264, MIT Press. (Russian original 1956).

Kolmogorov, A. (1963b) On tables of random numbers, *Sankhya* **25**, 369–376.

Kolmogorov, A. (1965) Three approaches to the quantitative definition of information, *Problems of Information Transmission* **1**, 1–7.

Kolmogorov, A. (1968) Logical basis for information theory and probability theory, *IEEE Transactions on Information Theory* **14**, 662–664.

Kolmogorov, A. (1983) Combinatorial basis of information theory and probability theory, *Russian Mathematical Surveys* **38**, 29–40.

Kolmogorov, A. (1984) On logical foundations of probability theory, *Lecture Notes in Mathematics* **1021**, 1–5.

Kolmogorov, A. (1986) *Probability Theory and Mathematical Statistics* (in Russian), Nauka, Moscow.

Kolmogorov, A. and M. Leontovich (1933) Zur Berechnung der mittleren Brownschen Fläche, *Physikalische Zeitschrift der Sowjetunion* **4**, 1–13.

König, D. and A. Szücs (1913) Mouvement d'un point abandonné à l'intérieur d'un cube, *Rendiconti del Circolo Matematico di Palermo* **36**, 79–90.

Koopman, B. (1930) Birkhoff on dynamical systems, *Bulletin of the American Mathematical Society* **26**, 162–166.

Koopman, B. (1931) Hamiltonian systems and transformations in Hilbert space, *Proceedings of the National Academy of Sciences* **17**, 315–318.

Kramers, H. A. and W. Heisenberg (1925) Ueber die Streuung von Strahlung durch Atome, *Zeitschrift für Physik* **31**, 681–708.

Kramers, H. A. and H. Holst (1925) *Das Atom und die Bohrsche Theorie seines Baues*, Springer, Berlin.

von Kries, J. (1886) *Die Principien der Wahrscheinlichkeits-Rechnung*, Mohr, Freiburg i. B.

Kronecker, L. (1884) Näherungsweise ganzzahlige Auflösung linearer Gleichungen, *Monatsberichte der Königlichen Preussischen Akademie der Wissenschaften zu Berlin vom Jahre 1884*, 1071–1080.

Krönig, A. (1856) Grundzüge einer Theorie der Gase, *Annalen der Physik und Chemie* **99**, 315–322.

Krüger, L. et al. eds. (1987) *The Probabilistic Revolution*, 2 vols., MIT Press.

Krylov, N. S. (1979) *Works on the Foundations of Statistical Physics*, Princeton University Press.

Kuzmin, R. (1928) Sur un problème de Gauss, *Atti del Congresso Internazionale dei Matematici*, Bologna 1928, vol. 5, 83–89. (Published 1932.)

Laemmel, R. (1904) *Untersuchungen über die Ermittlung der Wahrscheinlichkeiten*, Dissertation Zurich, as partly reprinted in Schneider (1988).

van Lambalgen, M. (1987) *Random Sequences*, Academisch Proefschrift, University of Amsterdam.

van Lambalgen, M. (1989) Algorithmic information theory, *Journal of Symbolic Logic* **54**, 1389–1400.

Langevin, P. (1908) Sur la théorie du mouvement brownien, *Comptes rendus* **146**, 533–539.

Laplace, P. S. (1951) *A Philosophical Essay on Probabilities*, Dover, New York. French original 1814.

Lebesgue, H. (1904) *Leçons sur l'intégration et la recherche des fonctions primitives*, Gauthier-Villars, Paris.

Lebesgue, H. (1917) Sur certaines démonstrations d'existence, *Bulletin de la Société mathématique de France* **45**, 132–144.

Leontovich, M. (1933) Zur Statistik der kontinuierlichen Systeme und des zeitlichen Verlaufes der physikalischen Vorgänge, *Physikalische Zeitschrift der Sowjetunion* **3**, 35–63.

Lévy, P. (1925) *Calcul des probabilités*, Gauthier-Villars, Paris.

Lévy, P. (1929) Sur les lois de probabilité dont dépendent les quotients complets et incomplets d'une fraction continue, *Bulletin de la Société mathématique de France* **57**, 178–194.

Lévy, P. (1934) Sur les intégrales dont les éléments sont des variables aléatoires indépendents, *Annali della R. Scuola Normale Superiore di Pisa* **3**, 337–366.

Lévy, P. (1937) *Théorie de l'addition des variables aleatoires*, Gauthier-Villars, Paris.

Lévy, P. (1972) Commentaire sur la théorie des probabilités dénombrables, in Borel (1972), vol. 1, 221–226. Originally published in 1940.

Lindeberg, J. (1922) Eine neue Herleitung des Exponentialgesetzes in der Wahrscheinlichkeitsrechnung, *Mathematische Zeitschrift* **15**, 211–225.

Loève, M. (1978) Calcul des probabilités, in J. Dieudonné ed., *Abrégé d'histoire des mathématiques 1700–1900*, vol. 2, 277–313, Hermann, Paris.

Loschmidt, J. (1876) Ueber den Zustand des Wärmegleichgewichtes eines Systems von Körpern mit Rücksicht auf die Schwerkraft. I., *Sitzungsberichte der Kaiserlichen Akademie der Wissenschaften (Wien), Abteilung II*, **73**, 128–142.

Lundberg, F. (1903) *Approximerad framställning av sannolikhetsfunktionen. Återförsäkring av kollektivrisker*, Thesis, Uppsala.

Maistrov, L. (1974) *Probability Theory: A Historical Sketch*, Academic Press, New York.

Markov [Markoff], A. A. (1912) Ausdehnung der Sätze über die Grenzwerte in der Wahrscheinlichkeitsrechnung auf eine Summe verketteter Grössen. Translation of original Russian article of 1908, in Markov's *Wahrscheinlichkeitsrechnung*, 272–298, Teubner. Leipzig.

Martin-Löf, P. (1966) The definition of random sequences, *Information and Control* **9**, 602–619.

Martin-Löf, P. (1969) The literature on von Mises' collectivs revisited, *Theoria* **35**, 12–37.

Martin-Löf, P. (1970) *Statistiska modeller*, Seminar notes from the academic year 1969–1970, University of Stockholm, Department of Mathematics.

Martin-Löf, P. (1970a) On the notion of randomness, in A. Kino et al. eds., *Intuitionism and Proof Theory*, 73–78, North-Holland, Amsterdam.

Martin-Löf, P. (1971) Complexity oscillations in infinite binary sequences, *Zeitschrift für Wahrscheinlichkeitstheorie und verwandte Gebiete* **19**, 225–230.

Matthews, P. T. (1987) Dirac and the foundation of quantum mechanics, in B. Kursunoglu and E. Winger, eds. *Reminiscences About a Great Physicist: Paul Adrien Maurice Dirac*, 199–224, Cambridge University Press.

Maxwell, J. C. (1860) Illustrations of the dynamical theory of gases, as republished in *The Scientific Papers of James Clerk Maxwell*, vol. 1, 377–409.

Maxwell, J. C. (1867) On the dynamical theory of gases, *Scientific Papers*, vol. 2, 26–78.

Maxwell, J. C. (1873) Does the progress of Physical Science tend to give any advantage to the opinion of Necessity (or Determinism) over that of the Contingency of Events and the Freedom of the Will? published in L. Campbell and W. Garnett, *The Life of James Clerk Maxwell*, 434–444, London 1882.

Maxwell. J. C. (1875) On the dynamical evidence of the molecular constitution of bodies, *Scientific Papers*, vol. 2, 418–438.

Maxwell, J. C. (1879) On Boltzmanns's theorem on the average distribution of energy on a system of material points, *Scientific Papers*, vol. 2, 713–741.

Medolaghi, P. (1907) La logica matematica ed il calcolo delle probabilità, *Bollettino della Associazione Italiana per l'Incremento della Scienza degli Attuari* **18**, 20–40.

Mehra, J. and H. Rechenberg (1982-) *The Historical Development of Quantum Theory*, Springer, New York.

von Mises, R. (1912) Ueber die Grundbegriffe der Kollektivmasslehre, *Jahresbericht der Deutschen Mathematiker-Vereinigung* **21**, 9–20.

von Mises, R. (1918) Ueber die "Ganzzahligkeit" der Atomgewichte und verwandte Fragen, *Physikalische Zeitschrift* **19**, 490–500.

von Mises, R. (1919a) Fundamentalsätze der Wahrscheinlichkeitsrechnung, *Mathematische Zeitschrift* **4**, 1–97.

von Mises, R. (1919b) Grundlagen der Wahrscheinlichkeitsrechnung, *Mathematische Zeitschrift* **5**, 52–99.

von Mises, R. (1920) Ausschaltung der Ergodenhypothese in der physikalischen Statistik, *Physikalische Zeitschrift* **21**, 225–232 and 256–262.

von Mises R. (1921a) Ueber die Wahrscheinlichkeit seltener Ereignisse, *Zeitschrift für Angewandte Mathematik und Mechanik* **1**, 121–124.

von Mises, R. (1921b) Ueber die gegenwärtige Krise der Mechanik, *Zeitschrift für Angewandte Mathematik und Mechanik* **1**, 425–431.

von Mises, R. (1928) *Wahrscheinlichkeit, Statistik und Wahrheit*, Springer, Vienna.

von Mises, R. (1930) Ueber kausale und statistische Gesetzmässigkeit in der Physik, *Die Naturwissenschaften* **18**, 145–153.

von Mises, R. (1931) *Wahrscheinlichkeitsrechnung und ihre Anwendung in der Statistik und theoretischen Physik*, Deuticke, Leipzig and Vienna.

von Mises, R. (1933) Ueber Zahlenfolgen die ein kollektiv-ähnliches Verhalten zeigen, *Mathematische Annalen* **108**, 757–772.

von Mises, R. (1934) Théorie des probabilités, fondements et applications, *Annales de l'Institut Henri Poincaré* **3**, 137–190.

von Mises, R. (1939) *Kleines Lehrbuch des Positivismus*, Den Haag.

von Mises, R. (1951) *Positivism*, Harvard University Press.

von Mises, R. (1964) *Mathematical Theory of Probability and Statistics*, Academic Press, New York.

von Mises, R. (1964a) *Selected Papers of Richard von Mises*, vol. 2, American Mathematical Society, Providence, Rhode Island.

Moore, G. (1982) *Zermelo's Axiom of Choice*, Springer, New York.

Nelson, E. (1987) *Radically Elementary Probability Theory*, Princeton University Press.

Neugebauer, O. (1975) *A History of Ancient Mathematical Astronomy*, 3 vols., Springer, Berlin.

von Neumann, J. (1928) Zur Theorie der Gesellschaftsspiele, *Mathematische Annalen* **100**, 295–320.

von Neumann, J. (1929) Beweis des Ergodensatzes und des H-Theorems in der neuen Mechanik, *Zeitschrift für Physik* **57**, 30–70.

von Neumann, J. (1932) *Mathematische Grundlagen der Quantenmechanik*, Springer, Berlin.

von Neumann, J. (1932a) A proof of the quasi-ergodic hypothesis, *Proceedings of the National Academy of Sciences* **18**, 70–82.

von Neumann, J. (1932b) Zur Operatorenmethode in der klassischen Mechanik, *Annals of Mathematics* **33**, 587–642, with additions, ibid., 789–791.

von Neumann, J. (1932c) Physical applications of the ergodic hypothesis, *Proceedings of the National Academy of Sciences* **18**, 263–266.

Newton, I. (1665) In *The Mathematical Papers of Isaac Newton*, ed. Whiteside, vol. 1, p. 60, Cambridge 1967.

Nicole Oresme (ed. Grant 1966) *De proportionibus proportionum* and *Ad pauca respicientes*, The University of Wisconsin Press, Madison.

Nicole Oresme (ed. Menut & Denomy 1968) *Le livre du ciel et du monde*, The University of Wisconsin Press, Madison.

Nicole Oresme (ed. Grant 1971) *Nicole Oresme and the Kinematics of Circular Motion*, The University of Wisconsin Press, Madison.

Nye, M. (1972) *Molecular Reality*, Macdonald, London.

Pais, A. (1977) Radioactivity's two early puzzles, *Reviews of Modern Physics* **49**, 925–938.

Pais, A. (1982) *Subtle is the Lord*, Oxford University Press.

Pais, A. (1986) *Inward Bound*, Oxford University Press.

Pauli, W. (1926) Ueber das Wasserstoffspektrum vom Standpunkt der neuen Quantenmechanik, *Zeitschrift für Physik* **36**, 336–363.

Pauli, W. (1927) Ueber Gasentartung und Paramagnetismus, *Zeitschrift für Physik* **41**, 81–102.

Pauli, W. (1979) *Scientific Correspondence*, vol 1, ed. A. Hermann et al., Springer, New York.

Pearson, K. (1905) The problem of the random walk, *Nature* **77**, p. 294.

Phillips, E. R. (1978) Nicolai Nicolaevich Luzin and the Moscow school of the theory of functions, *Historia Mathematica* **5**, 275–305.

Plancherel, M. (1913) Beweis der Unmöglichkeit ergodischer mechanischer Systeme, *Annalen der Physik* **42**, 1061–1063.

Planck, M. (1900) Zur Theorie des Gesetzes der Energieverteilung im Normalspektrum, *Verhandlungen der Deutschen Physikalischen Gesellschaft* **2**, 237–245.

Planck, M. (1917) Ueber einen Satz der statistischen Dynamik und seine Erweiterung in der Quantentheorie, *Sitzungsberichte der Preussischen Akademie der Wissenschaften*, 324–341.

von Plato, J. (1981) Nicole Oresme and the ergodicity of rotations, *Acta Philosophica Fennica* **32**, 190–197.

von Plato, J. (1982) The significance of the ergodic decomposition of stationary measures for the interpretation of probability, *Synthese* **53**, 419–432.

von Plato, J. (1991) Finite partial exchangeability, *Statistics & Probability Letters* **11**, 99–102.

Poincaré, H. (1890) Sur le problème des trois corps et les équations de la dynamique, *Acta mathematica* **13**, 1–270.

Poincaré, H. (1893) Le mécanisme et l'expérience, *Revue de métaphysique et de morale* **1**, 534–537. English: Brush (1966), 203–207.

Poincaré, H. (1894) Sur la théorie cinétique des gaz, *Revue générale des Sciences pures et appliquées* **5**, 513–521. Also in *Oeuvres de Henri Poincaré*, vol. 10, 246–263.

Poincaré, H. (1896) *Calcul des probabilités*, Gauthier Villars, Paris. 2nd ed. 1912.

Poincaré, H. (1902) *La Science et l'Hypothèse*, Flammarion, Paris.

Poincaré, H. (1905) Sur la méthode horistique de Gyldén, *Acta mathematica* **29**, 235–272.

Poincaré, H. (1908) *Science et méthode*, Flammarion, Paris.

Poincaré, H. (1912) (See 1896.)

Polya, G. (1920) Ueber den zentralen Grenzwertsatz der Wahrscheinlichkeitsrechnung und das Momentenproblem, *Mathematische Zeitschrift* **8**, 171–181.

Polya, G. (1941) Heuristic reasoning and the theory of probability, *American Mathematical Monthly* **48**, 450–465.

Popper, K. (1935) *Logik der Forschung*, Springer, Vienna.

Popper, K. (1959) *The Logic of Scientific Discovery*, Hutchinson, London.

Przibram, K. ed. (1963) *Briefe zur Wellenmechanik*, Springer, Vienna.

Ramsey, F. (1931) Truth and probability, in Ramsey's *The Foundations of Mathematics and other Logical Essays*, ed. R. Braithwaite, 156–198, Kegan Paul, London.

Ramsey, F. (1991) *Notes on Philosophy, Probability and Mathematics*, ed. M. C. Galavotti, Bibliopolis, Naples.

Regazzini, E. (1987) Probability theory in Italy between the two world wars. A brief historical review, *Metron* **45**, 5–42.

Reichenbach, H. (1915) *Der Begriff der Wahrscheinlichkeit für die mathematische Darstellung der Wirklichkeit*, Barth, Leipzig.

Reichenbach, H. (1929) Stetige Wahrscheinlichkeitsfolgen, *Zeitschrift für Physik* **53**, 274–307.

Reichenbach, H. (1932) Axiomatik der Wahrscheinlichkeitsrechnung, *Mathematische Zeitschrift* **34**, 568–619.

Reichenbach, H. (1935) *Wahrscheinlichkeitslehre*, Sijthoff's, Leiden.

Rosenthal, A. (1913) Beweis der Unmöglichkeit ergodischer Gassysteme, *Annalen der Physik* **42**, 796–806.

Rosenthal, A. (1914) Aufbau der Gastheorie mit Hilfe der Quasiergodenhypothese, *Annalen der Physik* **43**, 894–904.

Rutherford, E. (1900) A radio-active substance emitted from Thorium compounds, *Philosophical Magazine* **49**, 1–14.

Rutherford, E. (1911) The scattering of α and β particles by matter and the structure of the atom, *Philosophical Magazine* **21**, 669–688.

Rutherford, E. and F. Soddy (1903) Radioactive change, *Philosophical Magazine* **5**, 576–591.

Savage. L. J. (1954) *The Foundations of Statistics*, Wiley, New York.

Schneider, I. (1988) *Die Entwicklung der Wahrscheinlichkeitstheorie von den Anfängen bis 1933. Einführungen und Texte*, Wissenschaftliche Buchgesellschaft, Darmstadt.

Schrödinger, E. (1922) Was ist ein Naturgesetz?, *GA 4*, 495–498.

Schrödinger, E. (1924) Anmerkungen zum Kausalproblem, *GA 4*, 65–70.

Schrödinger, E. (1926a) Quantisierung als Eigenwertproblem (Erste Mitteilung), *GA 3*, 82–97.

Schrödinger, E. (1926b) Quantisierung als Eigenwertproblem (Zweite Mitteilung), *GA 3*, 98–136.

Schrödinger, E. (1926c) Ueber das Verhältnis der Heisenberg-Born-Jordanschen Quantenmechanik zu der meinen, *GA 3*, 143–165.

Schrödinger, E. (1926d) Quantisierung als Eigenwertproblem (Dritte Mitteilung: Störungstheorie, mit Anwendung auf den Starkeffekt der Balmerlinien), *GA 3*, 166–219.

Schrödinger, E. (1926e) Der stetige Uebergang von Mikro-zu Makromechanik, *GA 3*, 137–142.

Schrödinger, E. (1926f) Quantisierung als Eigenwertproblem (Vierte Mitteilung), *GA 3*, 220–250.

Schrödinger, E. (1927) Der Energieimpulssatz der Materiewellen, *GA 3*, 259–266.

Schrödinger, E. (1929) Antrittsrede des Hrn. Schrödinger, *GA 4*, 303–305.

Schrödinger, E. (1931) Ueber die Umkehrung der Naturgesetze, *GA 1*, 412–422.

Schrödinger, E. (1952) Are there quantum jumps? Parts I and II, *GA 4*, 478–492 and 493–502.

Schrödinger, E. (1984) *Gesammelte Abhandlungen*, 4 vols.

von Schweidler, E. (1905) Ueber Schwankungen der radioaktiven Umwandlung, *Premier Congrès International pour l'Etude de la Radiologie et de l'Ionization*, pp. 1–3, Imprimerie scientifique L. Severeyns, Brussels.

Shimony, A. (1955) Coherence and the axioms of confirmation, *Journal of Symbolic Logic* **20**, 1–28.

Shiryaev, A. N. (1989) Kolmogorov: life and creative activities, *The Annals of Probability* **17**, 866–944.

Sierpinski, W. (1909) Un théorème sur les nombres irrationels, *Bulletin International de l'Académie des Sciences et des Lettres, Cracovie*, Sér. A, 725–727. Reprinted in Hlawka and Binder (1986), 241–243.

Sierpinski, W. (1910) Sur la valeur asymptotique d'une certaine somme, *Bulletin International de l'Académie des Sciences et des Lettres, Cracovie*, Sér. A, 9–11. Reprinted in Hlawka and Binder (1986), 244–245.

Sierpinski, W. (1917) Une démonstration élémentaire du théorème de M. Borel sur les nombres absolument normaux et détermination effective d'un tel nombre, *Bulletin de la Société mathématique de France* **45**, 125–132.

von Smoluchowski, M. (1904) Ueber Unregelmässigkeiten in der Verteilung von Gasmolekülen und deren Einfluss auf Entropie und Zustandgleichung, in *Festschrift Ludwig Boltzmann zum 60. Geburtstag*, 626–641, Barth, Leipzig.

von Smoluchowski, M. (1906) Zur kinetischen Theorie der Brownschen Molekularbewegung und der Suspensionen, *Annalen der Physik* **21**, 756–780.

von Smoluchowski, M. (1912) Experimentell nachweisbare, der üblichen Thermodynamik widersprechende Molekularphänomene, *Physikalische Zeitschrift* **13**, 1069–1080.

von Smoluchowski, M. (1915) Ueber Brownsche Molekularbewegung unter Einwirkung äusserer Kräfte und deren Zusammenhang mit der verallgemeinerten Diffusionsgleichung, *Annalen der Physik* **48**, 1103–1112.

von Smoluchowski (1916) Drei Vorträge über Diffusion, Brownsche Molekularbewegung und Koagulation von Kolloidteilchen, *Physikalische Zeitschrift* **17**, 557–571 and 585–599.

von Smoluchowski, M. (1918) Ueber den Begriff des Zufalls und den Ursprung der Wahrscheinlichkeitsgesetze in der Physik, *Die Naturwissenschaften* **6**, 253–263.

Steinhaus, H. (1923) Les probabilités dénombrables et leur rapport à la théorie de la mesure, *Fundamenta Mathematicae* **4**, 286–310.

Steinhaus, H. (1938) La théorie et les applications des fonctions indépendents au sens stochastique, *Actualités scientifiques et industrielles* **738**, Hermann, Paris.

Struik, D. (1934) On the foundations of the theory of probabilities, *Philosophy of Science* **1**, 50–70.

Teske, A. (1977) *Marian Smoluchowski, Leben und Werk*, Polska Akademia Nauk.

Thomson, W. (1874) The kinetic theory of the dissipation of energy, as reprinted in Brush (1966) 176–187.

Tolman, R. C. (1938) *The Principles of Statistical Mechanics*, Oxford University Press.

Tomonaga, S. (1962–66) *Quantum Mechanics*, 2 vols., North-Holland, Amsterdam.

Tornier, E. (1929) Wahrscheinlichkeitsrechnung und Zahlentheorie, *Journal für die Reine und Angewandte Mathematik* **160**, 177–198.

Tornier, E. (1933) Grundlagen der Wahrscheinlichkeitsrechnung, *Acta mathematica* **60**, 239–380.

Tornier, E. and W. Feller (1930) Mass- und Inhaltstheorie des Baire'schen Nullraumes, *Mathematische Annalen* **107**, 165–187.

Troelstra, A. (1982) On the origin and development of Brouwer's concept of choice sequence, in A. Troelstra and D. van Dalen (eds.), *The L. E. J. Brouwer Centenary Symposium*, 465–486, North-Holland, Amsterdam.

Ulam, St. (1932) Zum Massbegriffe in Produkträumen, *Verhandlungen des Internationalen Mathematiker-Kongresses, Zürich*, vol. 2, 118–119.

Ville, J. (1939) *Etude critique de la notion de collectif*, Gauthier-Villars, Paris.

van der Waerden, B. L. ed. (1967) *Sources of Quantum Mechanics*, North-Holland, Amsterdam.

Wald, A. (1937) Die Widerspruchsfreiheit des Kollektivbegriffes in der Wahrscheinlichkeitsrechnung, *Ergebnisse eines mathematischen Kolloquiums* **8**, 38–72.

Weyl, H. (1910) Ueber die Gibbssche Erscheinung und verwandte Konvergenzphänomene, *Rendiconti del Circolo Matematico di Palermo* **30**, 377–407.

Weyl, H. (1914a) Ueber ein Problem aus dem Gebiete der Diophantinschen Approximationen, *Nachrichten der Königlichen Gesellschaft der Wissenschaften zu Göttingen. Mathematisch-physikalische Klasse*, 234–244.

Weyl, H. (1914b) Sur une application de la théorie des nombres à la mécanique statistique et la théorie des perturbations, *L'Enseignement mathématique* **16**, 455–467.

Weyl, H. (1916) Ueber die Gleichverteilung von Zahlen mod. Eins, *Mathematische Annalen* **77**, 313–352.

Weyl. H. (1919) *Raum, Zeit, Materie*, 3rd ed., Springer, Berlin.

Weyl, H. (1920) Das Verhältnis der kausalen zur statistischen Betrachtungsweise in der Physik, *GA 2*, 113–122. Originally in the *Schweizerische Medizinische Wochenschrift*.

Weyl, H. (1926) *Philosophie der Mathematik und Naturwissenschaft*, Oldenbourg, München u. Berlin

Weyl, H. (1927) Quantenmechanik und Gruppentheorie, *Zeitschrift für Physik* **46**, 1–46.

Weyl, H. (1938) Mean motion, *American Journal of Mathematics* **60**, 889–896.

Weyl, H. (1939) Mean motion II, *American Journal of Mathematics* **61**, 143–148.

Weyl, H. (1950) Ramifications, old and new, of the eigenvalue problem, *Bulletin of the American Mathematical Society* **56**, 115–139.

Weyl, H. (1954) Erkenntnis und Besinnung (Ein Lebensrückblick), *GA 4*, 631–649. Originally in *Studia Philosophica, Jahrbuch der Schweizerischen Philosophischen Gesellschaft*.

Weyl. H. (1968) *Gesammelte Abhandlungen*, 4 vols., Springer, Berlin.

Wiener, N. (1921) The average of an analytic functional and the Brownian movement, *Proceedings of the National Academy of Sciences* **7**, 294–298.

Wiener, N. (1924) Un problème de probabilités dénombrables, *Bulletin de la Societé mathématique de France* **52**, 569–578.

Wightman, A. (1976) Hilbert's sixth problem: mathematical treatment of the axioms of physics, in *Mathematical Developments Arising from Hilbert's Problems*, 147–240 (Proceedings of Symposia in Pure Mathematics **28**).

Wiman, A. (1900) Ueber eine Wahrscheinlichkeitsaufgabe bei Kettenbruchentwicklungen, *Kongliga Vetenskaps-Akademiens Förhandlingar*, 829–841.

Wiman, A. (1901) *Bemerkungen über eine von Gyldén aufgeworfene Wahrscheinlichkeitsfrage*, Håkan Ohlssons boktryckeri, Lund.

Yaglom, A. (1987) Einstein's 1914 paper on the theory of irregularly fluctuating series of observations, *IEEE ASSP Magazine* **4**, 7–11.

Zabell, S. (1982) W.E. Johnson's "sufficientness" postulate, *The Annals of Statistics* **10**, 1091–1099.

Zermelo, E. (1896a) Ueber einen Satz der Dynamik und die mechanische Wärmetheorie, *Annalen der Physik* **57**, 485–494.

Zermelo, E. (1896b) Ueber mechanische Erklärungen irreversibler Vorgänge, *Annalen der Physik* **59**, 793–801.

Zermelo, E. (1900) Ueber die Anwendung der Wahrscheinlichkeitsrechnung auf dynamische Systeme, *Physikalische Zeitschrift* **1**, 317–320.

Zermelo, E. (1904) Beweis, dass jede Menge wohlgeordnet sein kann (Aus einem and Herrn Hilbert gerichteten Briefe), *Mathematische Annalen* **59**, 514–516.

Zermelo, E. (1906) [Review of Gibbs' Statistical Mechanics (1902)], *Jahresbericht der Deutschen Mathematiker-Vereinigung* **15**, 232–242.

Zygmund, A. (1935) *Trigonometrical Series*, (Monografje Matematyczne, t. V), Warsaw.

# Index of Names

# Index of Subjects

www.ingramcontent.com/pod-product-compliance
Ingram Content Group UK Ltd.
Pitfield, Milton Keynes, MK11 3LW, UK
UKHW040811180125
453697UK00004B/32